Advanced Hydroinformatics

Special Publications 78

ADVANCED HYDROINFORMATICS

Machine Learning and Optimization for Water Resources

Gerald A. Corzo Perez
Dimitri P. Solomatine

Editors

This Work is a co-publication of
the American Geophysical Union and John Wiley and Sons, Inc.

AGU
ADVANCING EARTH AND SPACE SCIENCES

WILEY

This edition first published 2024

Published under the aegis of the AGU Publications Committee

Matthew Giampoala, Vice President, Publications
Carol Frost, Chair, Publications Committee
For details about the American Geophysical Union visit us at www.agu.org.

The right of Gerald A. Corzo Perez and Dimitri P. Solomatine to be identified as the editors of this work has been asserted in accordance with law.

Registered Office
John Wiley & Sons, Inc., 111 River Street, Hoboken, NJ 07030, USA

Editorial Office
111 River Street, Hoboken, NJ 07030, USA

For details of our global editorial offices, customer services, and more information about Wiley products visit us at www.wiley.com.

Wiley also publishes its books in a variety of electronic formats and by print-on-demand. Some content that appears in standard print versions of this book may not be available in other formats.

Library of Congress Cataloging-in-Publication Data

Names: Corzo Perez, Gerald Augusto, editor. | Solomatine, Dimitri P., editor. | John Wiley & Sons, publisher.
Title: Advanced hydroinformatics : machine learning and optimization for water resources / Gerald A. Corzo Perez, Dimitri P. Solomatine.
Description: Hoboken, NJ : Wiley, 2024. | Includes index.
Identifiers: LCCN 2023039032 (print) | LCCN 2023039033 (ebook) | ISBN 9781119639312 (hardback) | ISBN 9781119639329 (adobe pdf) | ISBN 9781119639343 (epub)
Subjects: LCSH: Hydrology–Data processing. | Hydrologic models.
Classification: LCC GB656.2.H9 A36 2024 (print) | LCC GB656.2.H9 (ebook) | DDC 551.480285/631–dc23/eng/20231023
LC record available at https://lccn.loc.gov/2023039032
LC ebook record available at https://lccn.loc.gov/2023039033

Cover design: Wiley
Cover image: © Alexander Nikitin/Getty Images; fotograzia/Getty Images

Set in 10/12pt TimesNewRomanMTStd by Straive, Chennai, India

CONTENTS

LIST OF CONTRIBUTORS

Nicholas G. Aumen
Southeast Region
United States Geological Survey
Boynton Beach, Florida, USA

Biswa Bhattacharya
IHE Delft Institute for Water Education
Delft, The Netherlands

Rolando Célleri
Department of Water Resources and Environmental Sciences, and Faculty of Engineering
University of Cuenca
Cuenca, Ecuador

Gerald A. Corzo Perez
IHE Delft Institute for Water Education
Delft, The Netherlands

Vitali Diaz
IHE Delft Institute for Water Education
Delft, The Netherlands; and
Delft University of Technology
Delft, The Netherlands

Santiago Duarte
IHE Delft Institute for Water Education
Delft, The Netherlands; and
Delft University of Technology
Delft, The Netherlands

Andreas Efstratiadis
Department of Water Resources and Environmental Engineering
National Technical University of Athens
Zografou, Greece

Mostafa Farrag
Department of Water Quality and Ecology Software
Deltares, Delft, The Netherlands; and
Hydrology Section
GFZ German Research Centre for Geosciences
Potsdam, Germany; and
Institute for Environmental Sciences and Geography
University of Potsdam
Potsdam, Germany

Jan Feyen
Faculty of Bioscience Engineering
Catholic University of Leuven
Leuven, Belgium

Courtney S. I. Forde
Caribbean Institute for Meteorology and Hydrology
Saint Michael, Barbados

Haiting Gu
Institute of Hydrology and Water Resources
Zhejiang University
Hangzhou, China

Daniel R. Hitchcock
Department of Agricultural Sciences
Clemson University
Clemson, South Carolina, USA

Jiping Jiang
School of Environmental Science and Engineering
Southern University of Science and Technology
Shenzhen, China; and
State Key Laboratory of Urban Water Resource and Environment
School of Environment
Harbin Institute of Technology
Harbin, China

Andreja Jonoski
Department of Hydroinformatics and Socio-Technical Innovation
IHE Delft Institute for Water Education
Delft, The Netherlands

Panagiotis Kossieris
Department of Water Resources and Environmental Engineering
National Technical University of Athens
Zografou, Greece

Li Liu
Institute of Hydrology and Water Resources
Zhejiang University
Hangzhou, China

Di Ma
Institute of Hydrology and Water Resources
Zhejiang University
Hangzhou, China

Paul Muñoz
Department of Water Resources and Environmental Sciences, and Faculty of Engineering
University of Cuenca
Cuenca, Ecuador

Suli Pan
Institute of Hydrology and Water Resources
Zhejiang University
Hangzhou, China

Tianrui Pang
State Key Laboratory of Urban Water Resource and Environment
School of Environment
Harbin Institute of Technology
Harbin, China

Fidel Perez
Mother and Teacher Pontifical Catholic University
Santo Domingo, Dominican Republic

Ioana Popescu
Department of Hydroinformatics and Socio-Technical Innovation
IHE Delft Institute for Water Education
Delft, The Netherlands

Vidya S. Samadi
Department of Agricultural Sciences
Clemson University
Clemson, South Carolina, USA

Germán Santos
Colombian School of Engineering Julio Garavito
Bogotá, Colombia

Bellie Sivakumar
Department of Civil Engineering
Indian Institute of Technology Bombay
Mumbai, India

Dimitri P. Solomatine
IHE Delft Institute for Water Education
Delft, The Netherlands; and
Water Resources Section
Delft University of Technology
Delft, The Netherlands; and
Water Problems Institute of the Russian Academy of Sciences
Moscow, Russia

Eric D. Swain
Caribbean-Florida Water Science Center
United States Geological Survey
Lutz, Florida, USA

Sadgeh Sadeghi Tabas
Department of Civil Engineering
Clemson University
Clemson, South Carolina, USA

Carlos Tami
Colombian School of Engineering Julio Garavito
Bogotá, Colombia

Sijie Tang
School of Environmental Science and Engineering
Southern University of Science and Technology
Shenzhen, China

Ioannis Tsoukalas
Department of Water Resources and Environmental Engineering
National Technical University of Athens
Zografou, Greece

Jose Valles
IHE Delft Institute for Water Education
Delft, The Netherlands

Henny A. J. Van Lanen
Hydrology and Quantitative Water Management Group
Wageningen University
Wageningen, The Netherlands

Daniel A. Vázquez
IHE Delft Institute for Water Education
Delft, The Netherlands

Han Wang
China Institute of Water Resources and Hydropower Research
Beijing, China

Catherine A. M. E. Wilson
Hydro-environmental Research Centre
School of Engineering
Cardiff University
Cardiff, United Kingdom

Na Wu
College of Environmental Science and Engineering
Tongji University
Shanghai, China

Jingkai Xie
Institute of Hydrology and Water Resources
Zhejiang University
Hangzhou, China

Yue-Ping Xu
Institute of Hydrology and Water Resources
Zhejiang University
Hangzhou, China

Yunqing Xuan
Zienkiewicz Centre for Computational Engineering
Swansea University
Swansea, United Kingdom

Yi Zheng
School of Environmental Science and Engineering
Southern University of Science and Technology
Shenzhen, China

Xingyu Zhou
Shanghai Investigate, Design & Research Institute Co., Ltd.
Shanghai, China

PREFACE

Hydroinformatics deals with advanced information technology, data analytics, modeling, artificial intelligence (AI), and optimization applied to problems of aquatic environments for the purpose of informing management. Many of these technologies have become standard tools that support water management decisions around the world. However, the technologies continue to develop and new ones are emerging, which allows them to be applied to more complex and interesting problems. There are multiple examples of environmental and hydrological problems being dealt with not only by employing physically based (process) models but also with advanced data analysis tools and machine learning models.

The rapid development of machine learning and AI brings new possibilities for hydroinformatics research and practice. Nowadays, complex issues are analyzed by identifying and explaining patterns and anomalies of measured or simulated data. With an increasing amount of data collected about the environment, physically based models are increasingly complemented (and sometimes even replaced) by data-driven models. Although data-driven models lack the ability of physically based models to explain the physics of underlying processes, they are able to discover the hidden patterns in data and often can be more accurate in forecasting. Thus, they play an important role. Pattern recognition has been one of the main tasks solved by machine learning, and lately has been given an additional push by the development and use of deep learning, an important class of machine learning algorithms and of AI in general.

Data analytics plays an important role in water resources when data are multidimensional, and spatial and time dimensions have to be dealt with in a coordinated fashion. In relation to water resources, both dimensions were always important, but recently the need to handle huge amounts of remote sensing data (big data) has become more pronounced. These developments have motivated new research efforts in the context of predicting hydrological extremes and called for testing novel approaches of spatiotemporal data analysis and machine learning. In many problems, machine learning applied to hydrological extremes is studied in connection with uncertainties.

An issue in water resources management is optimal planning and operation under uncertainties, and this is where the role of AI-driven approaches is also becoming more important. A traditional optimization approach typically cannot help much since such optimization is model based and objective functions cannot be analytically expressed. Algorithms developed under the framework of computational intelligence have been the focus of hydroinformatics for three decades but the new problems and the increased data availability lead to the necessity of testing new approaches and their critical analysis.

This book presents research results and experiences of applying hydroinformatics and, in particular, artificial intelligence and optimization technologies for water-related problems. It targets hydrologists, water resources engineers, modelers, forecasters, and hydroinformatics specialists interested in the latest experiences of applying machine learning and optimization techniques to various water resources problems.

The chapters in the book are grouped into three sections: modeling hydrological systems, forecasting water resources, and knowledge discovery and optimization.

Part I deals with advances in modeling hydrological systems concentrating on distributed representations. Chapters consider model identifiability by driving calibration with stochastic inputs, use of heterogeneous precipitation data sources, fuzzy committees of distributed models, and using machine learning to represent dynamics of rainfall runoff.

Part II covers various aspects of forecasting water resources, which is an area of great importance for planning and providing warnings in case of extreme events. The main challenge here is to ensure an extended horizon of forecasts considering the associated uncertainties. Adequate incorporation of physics in machine learning models via variable selection process is the key in ensuring interpretability and acceptance of the final results. Different types of data analysis methods and models are presented from wavelet decomposition and noise filter to water level and flow predictions using deep learning. Well-established (and accurate) neural networks (multilayer perceptrons) are not forgotten, as demonstrated by an example of the operational hydrological forecasting system used in El Salvador.

Part III covers relatively new areas of knowledge discovery and optimization based on using machine learning techniques. The chapters demonstrate how new vast data sources (big data) ranging from modeling results to spatiotemporal data analytics to social media provide new opportunities for discovering unseen patterns important for deeper understanding of water-related processes. One chapter demonstrates how information from news media about water extremes combined with sentiment analysis enables the discovery of unexpected knowledge patterns.

Another chapter shows how drought analysis from a 3-D perspective of clustering extreme events can be used to characterize how phenomena develop in space and time. A further contribution demonstrates how so-called complex networks are helping in discovering teleconnection patterns in water quality dynamics. Water management decisions are becoming more and more challenging, thus ideal for developing and testing various model-based multi-objective optimization approaches under uncertainty, an issue also dealt with in this part of the book.

The research results and experiences presented here demonstrate how machine learning and, more generally, artificial intelligence, can advance data analytics, accuracy of modeling and forecasting, and knowledge discovery for better water management under uncertainty. We hope this book will also provide inspiration for further advancing research in hydroinformatics.

We would like to acknowledge the valuable contributions of all the authors as well as their dedication and patience in multiple interactions with reviewers, whom we would also like to thank. This book includes contributions from researchers from many countries around the world, who present a broad scope of interesting problems of different geographical and hydrometeorological nature and scale. We are also thankful to our home, IHE Delft Institute for Water Education in the Netherlands, where hydroinformatics began 35 years ago, and where a number of our contributors studied. Our gratitude also goes to staff at AGU and Wiley who have played an important role by continuously supporting the preparation of this book.

Gerald A. Corzo Perez
IHE Delft Institute for Water Education
The Netherlands

Dimitri P. Solomatine
IHE Delft Institute for Water Education, and
Delft University of Technology
The Netherlands

1

HYDROINFORMATICS AND APPLICATIONS OF ARTIFICIAL INTELLIGENCE AND MACHINE LEARNING IN WATER-RELATED PROBLEMS

Gerald A. Corzo Perez[1] and Dimitri P. Solomatine[1,2,3]

[1] *IHE Delft Institute for Water Education, Delft, The Netherlands*
[2] *Water Resources Section, Delft University of Technology, Delft, The Netherlands*
[3] *Water Problems Institute of the Russian Academy of Sciences, Moscow, Russia*

In recent years, there has been a surge of interest in machine learning (ML) and artificial intelligence (AI) due to the effectiveness of deep learning algorithms and the increasing availability of large data sets. This chapter provides a brief overview of the applications of AI and ML techniques in hydroinformatics, a field that deals with advanced information technology, data analytics, and modeling for aquatic environment management. Data-driven models are becoming more common in water management as they can reveal hidden patterns in data and offer improved accuracy in certain situations. This chapter highlights the importance of spatiotemporal data analysis, pattern recognition, and optimization approaches in water resources management under uncertainty. It does not offer a comprehensive review of all methods but rather focuses on selected ML techniques widely used in water-related problems. Additionally, the

Advanced Hydroinformatics: Machine Learning and Optimization for Water Resources, Special Publications 78, First Edition. Edited by Gerald A. Corzo Perez and Dimitri P. Solomantine.

DOI: 10.1002/9781119639268.ch1

chapter discusses the challenges associated with using ML models, such as black-box criticisms, and the potential of hybrid models that combine the strengths of ML and physically based process models for more robust solutions in hydroinformatics.

1.1. Introduction

Hydroinformatics deals with advanced information technology, data analytics, modeling, artificial intelligence (AI), and optimization applied to problems of aquatic environment for the purpose of informing management. Many of these technologies have become standard tools that support water management decisions around the world. However, the technologies are developing further, new ones are emerging, and this allows for applying them to more complex and interesting problems. One can find multiple examples when environmental and hydrological problems have been dealt with not only by employing physically based (process) models, but also advanced data analysis tools and machine learning models have been used. Using AI techniques in geosciences has a long history. Hydroinformatics, formulated by Abbott (1991) 30 yr ago, has been defined as a union of computational hydraulics (CH) and AI (so that $HI = CH \cup AI$), and during the last three decades we have been witnessing a much wider use of AI, with a large number of successful practical applications. The first stage of such development has been covered, for example, in the edited volume *Practical hydroinformatics: Computational intelligence and technological developments in water applications* (Abrahart et al., 2008), and in dozens of other books and hundreds of research papers covering these new developments.

Currently, we see a new wave of interest in machine learning (ML) and AI, which is partly explained by the demonstrable effectiveness of the new generation of deep learning algorithms and availability of large data sets (see, e.g., Nearing et al., 2021), and this brings new possibilities for hydroinformatics research and practice. With an increasing amount of data collected about the environment, physically based models are more and more complemented and sometimes even replaced by data-driven models. Lacking the ability of physically based models to explain the physics of underlying processes, data-driven models are however able to discover the hidden patterns in data and often can be more accurate, and play an important supporting role, in water management. Pattern recognition (e.g., automatic identification of flooded areas on satellite images) has been one of the main tasks solved by machine learning, and lately has been given an additional push by the development and use of deep learning, an important class of machine learning algorithms, and of AI in general. Data analytics plays an important role in water resources when data are multidimensional, and spatial and time dimensions have to be dealt with in a coordinated fashion. In relation to water resources, both dimensions were always important, but recently the need to handle huge amounts of remote sensing data ("big data") has become more pronounced. These developments have motivated new research efforts in the context of predicting hydrological extremes and call for testing novel approaches of spatiotemporal data analysis and machine learning. Due to much easier access to supercomputing facilities, there are increased possibilities to study the models uncertainty (typically using Monte Carlo frameworks), and machine learning can also play a role in building predictive models of such uncertainties. An issue in water resources management is optimal planning and operation under uncertainties, and this is where the role of AI-driven approaches is also becoming more important. Classical optimization approaches (gradient-based nonlinear optimization) typically cannot help much, since such optimization is model based, and objective functions (and their gradients) cannot be analytically expressed. Optimization approaches developed under the framework of computational intelligence (various types of randomized search, e.g. evolutionary approaches) have been the focus of hydroinformatics for three decades, but the new problems and the increased data availability lead to the necessity of testing new approaches and their critical analysis.

This chapter aims at presenting a brief overview of AI- and ML-related building processes and methods widely used for water-related problems, in the context of the chapters presented in this volume. AI is a concept that covers a wide area of science and technology, however, quite often it is used interchangeably with ML, which is in fact a narrower notion. One may find in literature quite a large number of AI- and ML-related subareas: big data, data mining, pattern recognition (PR), natural language processing (NLP), neural networks, deep learning, and so on. We will not go into a discussion about terminology and differences in AI and ML; for the purpose of this chapter and the issues covered in the book, it would be right to use a somewhat narrower term, that is, machine learning.

ML techniques have been widely used in water resources during the last decades, however, at the same time, one may observe also inadequate use of ML-related modeling procedures, unjustified selection of algorithms, and even lack of understanding of why a model provides good or poor performance in mathematical and statistical sense. There is also well-known criticism of ML and statistical techniques by practitioners who are used to employing physically based (process) models; they are pointing out that a water resources problem interpretation is hidden in the so-called black box of a ML model. There is indeed a challenge of posing the problem in the right way: how domain knowledge can drive selection, building, and tuning a ML model. Lack of data and its uncertainty also makes it difficult for practitioners to feel confident about ML models.

On the other hand, the strength of ML is in its ability to represent the relationships between inputs and outputs, provided enough data are available. Although the relatively recent advances in deep learning have opened the door to the new ways of using spatiotemporal data, and at the same time motivating new algorithm developments from spatial patterns and, in general, all types of computer vision algorithms, not all problems can be tackled by ML. Input and output relations can be so complex that ML techniques may not be able to find the hidden patterns, and in such cases hybrid models, combining power of ML and process models (so-called physics-aware AI; see, e.g., Jiang et al., 2020) would be needed. Such hybrid approaches are given now increased attention in hydroinformatics.

This chapter is not intended to provide a comprehensive review of methods (which are covered in hundreds of books and in the referred literature herein), but rather focuses on some important elements of ML model building, and presents basics of several selected ML techniques quite widely used in solving water-related problems, allowing for "feeling the flavor" of ML.

1.2. Key Principles of ML/Hydroinformatics

1.2.1. AI and ML Definitions

There is a large number of evolving definitions of AI, and this can be explained by its permanent evolution and shifts in priorities, and the advances in the used mathematical instruments. Many literature sources point out that for the first time the term *AI* was used in 1956 at the Dartmouth Conference, were John McCarthy, Alan Turing, and other founding fathers of AI, help to coin the term *artificial intelligence*. One of the definitions reads: "AI is the field devoted to building artificial animals (or at least artificial creatures that, in suitable contexts, *appear* to be animals) and, for many, artificial persons (or at least artificial creatures that, in suitable contexts, *appear* to be persons)" (*Stanford Encyclopedia of Philosophy*, 2018). On the other hand, Wikipedia defines it as the "intelligence demonstrated by machines, unlike the natural intelligence displayed by humans and animals" (Artificial Intelligence, 2022). Yet another definition (sometimes referred to as being given by IBM) states, that "AI leverages computers and machines to mimic the problem-solving and decision-making capabilities of the human mind." All these definitions differ in details, but are very similar in the main idea: a machine (a programmed computer) is supposed to imitate some behavior of a living creature.

An old debate regarding whether humans will be replaced by machines has been reinitiated in various public media in the view of the latest developments in AI, especially generative AI, as implemented, for example in platforms like ChatGPT. Indeed, AI has evolved into different types, related to an extent to which it may take over some of humans' activities. The first ideas of what could be achieved are purely reactive, which is highly related to the beginnings of computer science, where, AI does not have any memory, which basically means no initial data base or information of processes. This concept can be applied to solving narrow specialized tasks. For example, a forecast is performed based only on the current situation, limited historical samples, and known variables. Further development can lead to building up memory, by collecting previous experience and more complex and voluminous data and continue adding it to the memory. Such AI systems have enough memory or experience to support humans in performing various tasks, but their ability is still limited and they are still seen as a helping hand. For example, it can provide adaptive forecasts depending on the context, such as previous performance, climatic conditions, type of a river basin, and others. An even higher level of AI can be explained as a theory of mind (Premack & Woodruff, 1978) where AI can understand thoughts and emotions and interact socially. This type of concept needs an integration of many components of AI, development of more sophisticated mathematical apparatus, so such developments are still at a rudimentary level. At the top level, it is possible to consider how these systems can become aware of life and even become self-aware. This concept links to the idea that AI machines can create new knowledge and, at the same time, build internal system concepts that link intelligence, sentience, and consciousness.

Advances of AI have been numerous and applied in various areas. We should admit however, that in water resources, only a few of such developments have been used, and these relate to application of specific machine learning techniques.

Figure 1.1 presents a schematization of some of the key techniques of ML, with references to decades when these methods started to develop. Due to a wide application of ANN, this architecture is presented in more detail. One of the relatively new developments is natural language processing (NLP); it uses deep learning (DL) to train models that help

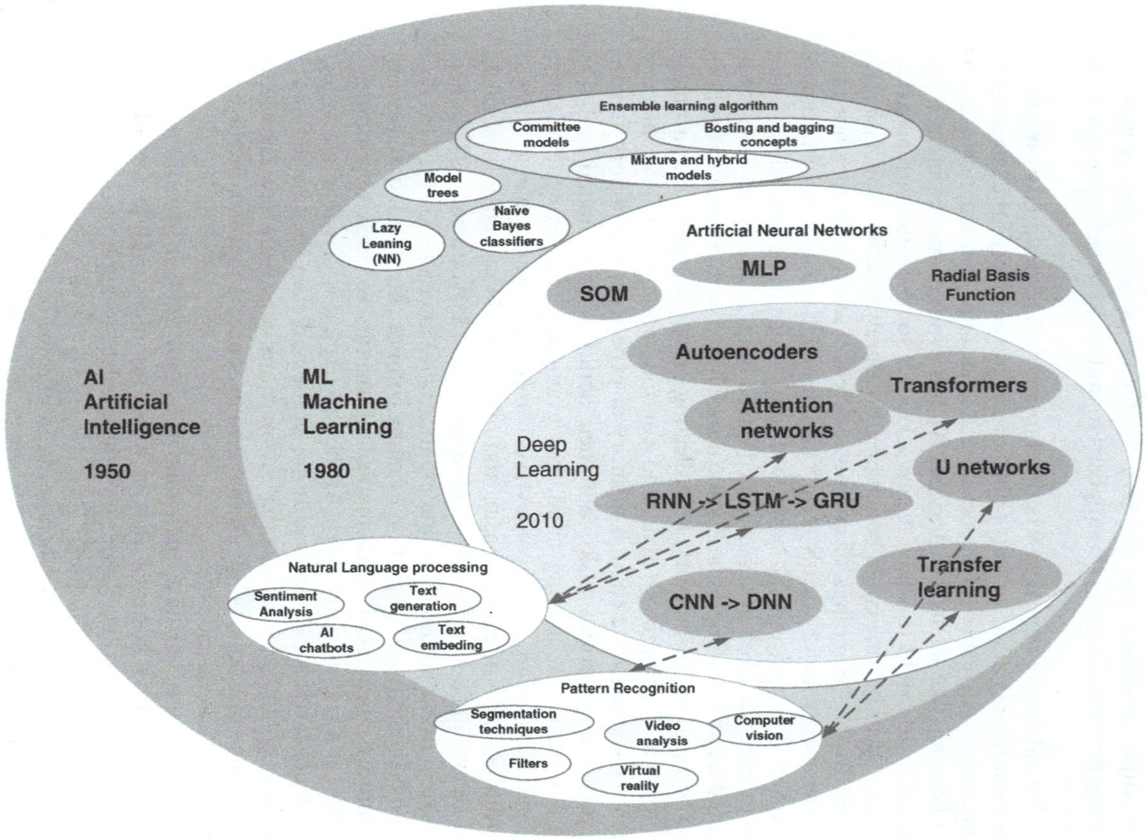

Figure 1.1 Evolution of AI topics: From artificial intelligence to machine learning and deep learning in hydroinformatics.

interpret text and reinforcement learning concepts to use DL. Pattern recognition uses convolutions and DL, which develop pattern recognition to extract features. Finally, metaheuristics provide the basis for new models of DL. The following are some of the concepts used in this chapter:

1. ML (machine learning). Mathematical models that aim to represent groups and/or input-output relationships from data
2. NLP (natural language processing). The use of language elements, in general, text encoded into numbers and its analysis, mainly from the transformation of text and processing it to solve, replicate semantics, and understand them
3. Pattern recognition. ML can be characterized as a subarea that explores how data and their attributes (variables or features) can be detected. Many ML algorithms do implicitly detect patterns and therefore these areas are interrelated. Computer vision is an important area of their application. It is worth noting that a number of important pattern recognition mathematical apparatus and algorithms are not explicitly positioned in the machine learning realm, for example, procedures of denoising and filtering, segmentation of images, 3D virtual reality patterns, vector fields flow, but they for sure contribute to solving the pattern recognition problems.

1.2.2. Machine Learning (ML)

There are various ways of contextualizing ML. From the perspective of computer science, the concept of ML can be seen as aiming at changing the programming paradigm (Fig. 1.2). Aim here is to develop a computer program that will not require significant analysis to understand how to create an algorithm to obtain certain responses; instead, a ML algorithm, theoretically, can learn from inputs and responses (outputs).

In many applications, however, ML is not seen as a tool to generate computer programs, but instead is expected to help in building input-output models by learning from data, in other words, data-driven models (see Fig. 1.6). Their use is quite varied and most of the time is justified by the idea that a system might be very complex and we may not observe all the internal states of a modeled system or process (e.g., in hydrological modeling this may be soil moisture). This implies that if we have a complex system, with only a limited understanding of the driving variables of a natural process (or any process in general), and we can measure the consequences of events (i.e., outputs resulting from particular

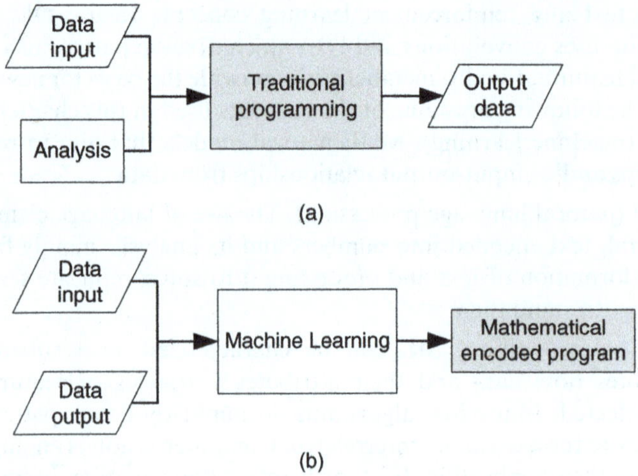

Figure 1.2 Differences between traditional programming and ML, as seen in computer science: (a) Computer science algorithm development; (b) Machine Learning algorithm development.

inputs), then with this information, it is possible to generate ML models (Fig. 1.3).

In most cases, the ML engines (e.g., artificial neural networks) work with numerical (real-valued) data, so are, in fact, nonlinear regression models. If data are nonnumerical (e.g., classes, images, or words), they have to be first transformed (encoded) into numerical form, and then processed.

In relation to Earth sciences, wide adoption of ML has not been fast, to say the least, since many scientists were pointing out that there is no clear justification for using these algorithms. Their reasoning was that the models, as descriptors of reality, should be based on scientific understanding of processes (e.g., physics), and not on a statistical encapsulation of data sets. Water resources are not an exception in this sense, and early applications of ML have been criticized, as they end up reproducing natural problems that do not need to be reproduced abstractly with ML, which was arguably resulting often in building a blind representation of a well-known problem. However, during the last two to three decades, there have been many examples of successful applications of ML reported and implemented in decision support systems. It has been shown that ML methods are often more accurate than the traditional hydrologic models in forecasting (see, e.g., Nearing et al., 2021; Arsenault et al., 2023). ML also helps to replace complex slow-running physically based models: a ML model is

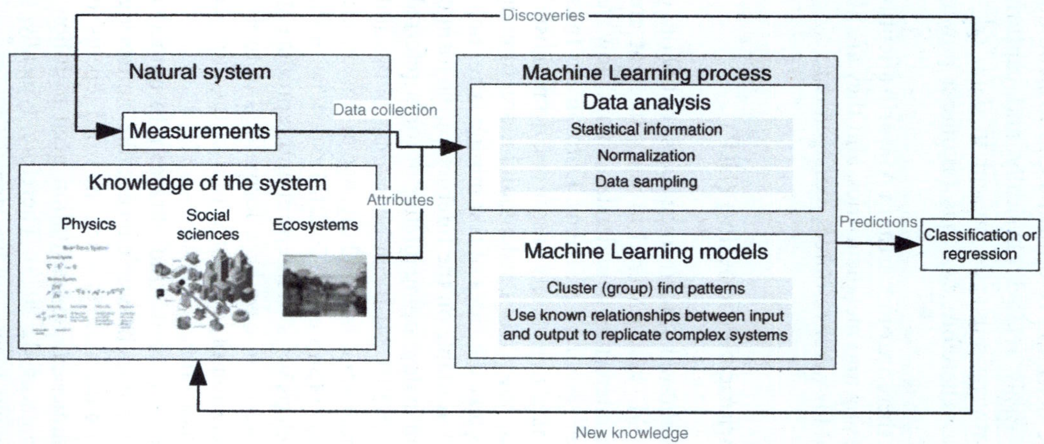

Figure 1.3 Encapsulation of natural systems and processes in ML models, with feedbacks.

trained on data generated by a process model, and such fast metamodel (surrogate) would replace a much slower process model in operational systems and therefore be used in real-time forecasting to provide warnings in an efficient manner. ML-based pattern recognition algorithms can also help to automate the detection of critical scenarios, combining variables that might not be easily related physically and capturing nontrivial relationships and patterns implicitly present in data, reproducing thus complex phenomena. Therefore, ML has become a powerful analytical and predictive tool.

Aside from ML, there are other areas in AI worth attention, that is, natural language processing and metaheuristics, and they are also considered due to their potential for water resources management.

1.2.3. Natural Language Processing (NLP)

The Internet has allowed us to arrange access to billions of documents, images, and audio and video material in very different areas of human activities. It would be interesting to understand if and how we can use these data for solving water resources problems. Data on the Internet are often not structured and linked to a variety of sources, from websites of organizations, to social media, news, blogs, videos, and more. In many cases useful data are presented as text. The idea of text mining is not new; however, the large amount of data available on the Internet, in the form of text format, has generated a boom in developing intelligent tools, referred to as natural language processing (NLP). NLP can be defined as the ability of a computer program to understand human language as it is spoken and written, that is natural language (Sun et al., 2022). It starts with the idea of processing text and develops ways to interpret and reproduce it. The ways of understanding how we write have been formalized in tools for sentiment analysis of text, generation of text, correction of text, text extraction, and concept of artificial assistants.

NLP converts letters, words, and phrases into numerical representation. This is done sometimes in simple terms, like numbering each word in a phrase and repeating the number when the word repeats itself. However, the results of this numerical representation need to follow the basics of the language (Khurana et al., 2023). Therefore, typically, the process of interpreting the language focuses on five steps:

1. Lexical (morphological) analysis: In essence, it is breaking text into paragraphs, phrases, and words. Furthermore, it is possible to understand at the level of individual words, the morphemes as the smallest units of a word. Last, lexical analysis identifies the

morphemes and allows us to characterize the word and understand its meaning knowing its root form. The final objective of this step is to help to identify words, which are normally referred to as tokens, since the original word in fact possesses some information, and, for programming, it is a sequence of characters, which represent a unit of information.

2. Syntax analysis: it allows for checking the grammar and with this the way words are arranged in a sentence. As a consequence, this order allows us to find how words should be normally arranged. Using this information, it is possible to build relationships between them. Knowing this, it is possible to assess the parts of a sentence (POS) and tag this information based on the structure found.
3. Semantic analysis: This step aims at finding the meaning of the statement, how the phrase reads literally. This understanding provides the basis for rejecting syntactically valid, but illogical statements.
4. Discourse integration: The context in which a phrase is used can be very important, so this step aims at establishing links between the different sentences, especially the immediately preceding one.
5. Pragmatic analysis, this concept uses a set of rules that describe cooperative dialogs, as in social content. What can be found in social media and common interactions can become a rule and with this we can comprehend the way the communication takes place.

NLP sentiment analysis in marketing, for example, has developed tools that allow us to provide basic knowledge extraction or information provider from how people feel about a product, and, with this, has opened numerous possibilities in decision making and understanding of people's behavior. In water resources, several exploratory projects have been carried out at the IHE Delft Institute for Water Education aiming at understanding how people write about their water bodies in social media and newspapers (see Fig. 1.4). The concept presented here is based on media information on the Web on how people express (share) feelings in the form of text in the media. This concept applies to society in general and not only to citizens per se, which will include how policy makers and news also influence the media and therefore citizens. This cycle has two parts, since people tend to be susceptible to changing their feelings toward other people based on media information. With this, the media is, at the same time, the engine that moves people, but it also records all actions on citizens' publications.

Application of NLP technologies resulted in development of AI chatbots. Conventional chatbots have evolved since 1966 (Weizenbaum,

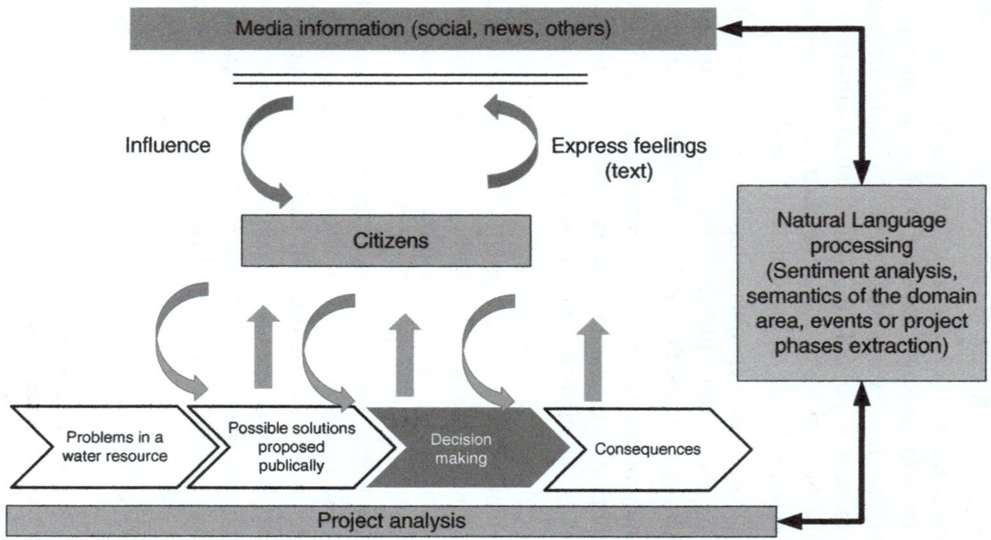

Figure 1.4 Interactions of digital information sources and citizens, and their relation to decision making in a project using NLP.

1966) when the first concept of a chat algorithm was conceived as chatterbot program ELIZA, and was defined as a program designed to interact with people by simulating human conversation. Since then, there have been plenty of tools and publications using sequential, functional database and object-oriented programming concepts. Most of these conventional programming techniques were too costly in development in terms of time and are not flexible enough to adapt to new data and new concepts. However, in the last 5 to 7 yr, technology using AI to improve the way we create chatbots has advanced considerably, and chatbots have been made widely available by their inclusion into standard smartphone software. Such chatbots serve as front-end engines to interpret human queries and then access knowledge bases to generate answers. These knowledge bases can encapsulate vast amounts of data stored on Internet servers worldwide, and they can be seen as ML systems trained on these data. Examples are Google Assistant (Google), Alexa (Amazon), and SIRI (Apple). In 2022–2023, the new noticeable implementations have been released: ChatGPT (OpenAI, a subsidiary of Microsoft), Bard (Google), and Ernie (Baidu).

1.2.4. Pattern (Image) Recognition

The area of image analysis and interpretation has been a focus in computer science and ML for several decades, and it is often referred to as pattern recognition. In this context, a pattern is an image that needs to be identified, interpreted, and classified. For example, recent research in atmospheric and hydrologic sciences has focused on how patterns in spatial data can be identified, likened to typical hydrometeorological events, and later used to predicts such events.

In water resources, pattern recognition techniques have been applied mainly to remote sensing (spatial data), finding flood patterns after extreme rainfall, identifying river networks and catchment land-use cover, learning from patterns of remote sensing bands. For example, in remote sensing, numerous pixels in a crop area are surveyed on site and then mapped to image bands. Having each sample as a training element to learn what the image band combinations are can be interpreted as this is quite important for obtaining the right spatial information about land cover and agricultural practices. As an example, recent works on determining the spatial extents of drought anomalies and the spatiotemporal patterns of rainfall that lead to floods can be mentioned (Corzo et al., 2018; Diaz et al., 2020; Khoshnazar et al., 2021; Varouchakis et al., 2021).

1.3. Model Building and Input Variable Selection in Machine Learning for Water-related Problems

ML procedures are described in detail in many books (e.g., Haykin, 1999). In relation to the use of ML in water-related problems, most papers and books present similar modeling frameworks (Fig. 1.5). For example, Elshorbagy et al. (2010) compiled information about methodologies used in various studies, presenting a general-purpose framework, which was tested on several relevant water-related cases. Similar frameworks can be found also in more recent publications (e.g., Potgieter & Dahlberg, 2022).

A large number of procedures for designing data-driven models (learning systems) have been proposed in different areas. Some of these procedures have been generalized in such a way that they can be applied to other modeling approaches (Mitchell, 2007; Pyle, 1999; Abrahart et al., 2008). Corzo Perez (2009) presented the concept of characterizing the way models can include physical-based knowledge. A typical procedure of model building, presented, for example, by Haykin (1999) and many other authors, is as follows:

1. Explore the problem and solution spaces, and state the problem.
2. What is the expected result and how will it be used?
3. Select the input and output variables (features).
4. Specify the appropriate modeling methods and choose the tools (software and algorithms).
5. Prepare and survey the data. Partition the date into the training, verification, and test subsets.
6. Build (train) the model, using training and validation subsets.
7. Test (validate) the model, using the test subset.
8. Apply the model and evaluate the results.

In reality, the process of modeling is not linear but continuous with feedback loops. For example, a lack of particular data may lead to a change in the modeling method selected. For these processes, there is a sort of checklist, or golden rules, widely accepted in the ML community, helping a modeler in the process of model building.

1. Clearly define the problem that the model will help to solve.
2. Specify the expected solution for the problem.
3. Evaluate if the delivered solution will be accepted and used in practice.
4. Learn the problem, collect the domain knowledge, analyze and understand it.

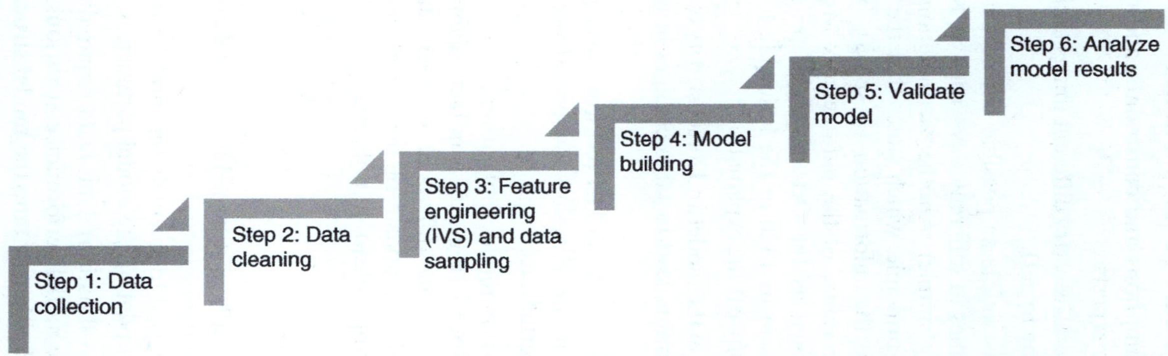

Figure 1.5 A typical sequence of steps in Machine Learning Modeling Framework.

5. Let the problem drive the modeling efforts, including selection of techniques, data preparation, and so on. Take the best tool for the job, not just a job you can do with the available tool.
6. Clearly define assumptions and constraints, and discuss them with domain knowledge experts.
7. Refine the model iteratively (try different things until the model seems as good as it is going to get).
8. Make the model as simple as possible but no simpler. This rule is formulated sometimes in different ways, such as KISS, for example (keep it sufficiently simple). Another formulation is the minimum description length principle, which states that the best model is the smallest (including the information to specify both the form of the model and the values of the parameters). More generally, this idea is widely known as the Occam's razor principle, formulated by William of Occam in 1320 in the following form: shave all the unneeded philosophy off the explanation.
9. Define instabilities in the model and its sensitivity (critical areas where small changes in inputs lead to large changes in output, or even to model crash).
10. Define uncertainties in the model (critical areas and ranges in the data set where the model produces low-confidence predictions). Try to reduce such uncertainties.
11. Draw conclusions from model application.

The model training process (Fig. 1.6) is in fact solving an optimization problem: minimization of the difference between the measured and predicted results through the update of the model parameters serve as decision variables in this optimization problem.

1.3.1. Data Partitioning

In ML, data are typically partitioned into three subsets, which ideally should be statistically similar.

1. Training data: Data used to calculate the model error with the aim of its minimization by updating the model parameters.
2. Validation (or cross-validation) data: Data required for the intermediate evaluation of the model performance at various steps of training. These data are used either to tune the model structure or to aid the early stopping of training to prevent overfitting. A common practice is also to use the so-called n-cross-validation procedure (n is often set

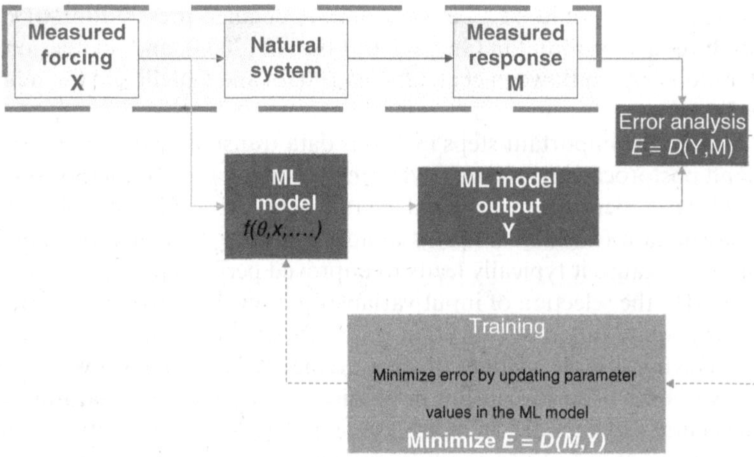

Figure 1.6 Training a Machine Learning model as an iterative optimization process. X is a vector of input variables (forcing); Θ is the parameters vector (decision variables); D is the error metrics to be minimized, the function measuring the difference between the predicted (Y) and the measured values (M).

to 10). In 10-fold cross-validation, 10 models are trained on 90% of the training set and validated on the remaining 10%. A similar performance of all models shows that the modeling process can be seen as successful, and any model can be taken as the final model, or the final model could be an ensemble of all these n trained models.

3. Testing (or verification) data: Data used for the final performance assessment of a model before its use. It requires a subset that has not been used during model training.

1.3.2. Input Variable Selection (IVS) (Feature Engineering)

It is important to ensure that a ML model would use the data and variables relevant for the objective of data analysis or modeling. Selecting the relevant variables (features) is referred to as feature engineering, or input variables selection (IVS). Input variables have to have relatedness to the output, and the simplest method here would be to use correlation analysis, or average mutual information (AMI). The correlation analysis reflects only linear relationships; therefore, when processes are highly nonlinear, AMI would be a better choice. A detailed presentation of IVS procedures can be found in Guyon and Elisseeff (2003), and, in relation to water modeling, in Bowden et al. (2005a, 2005b) and Galelli and Castelletti (2013).

One of the important steps in ML is data transformation, preprocessing and postprocessing of data (Pyle, 1999). For some of the methods, such as multilayer perceptron artificial neural networks (MLP ANNs), data transformation (normalization) is almost a must. Other methods benefit from this because it typically leads to improved performance.

In ML, the selection of input variables is of great importance, and this process is quite different from that normally used in process-based modeling. Having a large volume of measurements, and even knowing what input variables drive the output in physical terms, does not guarantee the high accuracy of a model. In most cases, it is necessary to identify (1) when and (2) to what extent these inputs contribute to the model output. Since information about the modeled phenomena is not always available in ML, most methods rely on statistics and information theory to determine the appropriate input variables for data-driven models.

Normally, the input selection process starts with all the knowledge (data) about the process that will be modeled, and then the selection space is narrowed based on a more detailed analysis. In contrast to hydrological process-based models, ML allows for the inclusion of any variable (or its combination), even of those that do not necessarily force the phenomenon (discharge) directly. For example, rainfall runoff models may use past discharges to forecast current or future ones; however, they are not the actual triggers of flood situations (in relation to hydrological modeling, see, e.g., discussion in Moreido et al., 2021).

To consider past information about the modeled phenomenon, often a model has to be fed with previous (lagged) values of an input variable, perhaps aggregated. In the context of hydrological forecasting, these are typically precipitation and discharges, optimally lagged. In this context, lag is defined as the number of time steps by which a time series is shifted from its current value, itself (when autocorrelated), or relative to the corresponding time values of another time series (when cross-correlated). Table 1.1 presents an example of building an input data matrix with the lagged precipitation (current time is assumed to be 01/01/2015).

Understanding of how much delayed inputs are correlated with output is used to make a comparative graph of autocorrelation or cross-correlated variables (Fig. 1.7). In principle, all lags can be important to represent different types of responses due to the memory of the system. However, every

Table 1.1 Example of the input matrix generation for a rainfall-runoff model (Ourthe river basin).

Dates	Precipitation (mm)	Lag prec 1 day	Lag prec 2 day	Lag prec 3 day	Lag prec 4 day	Discharge (m^3/s)
01/01/2015	4.37					4.37
02/01/2015	11.72	4.37				11.72
03/01/2015	33.57	11.72	4.37			33.57
04/01/2015	32.20	33.57	11.72	4.37		32.20
05/01/2015	**8.20**	**32.20**	**33.57**	**11.72**	**4.37**	8.20
06/01/2015	0.56	8.20	32.20	33.57	11.72	0.56
07/01/2015	26.41	0.56	8.20	32.20	33.57	26.41
08/01/2015	28.84	26.41	0.56	8.20	32.20	28.84
09/01/2015	0.23	28.84	26.41	0.56	8.20	0.23
10/01/2015	0.77	0.23	28.84	26.41	0.56	0.77
11/01/2015	10.79	0.77	0.23	28.84	26.41	10.79
12/01/2015	1.27	10.79	0.77	0.23	28.84	1.27
13/01/2015	32.60	1.27	10.79	0.77	0.23	32.60
14/01/2015	37.05	32.60	1.27	10.79	0.77	37.05
15/01/2015	1.15	37.05	32.60	1.27	10.79	1.15
16/01/2015	0.16	1.15	37.05	32.60	1.27	0.16
		0.16	1.15	37.05	32.60	
			0.16	1.15	37.05	
				0.16	1.15	
					0.16	

new variable represents an increase in the degrees of freedom in the problem and an unnecessary increase in a model complexity, and possibly, may lead to overfitting.

In this problem of rainfall-runoff, as stated before, the size of the basin plays an important role in what is to be expected from the complexity of the model. Although average precipitation is considered, it is often not enough to include only one or two lagged variables in the model. Precipitation events taking place close to the discharge measurement point would lead to an increase in discharge with a lag, which could be much smaller than the average one, and events far from this point would have a larger lag. This leads to another type of problem, which is spatiotemporal in nature.

The correlation coefficient is commonly used to determine mathematical linear relations between two samples of random variables or time series. In the case of building a rainfall runoff model, the variables are lagged precipitation and discharge. The lag time, leading to a high correlation, would be the one to adopt as the most probable candidate for the final model.

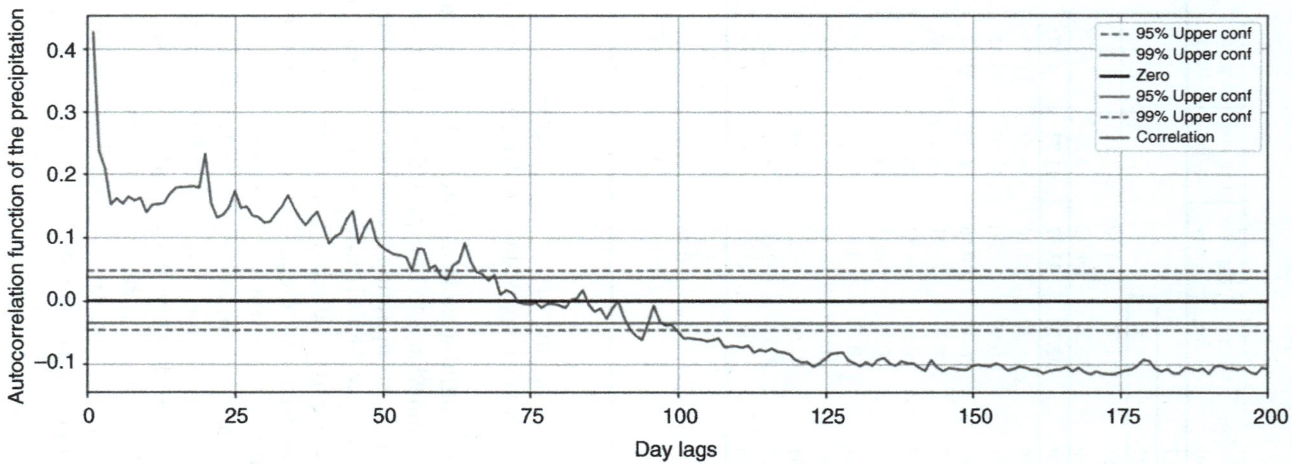

Figure 1.7 Autocorrelation for the discharge at Ourthe River basin (tributary of the Meuse).

Since the correlation coefficient can be misleading if not enough data are used, or there could be events with a different correlation structure, it could be useful to employ the two other analysis techniques: (1) to analyze the variation of the correlation coefficient with different sizes of data sets and (2) to perform the correlation analysis separately for certain events, for example, with the discharge in a particular range.

It should be noted that a number of ML models (e.g., recurrent neural networks; see below) are based on an architecture where the whole time series (or its part), representing an input variable, is fed into the model, and the choice of the previous (lagged) values and their weighting is done via an optimization process during the model training. In such cases, IVS is in choosing the right variables (rather than specific lagged values), without paying much attention to how they should be lagged in a model since this is done automatically.

1.3.3. Optimization

The third step shown in Figure 1.5 is the modeling process per se, which in principle is called learning and is an optimization process to choose the model parameters and the structure to fit the model output to the measured (target) values. Numerous single-objective and multiobjective optimization techniques are available for such purposes, and they are typically very specific to each of the techniques. It is not possible to state that one technique will always provide a better solution than another. This can be partly explained by the existence of random components in many algorithms.

1.3.4. Model Evaluation (Testing)

The final step in the training process is the evaluation of the model's performance. After the model is optimized and, as such, probably selected after some validation iterations are done, the error of the model on test data is used to assess the resulting model performance.

It is worth mentioning that in process-based modeling in situations of poor data availability, data partitioning into training, validation, and test subsets is not always carried out, so proper model testing may be difficult to arrange. However, in machine learning exercises, the procedures of data partitioning are normally followed.

The error measure (metrics) used in the training (learning) process of the ML model (typically based on the mean squared error) should not be seen as the only way of model performance estimation. Commonly, visual comparison, additional tests on extreme phenomena (crash tests), and a benchmark or reference model are used to assess the quality of the model as well. It is difficult to choose only one performance measure for a good model, so model training can be posed also as a problem of multiobjective optimization; on multiobjective machine learning, see, for example, Corzo and Solomatine (2006); and on multitask learning, a good introductory overview is given by Ruder (2017). In such cases, the modeling results should preferably be a matter of further discussions with stakeholders about the selection of a single model out of a Pareto-optimal set of models generated as a result of multiobjective optimization (training).

1.3.5. Uncertainties in the Process of Building a Model

Modeling leads to a generation of new knowledge and new insights into the modeled processes, and typically involves cycles of progressive refinement and improvement, and even reformulations of the initial modeling goals. Some of possible questions raised during modeling experiments, and the related cycles, are shown in Figure 1.8. Each of these questions is associated with an uncertain variable in the process. Uncertainty in data inevitably propagates to models' outputs. Changes in data partitioning also lead to different model parameterizations. Changes in the input variables set can lead to different model outputs and hence varying performance. Many ML models require random initializations of parameters, so that multiple runs may lead to different results. Finally, the idea of having one objective function in the optimization process, where the real problem can normally have multiple objectives, makes the model biased to a certain modeling objective. For example, a ML model built for high flow might not be optimal for low flows and vice versa. Since there is no perfect measure for performance, there is uncertainty associated with the choice of what is defined as optimal in the modeling process.

In ML theory and practice, all these problems have been studied, are quite well understood, and adequate techniques have been developed. For example, *n*-fold cross-validation is used to deal with uncertainties in data partitioning. Uncertainty is reduced if instead of using a single model, multiple (ensemble) models are built, based on different parameter (weights) randomizations, and then aggregated using (dynamic) weighting. The problem of multiregimes of the modeled system and seasonality is solved by modular or committee systems (Corzo & Solomatine, 2007b, 2007a; Kayastha et al., 2013). For problems related to the IVS, feature engineering offers a number of techniques (see Guyon & Elisseeff 2003; Bowden et al., 2005a, 2005b; Galleli & Castelletti, 2013). Robustness

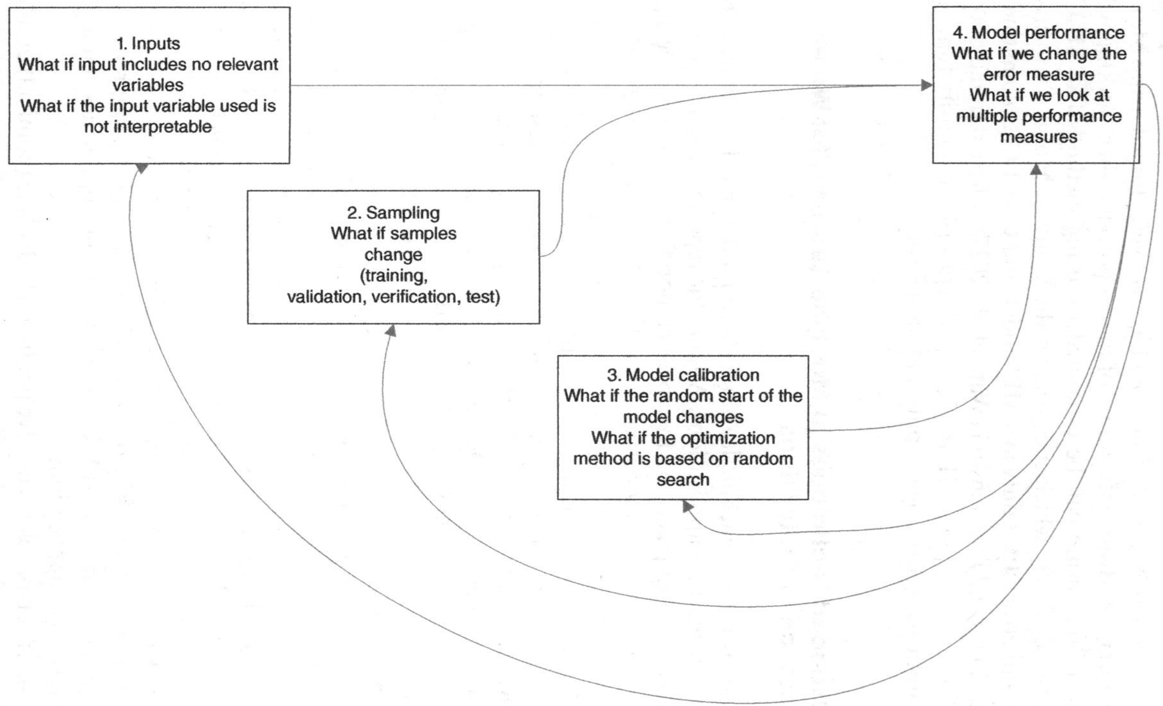

Figure 1.8 Possible cycles in ML model building.

of models can be increased also by encoding inputs and generating new features aggregating variables and thus reducing the input space dimension.

Uncertainty of the final model can be effectively studied using Monte Carlo techniques. More, uncertainty of model outputs when working with new data in the future can be estimated by using methods like MLUE (Shrestha et al., 2009), which encapsulates the results of Monte Carlo analysis of parametric uncertainty in a ML model, and UNEEC (Solomatine & Shrestha, 2009) and its extension (Wani et al., 2017), where residual uncertainty is encapsulated in a ML model, and its probability density function can be forecasted for the new inputs and model runs.

1.4. Advanced Techniques in Machine Learning for Water Resources Applications

A wide range of machine learning (ML) models has been extensively used in water resource applications. Among the reported research, a significant focus has been placed on five to seven popular techniques, including linear regression, multilayer perceptron (MLP) ANN, radial-basis function (RBF) ANN, support vector machines, M5 model and regression trees, random forests, and, more recently, LSTM (deep learning) networks.

Machine learning plays a crucial role in various water resources tasks, such as hydrological modeling, flood forecasting, drought prediction, and water resource management. Depending on the nature of the output, these application areas can be broadly classified into either classification or regression (numerical prediction) problems. In this chapter, we not only explore the popular ML techniques employed in water resources but also explicitly focus on the fascinating realm of deep learning. Deep learning, characterized by architectures based on recurrent multilayered neural networks, has garnered significant attention in recent years. Specifically, we delve into Long Short-Term Memory (LSTM) networks, Gated Recurrent Units (GRUs), and convolutional networks, which are widely used in water-related applications. By providing detailed explanations and insights into these techniques, this chapter aims to equip researchers and practitioners in the field of hydroinformatics with a comprehensive understanding of the various ML and deep learning approaches employed in water resources applications.

Classification problems involve predicting a discrete (nominal) output, such as whether a flood event will occur or not. In classification, ML models are trained to output a class. For this purpose, regression models (like ANN) can be used as well: the predicted numerical value is encoded as a class index. For example, in the case of flood forecasting, a ML numerical prediction model can be trained to predict the discharge values, but the final result will be presented as a binary class: "no flood" (if discharge is below a predefined threshold), and "flood" (otherwise).

There are also other ways of encoding classes: data are transformed, and the problem is formulated as having *N* binary outputs where *N* is the number of classes. For example, in case of two possible outcomes (like no flood and flood), output vectors (real-valued) would be either (0, 1) for flood conditions or (1, 0) for no-flood conditions. A regression ML model is trained to reproduce real-valued vectors. If, for the new inputs, this model gives an output vector close to (0, 1), for example (0.11, 0.85), then it is interpreted as flood, and if it is close to (1, 0), for example (0.15, 0.91), it is interpreted as no-flood.

However, in water resources, a much more widely spread type of problem is a regression (numerical prediction) problem, which aims to predict a continuous output, such as the discharge of a river or the water level in a reservoir. For example, in the case of water resource management, a ML model is employed to predict the reservoir inflows, water demand, and water availability, and then to use the predictions to optimize the water allocation.

This section investigates the use of machine learning techniques in water resources, emphasizing their relevance and efficacy. It begins by looking into artificial neural networks (ANNs) with a focus on the popular multilayer perceptron (MLP) ANN. MLP ANNs are popular in water resources research due to their capacity to handle complicated interactions and big data sets. The section gives a review of MLP ANN architecture and training procedures, emphasizing its applicability to water resource modeling. Following that, the subject shifts to regression and model trees, as well as bagging and boosting strategies for developing ensemble models. These methods provide modular solutions as well as statistical tools for increasing model stability, accuracy, and generalization. In regression and model trees, the combination of decision trees with linear models allows for flexible modeling of nonlinear connections, while bagging and boosting approaches improve performance by combining numerous models. The section also discusses the recent spike in interest in deep learning, particularly in relation to water resources. Deep learning, as represented by recurrent neural networks, exhibits exceptional ability in managing time series data and capturing complicated patterns. Long short-term memory (LSTM) networks, in particular, have shown considerable potential in hydrological modeling by overcoming long-term dependency concerns. The section emphasizes the significance of using deep learning methods into water management and forecasting. Overall, this in-depth examination of machine learning approaches in water resources highlights their potential and usefulness in furthering research and applications in the field.

1.4.1. Artificial Neural Networks (ANNs)

The most widely used ANN is a multilayer perceptron (MLP). An MLP ANN consists of multiple layers of interconnected nodes (neurons; Fig. 1.9) that process input vectors to produce a vector output (often a single value, however). The nodes in each layer first linearly combine inputs and then apply a nonlinear function (sigmoid or hyperbolic tangent) to transform the input further and pass the result to the next layer, whereas the final layer produces the network's output. MLPs are typically trained by backpropagation, a method that progressively adjusts the weights between nodes to minimize the difference between the network's output and the desired output; it can be seen as a specialized version of nonlinear gradient-based optimization. MLP may have several intermediate (hidden) layers, but the most widely used architecture uses only one. On details see, for example, Haykin (1999), and many other books and Internet resources.

Overall, MLP ANNs have proven to handle complex, nonlinear relationships and large amounts of data, and can make predictions with high accuracy. There are hundreds of successful applications of ANN in water resources (see, e.g., Abrahart et al., 2012), and a MLP ANN is often seen as the first choice of a ML model.

1.4.2. Regression and Model Trees

Following a modular approach to modeling, a data-driven model should comprise several submodels. The training set is split into subsets corresponding to a particular subprocess to be modeled, and then each module (Fig. 1.1) is trained on these nonintersecting subsets (actually, these subsets can be intersecting, leading to some versions of ensemble models, but this option is not considered here). When a new input vector is presented, it is first classified into one of the regions (corresponding to the subsets) for which the modules were trained, and then only one module is run to generate the prediction. A class of such methods employing consecutive progressive splits is typically referred to as *trees*. Examples are decision trees, regression trees (Breiman, 1984), which use zero-order models, that is, constants, in leaves; multiadaptive regression splines (MARS) (Friedman, 1991); and M5 model trees (Quinlan, 1992), which use linear

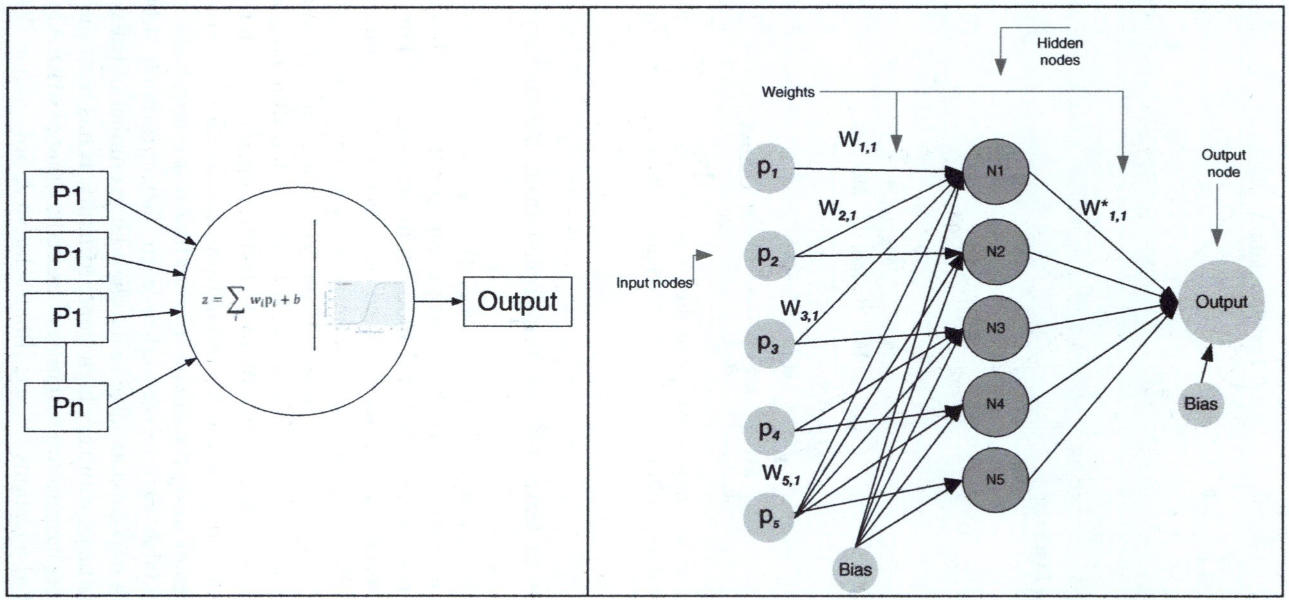

(a) (b)

Figure 1.9 A typical neuron and a single-output multilayer perceptron (MLP ANN): (a) A neuron in an MLP ANN; (b) multilayer perceptron diagram.

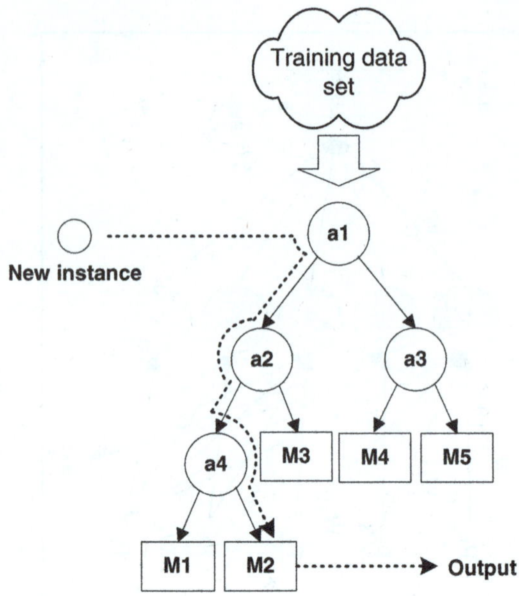

Figure 1.10 Training M5 model trees and their operation on a new unseen instance: a1–a3 are data partitioning rules; M1–M5 are multiple linear regression models.

regression models in leaves (Fig. 1.10, adopted from Bhattacharya & Solomatine, 2005).

Since for each data instance (input vector) only one local model is used for prediction, there is a problem with compatibility at the boundary between the regions for which the modules are responsible. For the two neighboring input vectors, the predicted outputs could be distinctive. A solution could be to update the local models to make them compatible at the boundaries, as is done in the M5 model tree algorithm through smoothing. Wang and Witten (1996) presented the M5 algorithm based on the original M5 algorithm but were able to deal with enumerated attributes, treat missing values, and use a different splitting termination condition. Several advantages of using the model tree are that it is a non-black-box model, understandable, easy to use and to learn, fast in training, robust when dealing with missing data, able to handle a large number of features, and able to tackle tasks with very high dimensionality. It has been shown that M5 model trees have accuracy similar or exceeding that of MLP ANNs (on water-related applications see, e.g., Bhattacharya & Solomatine, 2005; Solomatine & Xue, 2004).

In summary, M5 model trees are a type of machine learning algorithm that combines the strengths of regression trees and linear models, allowing for more flexibility in modeling nonlinear relationships.

1.4.3. Bagging and Boosting Techniques for Building Committee Models

Bagging

Two or more models whose outputs are combined, form a committee model. Statistical techniques widely used in preparing data for building such models are bagging and boosting. These methods are used to generate subsamples of data, which are used to train several models and then combine them in a committee.

The term *bagging* comes from bootstrap aggregating (bagging), which is an algorithm used to improve classification and regression models in terms of stability and classification accuracy. It also reduces variance and helps avoid overfitting. Bagging can be seen as a special case of the model averaging approach. Bagging was proposed by Leo Breiman (1984).

Given a standard training set D of size n, bagging generates m new training sets Di of size $n' \leq n$, by sampling examples from D uniformly and with replacement. By sampling with replacement, it is likely that some examples will be repeated in each Di. If $n' = n$, then for large n, the set Di is expected to have 63.2% of the unique examples of D, with the rest being duplicates. This kind of sample is known as a bootstrap sample. The m models are built using the above m bootstrap samples and combined by averaging the output (for regression) or voting (for classification). Since the method averages several predictors, it is not useful for improving linear models.

Random Forest

A very popular algorithm based on bagging that uses tree-like structuring of data is the random forest algorithm (Breiman, 2001). It is an ensemble method that combines multiple decision trees (in the original version, regression trees) to make predictions. The basic idea behind this method is to build a large number of decision (regression) trees using random subsets of the data and features, and then combine the predictions of these trees to make a final prediction.

The random forest algorithm works by first selecting a random subset of the data, called a bootstrap sample, to use as the training set for each decision tree. Then, for each tree, a random subset of the features is selected to use as the split variables at each node. The algorithm then grows a decision tree using this random subset of data and features. This process is repeated many times to build a large number of decision trees, each with its own random subset of data and features. Finally, the predictions of all the trees (constituting an ensemble) are combined to make a final prediction.

One of the advantages of random forest is that it can reduce overfitting by averaging out the predictions of many decision trees. The algorithm also has the ability to handle large numbers of input variables and nonlinear relationships between the input variables and the target variable, which is a limitation of linear regression models.

Boosting

Boosting is an approach leading also to a committee (ensemble) of ML models, but it is a sequential model, where each subsequent model is dependent on the outcome of the previous. Boosting assigns weak learners to a weighted subset of the original data set. Weak learners have little predictive ability and perform just marginally better than random guessing. Subsets that were previously misclassified are given more weight and hence the probability to be selected for the subsequent learner. As a result, the ensemble has good generalizing ability. The two widely used versions of boosting are gradient boosting (Friedman, 2001) and adaptive boosting, AdaBoost (on its realization for regression problems, AdaBoost.RT, see Shrestha & Solomatine, 2006). A popular implementation of the former is in XGBoost (extreme gradient boosting), a C++ library with APIs for several languages (XGBoost, 2023), which is extensively employed in various types of applications.

1.4.4. Deep Learning

In relation to water resources, lately, a lot of attention has been given to the so-called deep learning, a general term given to architectures based on recurrent multilayered neural networks. Especially popular are long short-term memory (LSTM), gated recurrent units (GRUs; a version of LSTM), and convolutional networks. Deep learning (DL) is commonly referred to models with large number of layers and complex nodes with memory, for example, able to handle time series "deep in time" or process images. DL has become a fashionable term, and currently is widely used, sometimes even (incorrectly) replacing a more general term ML.

1.4.5. Recurrent Neural Networks (RNNs) and LSTM

RNN

RNNs are designed to automatically learn and understand the mutual relationship between inputs, which are provided in sequential order. RNNs are often applied to NLP and speech recognition, since they have sequential structure fit to process long time series (Fig. 1.11). In a typical RNN, the output becomes a variable of state (h) used as input into the same network for the following time step in the sequence.

Long Short-Term Memory (LSTM) Networks

Traditional RNNs face difficulties in training models with long sequences, especially vanishing or exploding gradient problem. This problem was addressed in many studies, and one of the most popular techniques nowadays is the long-short-term-memory (LSTM) network, which was introduced by Hochreiter and Schmidhuber (1997). This and later developments allowed for much wider adoption of deep learning in time series forecasting.

Learning of such long-term dependencies between inputs and outputs is vital in hydrological modeling, because the lag time between hydrological processes may vary from days to years. For instance, the long lag time can be seen in the cases of glaciers, snow, and groundwater processes (Kratzert et al., 2019). Due to developments of new DL algorithms in the last several years, there has been clear revival of interest in using ML in water-related problems. These have demonstrated high accuracy, for example, in flood forecasting (e.g., Kratzert et al., 2019; Kao et al., 2020; Arsenault et al., 2023) and drought forecasting (Brust et al., 2021; Dikshit & Pradhan, 2021).

Kratzert et al. (2019) investigated the use of LSTM in rainfall runoff hydrology using Catchment Attributes and Meteorology for Large-Sample Studies (CAMELS) data sets for numerous catchments. The authors also studied the extent to which a single LSTM model could be regionalized, that is, applied to other catchments. As a result, the LSTMs predicted discharge with good accuracy compared with the reference model. The final conclusion claims that by using LSTM, the problem that is generally related to ML models of being black box models could be overcome, and explanation is in the following. In LSTM the cell state's behavior in response to the hydrological trends and patterns can be physically interpreted, for instance, by analyzing the melting state of the cell state in comparison with the change in temperature. Consequently, with low temperatures, there is a

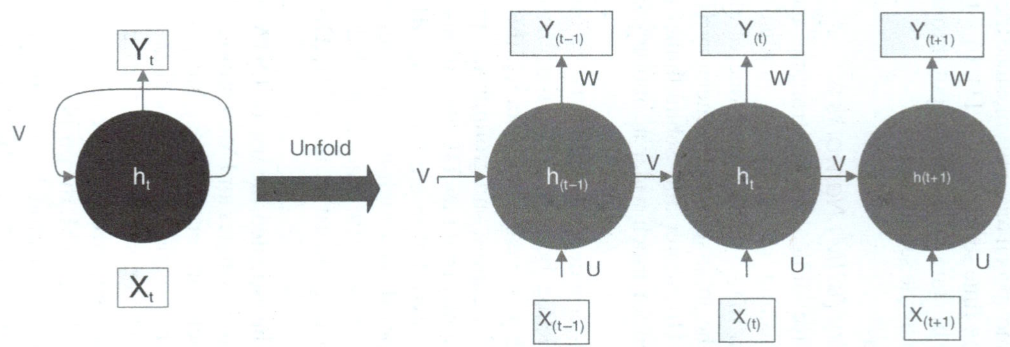

Figure 1.11 Recurrent neural network. Inputs from a time series (X) are fed into processing nodes, producing output (Y) at each time step. The state or memory variable h_t is equal to $Y_{(t)}$ in the basic algorithm.

certain increase in the cell state, and when the temperature exceeds certain degrees, the cell state starts to decrease (this imitates the snow accumulation and melting processes).

Examples of recent applications of DL in hydrometeorological problems, which are characterized by very large data sets, can be mentioned also. Schultz et al. (2021) evaluated numerical weather prediction models, which are quite important for analyzing all types of water resources problems. In this work, it is highlighted that DL is not yet widely used within the numerical weather prediction (NWP) context, mainly due to the lack of both the interpretability of the neural networks and the physical constraints. They mention also that many of the early experiments using simple neural networks did not capture the complexity of weather processes. Although these limitations are important, there are three main reasons for considering DL for weather prediction problems. First, large amounts of data are currently available. Second, new architectures of the neural networks have been developed that can capture well the time dependencies (RNNs), as well as spatial distributions and imagery, such as the convolutional neural network (CNN). Third, the computational power capacity needed for training such networks (graphical and specialized processors available on the cloud) has become easily available. All this brings the latest advanced ML techniques at the disposal of modelers and forecasters in water resources, leading thus to optimal water management.

For further information on the use of AI and ML in Earth system sciences, the reader is directed, for example, to the recorded lectures at the Summer School "Artificial Intelligence for Earth System Science" (AI4ESS, 2020), where some of the methods and impressive applications are presented.

1.5. Future Directions and Challenges

Use of modeling and forecasting tools is indispensable for effective water management. Traditionally, such models belonged to the class of the process-based (physically based) models, typically based on numerical solutions of differential equations of water motion. During the last three decades advances in machine learning techniques, increased computing power and data availability have led to changes in this landscape. Hydroinformatics is seen as the major area where machine learning such algorithms have been tested and applied to a large variety of water-related problems, and have proven to be tools effectively complementing, and sometimes replacing physically based models. An improved physical representation does not necessarily guarantee an increased accuracy and utility of traditional models, so alternative ways are sought. This chapter presented an overview of the main approach in ML, relevant for this volume, in most cases with references to the authors' own experience, and to the sources for further reading.

There is currently a serious change in the attitude toward the data-driven models within the practicing modelers community, and the current discussions in the hydrological literature is an indicator of this (see, e.g., Nearing et al., 2021; Beven, 2020). ML not only provides techniques for modeling, but also enhances ability of citizens to get access and better understand the modeling results via the natural language processing (NLP) and intelligent chatbots.

We see the future advances of modeling technologies in more effective combination of various types of models and their hybridization (e.g., in the line of physics-aware AI, see Jiang et al., 2020) and coevolution (as outlined by Razavi et al., 2022). Out of the experience of the authors, we see the following as the main future directions:

1. Integration of AI and ML into operational water management: Integrating AI and ML models into real-time decision support systems for efficient and adaptive water management is one potential approach. This entails creating models that can adapt to changing environmental circumstances in real time and improve water allocation and distribution.
2. Improved data availability and quality: ML models require access to high-quality, diversified data. Future work should concentrate on improving data-gathering methods, strengthening data-sharing channels, and creating approaches for dealing with missing or incomplete data. This will aid in the development of more accurate and robust ML models.
3. Uncertainty quantification and risk assessment: Addressing uncertainty and measuring risks associated with machine learning predictions are major problems. Future research should concentrate on creating approaches for quantifying uncertainties in ML-based predictions, as well as incorporating risk assessment frameworks for improved decision making under uncertainty.
4. Explainability and interpretability of ML models: ML models, particularly deep learning models, are often considered black boxes due to their complex structures. Enhancing the interpretability and explainability of ML models in water resource applications is an important

direction. This will enable stakeholders to understand the reasoning behind model predictions and build trust in their use.

5. Integration of domain knowledge and expert systems (hybrid models): Integration of domain knowledge and expert systems can boost ML models. To increase model accuracy and interpretability, future initiatives include constructing hybrid models that integrate ML methods with physics-based models or expert knowledge.
6. Ethical and responsible AI: As AI and ML play an increasing role in water resources management, addressing ethical considerations, data privacy, and bias becomes crucial. Future efforts should focus on developing guidelines, standards, and frameworks for the ethical and responsible use of AI in water-related applications.
7. Bridging the research-practice divide: It is critical to translate research results into practical applications and to develop cooperation among scholars, practitioners, and policymakers. Future directions should emphasize knowledge transfer, capacity building, and effective communication to ensure that ML techniques are effectively applied in real-world water management scenarios.

Finally, the chapter presented a thorough assessment of the future directions and difficulties in the field of hydroinformatics and machine learning for water-related problems. As we navigate the ever-changing world of water resource management, it is evident that embedding AI and machine learning technologies into operational systems provides a viable path toward better water allocation and distribution. However, in order to fully exploit these technologies' promise, data availability and quality must be enhanced, uncertainties measured, and dangers analyzed. Furthermore, the explainability and interpretability of ML models, domain knowledge integration, ethical issues, and effective knowledge transfer are critical topics that require study. We can support sustainable water management practices and assure the appropriate and successful use of AI and ML in tackling the ever-growing complexities of water-related concerns by embracing these difficulties and investigating the recommended future approaches.

References

Abbott, M. B. (1991). *Hydroinformatics: Information technology and the aquatic environment*. Avebury Technical.

Abrahart, R. J., Anctil, F., Coulibaly, P., Dawson, C. W., Mount, N. J., See, L. M., et al. (2012). Two decades of anarchy? Emerging themes and outstanding challenges for neural network river forecasting. *Progress in Physical Geography, 36*(4), 480–513. https://doi.org/ 10.1177/0309133312444943

Abrahart, R. J., See, L. M., & Solomatine, D. P. (Eds.) (2008). *Practical hydroinformatics: Computational intelligence and technological developments in water applications* (Vol. 68). Springer Science & Business Media.

AI4ESS (2020). *Artificial Intelligence for Earth System Science Summer School* 2020. https://www2.cisl.ucar.edu/events/summer-school/ai4ess/2020

Arsenault, R., Martel, J.-L., Brunet, F., Brissette, F., & Mai, J. (2023). Continuous streamflow prediction in ungauged basins: Long short-term memory neural networks clearly outperform traditional hydrological models. *Hydrology and Earth System Sciences, 27*(1), 139–157.

Artificial intelligence (2022, December 2). In Wikipedia. https://en.wikipedia.org/wiki/Artificial_intelligence

Beven, K. (2020). Deep learning, hydrological processes and the uniqueness of place. *Hydrological Processes, 34,* 3608–3613.

Bhattacharya, B., & Solomatine, D. P. (2005). Neural networks and M5 model trees in modeling water level-discharge relationship. *Neurocomputing, 63,* 381–396.

Bowden, G. J., Dandy, G. C., & Maier, H. R. (2005a). Input determination for neural network models in water resources applications. Part 1, Background and methodology. *Journal of Hydrology, 301,* 75–92.

Bowden, G. J., Dandy, G. C., & Maier, H. R. (2005b). Input determination for neural network models in water resources applications. Part 2, Case study: Forecasting salinity in a river. *Journal of Hydrology, 301,* 93–107.

Breiman, L. (1984). *Classification and regression trees.* Routledge. https://doi.org/10.1201/9781315139470

Breiman, L. (2001). Random forests. *Machine Learning, 45,* 5–32.

Brust, C., Kimball, J. S., Maneta, M. P., Jencso, K., & Reichle, R. H. (2021). Droughtcast: A machine learning forecast of the United States drought monitor. *Frontiers in Big Data, 4*. https://doi.org/ 10.3389/fdata.2021.773478

Corzo, G., & Solomatine, D. (2007a). Baseflow separation techniques for modular artificial neural network modelling in flow forecasting. *Hydrological Science Journal, 52*(3), 491–507.

Corzo, G., & Solomatine, D. (2007b). Knowledge-based modularization and global optimization of artificial neural network models in hydrological forecasting. *Neural Networks, 20*(4), 528–536.

Corzo, G. A., & Solomatine, D. P. (2006). *Optimization of base flow separation algorithm for modular data-driven hydrologic models.* Proceedings of the 7th International Conference on Hydroinformatics (HIC 2006). Nice, France.

Corzo, G. A., Diaz Mercado, V., & Laverde Barajas M. (2018). Spatiotemporal Hydrological Analysis. *International Journal of Hydrology, 2*(1), 25–26.

Corzo Perez, G. A. (2009). *Hybrid models for hydrological forecasting: Integration of data-driven and conceptual modeling techniques.* Taylor & Francis, CRC Press.

Diaz, V., Corzo Perez, G. A., van Lanen, H. A. J., Solomatine, D., & Varouchakis, E. A. (2020). An approach to characterize spatio-temporal drought dynamics. *Advances in Water Resources.*

Dikshit, A., & Pradhan, B. (2021). Explainable AI in drought forecasting. *Machine Learning With Applications, 6*, 100192.

Elshorbagy, A., Corzo, G., Srinivasulu, S., & Solomatine, D. P. (2010). Experimental investigation of the predictive capabilities of data-driven modeling techniques in hydrology. Part 1, Concepts and methodology. *Hydrology and Earth System Sciences.*

Friedman, J. H. (1991). Multivariate adaptive regression splines. *The Annals of Statistics, 19*(1), 1–67. http://www.jstor.org/stable/2241837

Friedman, J. H. (2001). Greedy function approximation: A gradient boosting machine. *The Annals of Statistics, 29*(5), 1189–1232.

Galelli, S., & Castelletti, A. (2013). Tree-based iterative input variable selection for hydrological modeling. *Water Resources Research., 49* (7), 4295–4310.

Guyon, I., & Elisseeff, A. (2003), An introduction to variable and feature selection. *Journal of Machine Learning Research, 3*, 1157–1182.

Haykin, S. (1999). *Neural networks and learning machines.* Pearson / Prentice Hall.

Hochreiter, S., & Schmidhuber, J. (1997). Long short-term memory. *Neural Computation, 9*(8), 1735–1780.

Jiang, S., Zheng, Y., & Solomatine, D. (2020). Improving AI system awareness of geoscience knowledge: Symbiotic integration of physical approaches and deep learning. *Geophysical Research Letters, 47*(13), e2020GL088229.

Kao, I-F., Zhou, Y., Chang, L-C., & Chang, F-J. (2020). Exploring a long short-term memory based encoder-decoder framework for multi-step-ahead flood forecasting. *Journal of Hydrology, 583*, 124631.

Kayastha, N., Ye, J., Fenicia, F., Kuzmin, V., & Solomatine, D. P. (2013). Fuzzy committees of specialized rainfall-runoff models: Further enhancements and tests. *Hydrology and Earth System Sciences, 17*, 4441–4451.

Khoshnazar, A., Corzo Perez, G. A., & Diaz, V. (2021). Spatiotemporal drought risk assessment considering resilience and heterogeneous vulnerability factors: Lempa transboundary river basin in the central American dry corridor. *Journal of Marine Science and Engineering, 9*(4).

Khurana, D., Koli, A., Khatter, K., et al. (2023). Natural language processing: State of the art, current trends and challenges. *Multimedia Tools and Applications, 82*, 3713–3744.

Kratzert, F., Klotz, D., Shalev, G., Klambauer, G., Hochreiter, S., & Nearing, G. (2019). Benchmarking a catchment-aware long short-term memory network (LSTM) for large-scale hydrological modeling. *Hydrology and Earth System Sciences Discuss, 2019*, 1–32.

Mitchell, T. M. (2007). *Machine learning* (Vol. 1). New York: McGraw-Hill.

Moreido, V., Gartsman, B., Solomatine, D. P., & Suchilina, Z. (2021). How well can machine learning models perform without hydrologists? Application of rational feature selection to improve hydrological forecasting. *Water, 13*(12), 1696.

Nearing, G. S., Kratzert, F., Sampson, A. K., Pelissier, C. S., Klotz, D., Frame, J. M., et al. (2021). What role does hydrological science play in the age of machine learning?. *Water Resources Research, 57*, e2020WR02809.

Potgieter, T., & Dahlberg, J. (2022). *Automated machine learning on AWS: Fast-track the development of your production-ready machine learning applications the AWS way.* PACkt Publishing

Premack, D., & Woodruff, G. (1978). Does the chimpanzee have a theory of mind? *Behavioral and Brain Sciences*, *1*(4), 515–526.

Pyle, D. (1999). *Data preparation for data mining*. MorganKaufmann, USA.

Quinlan, J. R. (1992). Learning with continuous classes. *Proceedings Australian Joint Conference on Artificial Intelligence* (pp. 343–348). Singapore: World Scientific.

Razavi, S., Hannah, D. M., Elshorbagy, A., Kumar, S., Marshall, L., Solomatine, D. P., et al. (2022). Coevolution of machine learning and process-based modeling to revolutionize Earth and environmental sciences: A perspective. *Hydrological Processes, 36*, e14596. https://doi.org/10.1002/hyp.14596

Ruder, S. (2017). *An overview of multi-task learning in deep neural networks*. https://www.ruder.io/multi-task

Schultz, M. G., Betancourt, C., Gong, B., Kleinert, F., Langguth, M., Leufen, L. H., et al. (2021). Can deep learning beat numerical weather prediction? *Philosophical Transactions of the Royal Society A, 379* (2194).

Shrestha, D. L., & Solomatine, D. P. (2006). Experiments with AdaBoost.RT, an improved boosting scheme for regression. *Neural Computation, 17,* 1678–1710.

Shrestha, D. L., Kayastha, N., & Solomatine, D. P. (2009). A novel approach to parameter uncertainty analysis of hydrological models using neural networks. *Hydrology and Earth System Sciences, 13,* 1235–1248.

Solomatine, D. P., & Shrestha, D. L. (2009). A novel method to estimate model uncertainty using machine learning techniques. *Water Resources Research, 45*, W00B11.

Solomatine, D. P., & Xue, Y. (2004). M5 model trees compared to neural networks: Application to flood forecasting in the upper reach of the Huai River in China. *Journal of Hydrologic Engineering, 9*(6), 491–501.

Stanford Encyclopedia of Philosophy (2018). Artificial intelligence. https://plato.stanford.edu/entries/artificial-intelligence/

Sun, T. X., Liu, X. Y., Qiu, X. P., & Huang, X. J. (2022). Paradigm shift in natural language processing. *Machine Intelligence Research, 19*(3), 169–183.

Varouchakis, E. A., Hristopulos, D. T., Karatzas, G. P., Perez, G. A. C., & Diaz, V. (2021). Spatiotemporal geostatistical analysis of precipitation combining ground and satellite observations. *Hydrology Research, 52*(3).

Wang, Y., & Witten, I. H. (1996). Induction of model trees for predicting continuous classes. In Artificial intelligence, Wikipedia (2023, March 30). https://en.wikipedia.org/wiki/Artificial_intelligence

Wani, O., Beckers, J., Weerts, A. H., & Solomatine, D. P. (2017). Residual uncertainty estimation using instance-based learning with applications to hydrologic forecasting, *Hydrology and Earth System Sciences, 21*, 4021–4036.

Weizenbaum, J. (1966). ELIZA: A computer program for the study of natural language communication between man and machine. *Communications of the ACM, 9*(1), 36–45.

XGBoost (2023). https://github.com/dmlc/xgboost

Part I

Modeling Hydrological Systems

2

IMPROVING MODEL IDENTIFIABILITY BY DRIVING CALIBRATION WITH STOCHASTIC INPUTS

Andreas Efstratiadis, Ioannis Tsoukalas, and Panagiotis Kossieris

Department of Water Resources and Environmental Engineering, National Technical University of Athens, Zografou, Greece

For a long time, the classical problem of identifying the optimal modeling structure and/or parameters followed the calibration-validation norm, originating from the iconic split-sample scheme by Vit Klemeš. A common feature of such approaches is their dependence on the length and representativeness of the available data. This introduces several questions since the inferred parameters are selected according to a specific subset (or subsets) of historical data, while the rest of data is used for validation. In this vein, we propose a conceptually simple approach driven by the well-known stochastic simulation paradigm, which builds upon the idea of calibrating models using alternative, yet probabilistically consistent, synthetic data. Decoupling this way, the available data now become the basis to generate stochastic inputs, as well as for model validation and parameter uncertainty assessment. This allows for embedding the stochasticity of real-world drivers (rainfall, evapotranspiration) and responses (runoff) and thus their hydrological uncertainty. Furthermore, it results

Advanced Hydroinformatics: Machine Learning and Optimization for Water Resources, Special Publications 78, First Edition. Edited by Gerald A. Corzo Perez and Dimitri P. Solomantine.

DOI: 10.1002/9781119639268.ch2

to stable and robust models, as calibration is performed using long enough time series that reproduce important properties that are associated with the changing climate (e.g., long-term persistence), which are generally hidden in the short historical samples. Identifying this way, the derived parameters are optimal not only for the historical data set, but for any alternative plausible realization of the modeled processes.

2.1. Introduction

Hydrological models are essential elements for supporting water engineering and management decisions. These typically adopt the concept of stationarity, which allows for handling the process descriptions and model parameters as time-invariant. In this respect, while model-based decisions refer to the intrinsically uncertain future, the identification of the underlying modeling scheme and the evaluation of its predictive capacity as well, are founded on past information about the catchment's response against a known set of meteorological inputs.

In a generic context, the identification of a particular hydrological model is subject to a sequence of user decisions, where the interplay of parameter values, process parameterization equations, and the overall system architecture should ensure the minimal uncertainty against the available information (Gharari et al., 2021). However, the typical problem posed in real-world applications is restricted to the parameter estimation of a given model structure. According to the common hydrological practice, this task is formalized by means of the well-known split-sample test (Klemeš, 1986) and its variants (e.g., Gowda et al., 2012; Arsenault et al., 2018; Nicolle et al., 2021). This requires separating the observed data in a calibration subset for parameter adjusting, and a second one for validation. Any important drop in model performance in validation with respect to calibration may raise flags about the model robustness, which is often due to overfitting against nonrepresentative calibration data. Therefore, key decisions are the separation strategy and the length of calibration, which should embed diverse hydroclimatic conditions, also including unusual events (Singh & Bárdossy, 2012), to ensure that all model features are well activated. The major role of data quantity and quality as well has been early recognized (Gupta & Sorooshian, 1985), next being subject of numerous studies (e.g., Seibert, 2003; Perrin et al., 2007; Juston et al., 2009;

Andréassian et al., 2009, 2012; Bennett et al., 2013; Jaafar & Bargaoui, 2020; Myers et al., 2021). In this vein, several researchers revealed the need for taking advantage of multisource information within calibration, by means of both hard (i.e., measurements) and soft data as well as expert judgment, in order to improve model realism and parameter plausibility (e.g., Seibert & McDonnell, 2002; Efstratiadis & Koutsoyiannis, 2010; Nalbantis et al., 2011; Finger et al., 2015; Mohammed et al., 2021).

An evident drawback of the split-sample rationale is that the inferred parameter set is only conditioned on a subset of the available data, thus losing important information that may be embedded in the validation period. This has triggered researchers to provide hybrid calibration schemes across sliding windows (Coron et al., 2012, 2014; Vormoor et al., 2018) or subperiods (Gharari et al., 2013; Thirel et al., 2015a), in order to detect parameter sets that exhibit consistent performance in time. Following a rather heretic approach, Singh and Bárdossy (2012) and Arsenault et al. (2018) have manifested to entirely disregard the validation procedure under certain conditions, and use instead the full data set for parameter identifiability. To our knowledge, this approach is quite common among practitioners, who usually employ the split-sample test in order to obtain a rough picture of the model robustness and then recalibrate the model on the entire data set. Perrin et al. (2008) have also demonstrated that in poorly gauged basins, where the available flow records are very short, it is preferable to apply already optimized parameters from areas exhibiting satisfactory hydrological similarity instead of employing calibration with too limited information. On the other hand, Razavi and Tolson (2013) highlighted the issue of computational cost when calibrating time-demanding models against long data, thus proposing surrogate-based approaches to approximate the model performance on the full-length data period, based on much shorter ones.

Nevertheless, the question of how the model will perform if applied to conditions outside those used for model calibration, remains an open issue, which is also induced by the changing climate premise (Blöschl & Montanari, 2010; Kling et al., 2014; Broderick et al., 2016; Fowler et al., 2016, 2018; Dakhlaoui et al., 2017, 2019; Stephens et al., 2020). For this purpose, Klemeš (1986) introduced the so-called Differential Split Sample Testing (DSST), in which the determination of calibration and validation subperiods is appropriately made to span over as much as dissimilar hydroclimatic data, which in turn allows for better evaluating model transferability across time. Yet, common experience with DSST studies indicated strong dependence of model parameters on the conditions dominating the calibration period, while it is rather questionable whether short historical data can provide sufficient analogues to represent the plausible range of highly uncertain projections for climate (and environmental, in a more general context) changes (e.g., Thirel et al., 2015b, and citations therein). Several studies found little improvement of the observed poor model performance under contrasting climate conditions, after employing longer calibration periods (e.g., Brigode et al., 2013; Coron et al., 2014; Duethmann et al., 2020). In this respect, the assumption of time-invariant parameters is disputed, and some researchers directly link parameter change with hydroclimatic variability (e.g., Merz et al., 2013; Zhao et al., 2020).

The dilemma of which calibration approach is the more suitable may not make sense if long enough data were available, provided that the latter are really representative of the full regime of the modeled catchment. In order to address this issue, a well-known option is offered by stochastic models, which allow for synthesizing hydrological data to capture a wide range of plausible hydroclimatic conditions. Taking advantage of the long experience with stochastic simulation approaches in water resources planning and management studies, we reveal a novel field of application of stochastics for calibrating hydrological models. The overall idea is very simple and follows the well-known calibration-validation paradigm, where calibration is employed with synthetic data and validation against the actual (i.e., observed) ones, which are also used as guidance for recognizing the stochastic behavior of the processes of interest. In this chapter, we introduce the so-called stochastic calibration approach, which is founded on the coupling of two totally different modeling philosophies (i.e., mechanistic and stochastic), and reveal its significant advantages over conventional parameter estimation methods. Emphasis is paid to the specifications of the stochastic model, which is the most critical component of the overall procedure. The method is demonstrated by means of a proof of concept involving the fitting of conceptual model to a real-world river basin in Greece.

2.2. Methodology

Let $\underline{y} = \mathcal{M}\left(\underline{x}|\theta\right)$ be a transformation of the driving meteorological processes across a river basin, $\underline{x}$ (rainfall, temperature, etc.), to its response processes, $\underline{y}$ (in the typical case, runoff at its outlet), through a deterministic hydrological model of given structure, $\mathcal{M}$, and for a given set of parameter values, $\boldsymbol{\theta}$. For convenience, we use the underline notation to denote a random quantity (variable or process), while the nonunderlined typeface denotes a realization of it. The identification of $\boldsymbol{\theta}$ is typically carried out through calibration, by considering a set of realized inputs and outputs, $\boldsymbol{x}_i$ and $\boldsymbol{y}_i$, respectively, and a performance function, $\mathcal{P}\left[\boldsymbol{y}_i, \mathcal{M}\left(\boldsymbol{x}_i | \boldsymbol{\theta}_i\right)\right]$, to evaluate the closeness of simulated to actual outputs (e.g., efficiency). We remark that as result of calibration, although the transformation function, $\mathcal{M}$, is deterministic, its parameters also can be considered as random, since they are inferred from the random processes $\underline{\boldsymbol{x}}$ and $\underline{\boldsymbol{y}}$.

In the classical split-sample scheme and its alternatives, the realized data are obtained from observed records, $\boldsymbol{x}_o$ and $\boldsymbol{y}_o$, a subset of which is used to calibrate the model and the rest for validating its predictive capacity against an independent data set. As already discussed, well-known shortcomings of such approaches are the potential lack of representativeness of historical data and their inability to represent long-term hydroclimatic changes (often, but not solely, associated with their length), which eventually results in poorly identified parameters. This problem is further intensified since we have to keep a significant part (e.g., half) of observations for validation purposes, thus loosing valuable information that may be contained in the associated data subset. Furthermore, when the observed records have different sizes, only their common part can be accounted for in calibration and validation, while the nonoverlapping period of data is ignored. This case occurs quite often, since rainfall data, which is a rather easily retrievable information, usually exceeds the period of runoff observations.

Here we manifest an alternative approach for parameter estimation, which follows the stochastic simulation paradigm (Efstratiadis, 2011), as depicted in Figure 2.1. Similar to the classical slit-sample scheme, this also involves a twofold strategy, yet by using long synthetic data within calibration (we apply symbols $\boldsymbol{x}_s$ and $\boldsymbol{y}_s$ for synthetic inputs and outputs, respectively), while the validation is employed against the full set of observed data, $\boldsymbol{x}_o$ and $\boldsymbol{y}_o$. Apart from the deterministic hydrological procedure, $\mathcal{M}$, and the performance function, $\mathcal{P}$, the computational procedure also requires a synthetic time series generator, $\mathcal{G}$. Ideally, this has to reproduce the true hydrological regime of the input and output processes, $\underline{\boldsymbol{x}}$ and $\underline{\boldsymbol{y}}$, respectively. This option is offered by stochastic models, which are generally built upon the information contained in the historical data. In this respect, the synthesis procedure can be expressed as a multivariate stochastic transformation of the form:

$$[\boldsymbol{x}_s, \boldsymbol{y}_s] = \mathcal{G}[\boldsymbol{x}_o, \boldsymbol{y}_o]. \tag{2.1}$$

The stochastic similarity of the synthetic data $[\boldsymbol{x}_s, \boldsymbol{y}_s]$ and the observed ones $[\boldsymbol{x}_o, \boldsymbol{y}_o]$ allows for taking advantage of substantially longer records for

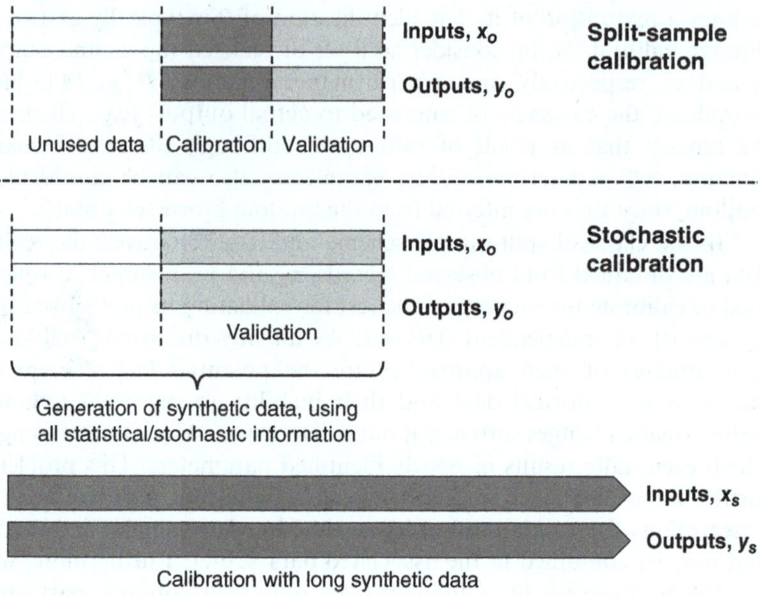

Figure 2.1 The classical calibration paradigm, based on the split-sample approach, contrasted with the stochastic calibration framework.

estimating the model parameters via calibration, thus paving the way to obtain more robust and stable parameters. The use of synthetic instead of real data is acceptable since, from a practical viewpoint, calibration is only an auxiliary stage of the overall modeling procedure. Furthermore, in order to assess the predictive capacity of the model, all available historical information is accounted for, thus the validation data set is also extended. Under this premise, the decision of which part of data should be used for calibration and which for validation makes no sense any more.

The use of stochastically generated data instead of historical samples that contain relatively few combinations of system inputs (rainfall), outputs (runoff), and states (storage conditions), and obviously miss long-term hydroclimatic fluctuations and rare events, also allows for addressing the major issue of representativeness of hydrological information from which the model parameters are inferred. It is recognized that the observed data partially only reflect the actual statistical and stochastic behavior of the input and output processes, while further uncertainties arise due to measurement and processing errors. In this respect, a well-designed synthesis procedure based on a good stochastic model can improve the information gained from the data available so far, toward at least three key directions: (1) providing wider dependency patterns among the inputs, $\underline{x}$, and the outputs, $\underline{y}$; (2) representing hydrometeorological extremes that are rarely appearing in short historical samples (Singh & Bárdossy, 2012); and (3) describing long-term hydroclimatic fluctuations, such as persistent wet/dry periods, which also may be induced by the changing climate.

Section 2.3 discusses in brief the specifications for the synthetic data generation approach within our stochastic calibration framework.

2.3. Insights to the Stochastic Simulation Procedure

The idea of substituting short historical records with synthetic ones is very popular in a wide range of hydrological design, planning, management, and forecasting applications, involving the use of deterministic models that are driven by synthetic inputs, provided by stochastic models (e.g., Koutsoyiannis & Economou, 2003). The coupling of deterministic and stochastic models allows for assessing probabilistic quantities, such as reliability, resilience, and risk, and expected values of long-term performance metrics, such as yield, costs, benefits, and deficits, on the basis of simulations with sufficiently large samples (e.g., with length of hundreds or thousands of years). On the other hand, less often is synthetic data used for evaluating the behavior of hydrological models in the long run (e.g., Nalbantis et al., 2011; Efstratiadis et al., 2015).

Cornerstone of all aforementioned approaches, belonging to the broader family of Monte Carlo simulation, is the time series generation model, which has to ensure a concise representation of the probabilistic and stochastic structure of the actual processes, $\underline{x}$ and $\underline{y}$. For a long time, this prerequisite has been determined as the preservation of a finite set of statistical characteristics of historical data within the synthetic ones by means of low-order moments (typically the mean, variance, skewness) and low-order auto- and cross-correlations (Matalas & Wallis, 1976). However, in the context of stochastic calibration, this approach is not sufficient, since these metrics describe only partially the marginal and dependence regime of the actual processes $\underline{x}$ and $\underline{y}$, while under some quite common conditions, unrealistic dependence patterns may appear in the synthetic data (Tsoukalas et al., 2018b).

In this respect, we impose the following specifications for the data synthesis procedure, to be implemented within a consistent and robust stochastic calibration framework.

1. Representation of the full probabilistic behavior of the input and output processes, through careful assignment of suitable distribution

models across processes and scales instead of a blind reproduction of the observed statistical characteristics per se (Tsoukalas et al., 2019, 2020). The distribution models should capture all well-known peculiarities of hydrometeorological processes, such as nonnormality, periodicity, intermittency, and time irreversibility (Efstratiadis et al., 2014; Tsoukalas et al., 2018a, 2019, 2020; Kossieris et al., 2019; Koutsoyiannis, 2020), to be eventually reflected in the calibrated parameters. If the hydrological model aims at simulating extreme events (e.g., floods), specific attention should also be paid to the tail properties of the distribution models, which govern both the magnitude and frequency of extreme events.

2. Representation of auto- and cross-dependencies, which are statistical descriptors of the major drivers and cause-effect phenomena across the hydrological cycle. For instance, the short-term dependence structure of runoff, traditionally referred to as *memory*, is associated with storage and recession mechanisms, which are explicitly represented in deterministic hydrological models. On the other hand, the auto-dependence structure of all involved processes at large time scales, also named long-term persistence or Hurst-Kolmogorov dynamics, is associated with the perpetually changing hydroclimate and the appearance of trends and irregularly fluctuating clusters of wet and dry periods across all scales (Koutsoyiannis, 2013). Finally, the cross-dependencies between the input and output processes determine the level of complexity of hydrological transformations; the larger these dependencies are, the more explainable (e.g., linear) the transformations.
3. Multiscale consistency, through effective coupling of stochastic models across multiple time scales, ensures the reproduction of the marginal and dependence behavior of the modeled processes not only at the temporal resolution of the hydrological simulation (e.g., daily, monthly), but also across coarser levels of aggregation (e.g., annual).

We emphasize that the key of success of stochastic calibration is the representativeness and realism of synthetic output data, y_s, which is introduced in the performance function, p, as proxy of the actual response of the hydrological system. It is essential to underline that the reproduction of a desirable probabilistic and stochastic structure within both input and output synthetic time series imposes the generation of long enough data (ideally, of infinite length), which eventually ensures that the hydrological model is trained against a much wider range of potential drivers, responses, and states with respect to historical records.

2.4. Proof of Concept

2.4.1. Data and Model Description

In order to demonstrate our framework, we apply a lumped conceptual model to represent the monthly runoff produced upstream of Evinos dam, western Greece, which is a key component of the water supply system of Athens (Koutsoyiannis et al., 2003). The river basin area is 351.9 km^2, and the available hydrological data (precipitation, potential evapotranspiration, runoff) extends over 50 hydrological years (from October 1970 to September 2020). Table 2.1 summarizes the key statistical information of historical rainfall and runoff data (mean, standard deviation, skewness, auto-correlation, and cross-correlation), to be reproduced within the synthetic time series that are used in stochastic calibration. The latter have been generated via the anySim package (Tsoukalas et al., 2020), and they extend over a time horizon of 2,000 yr. The synthesis scheme ensured the generation of dependency patterns (Fig. 2.2) and long-term fluctuations (Fig. 2.3) that are much wider than the ones found in the observed data. These are expected to represent the full hydroclimatic regime of the basin, which may be hidden, as a result of the limited information embedded in the observed data. In fact, only half of this information is accounted for within the split-sample calibration approach.

Figure 2.4 depicts an outline of the model and its governing equations. The model employs a typical bucket-type scheme comprising three interconnected tanks, which gets as inputs the monthly precipitation, P, and potential evapotranspiration, PET. The upper tank has infinite capacity and represents the major hydrological processes over the terrain, namely the direct (overland) runoff, Q_D, which is considered as a portion κ of precipitation, and the direct evapotranspiration, ET_D, which is fulfilled by the water excess that is temporarily stored in the ground and the vegetation. The remaining water infiltrates to an intermediate tank of storage capacity S_{max}, which employs the soil moisture accounting across the unsaturated zone. This implements three fluxes, namely the evapotranspiration losses through the soil, ET_S, which are controlled by the storage ratio, S/S_{max}, and the evapotranspiration deficit, $PET - ET_D$, the percolation to the groundwater tank, W, which is controlled by a recession parameter, λ, and the saturation runoff, Q_S, which is represented as the excess water that overflows from the tank. Finally, the lower tank has infinite capacity and employs the regulation across the groundwater zone. In particular, it receives the percolation and generates baseflow, Q_B, through another recession parameter, μ. In this respect, the runoff arriving at the basin outlet is the sum of three components, namely, $Q = Q_D + Q_S + Q_B$.

Table 2.1 Key summary statistical characteristics of observed rainfall and runoff data and their lag-0 cross-correlation coefficient at Evinos site, Greece, at monthly and annual time scales.

Rain (mm)	Oct	Nov	Dec	Jan	Feb	Mar	Apr	May	Jun	Jul	Aug	Sep	Annual
Mean	98.5	195.5	190.5	155.9	153.0	114.8	88.4	57.7	37.2	24.6	26.7	58.1	1196.6
St. deviation	66.0	98.8	116.2	108.3	94.4	68.6	56.7	35.4	24.8	28.7	23.9	50.7	270.5
Skewness	0.683	0.211	0.813	0.953	0.749	0.731	1.361	0.483	0.530	2.244	1.130	1.444	0.003
Lag-1 correl.	0.228	0.011	0.158	0.189	0.156	0.432	0.021	0.177	0.321	0.129	0.414	0.203	0.083
Runoff (mm)													
Mean	23.6	72.0	133.1	125.6	130.8	124.8	91.1	50.5	21.6	13.1	9.4	10.8	803.0
St. deviation	18.2	49.8	93.3	79.6	69.2	66.4	34.0	20.5	7.5	4.4	4.7	8.8	224.0
Skewness	1.296	1.081	1.415	0.838	0.281	1.050	0.439	0.324	0.857	0.842	2.738	3.948	−0.249
Lag-1 correl.	0.237	0.170	0.275	0.026	0.575	0.504	0.268	0.490	0.527	0.527	0.479	0.053	0.192
Lag-0 correl.	0.650	0.740	0.870	0.818	0.684	0.729	0.486	0.370	0.272	0.267	0.270	0.670	0.708

Note: Statistical characteristics = mean, standard deviation, skewness coefficient, lag-1 autocorrelation; rainfall and runoff data are in mm.

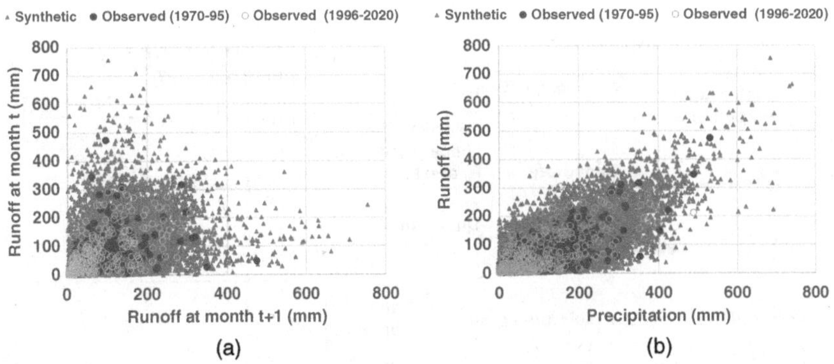

Figure 2.2 Dependency patterns among runoff data (a) between subsequent months and (b) between rainfall and runoff, derived from the observed data (split into two periods) and the synthetic ones (24,000 values).

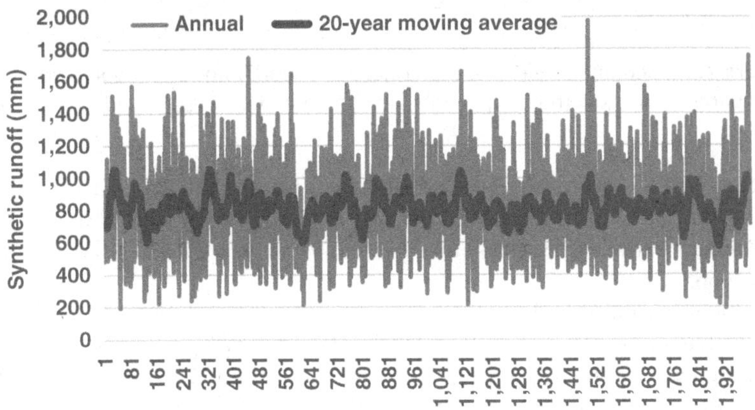

Figure 2.3 Synthetic runoff data of 2,000 yr length at Evinos basin, aggregated at annual and 20 yr moving average scales.

2.4.2. Results

Initially we apply the classical split-sample scheme by considering the first 25 yr for calibration and the next ones for validation, and by using the well-known Nash-Sutcliffe efficiency (NSE) metric to evaluate the model performance. The estimation of model parameters is handled as a global optimization problem, which is carried out via the Surrogate-Enhanced Evolutionary Annealing Simplex (SEEAS) algorithm (Tsoukalas et al.,

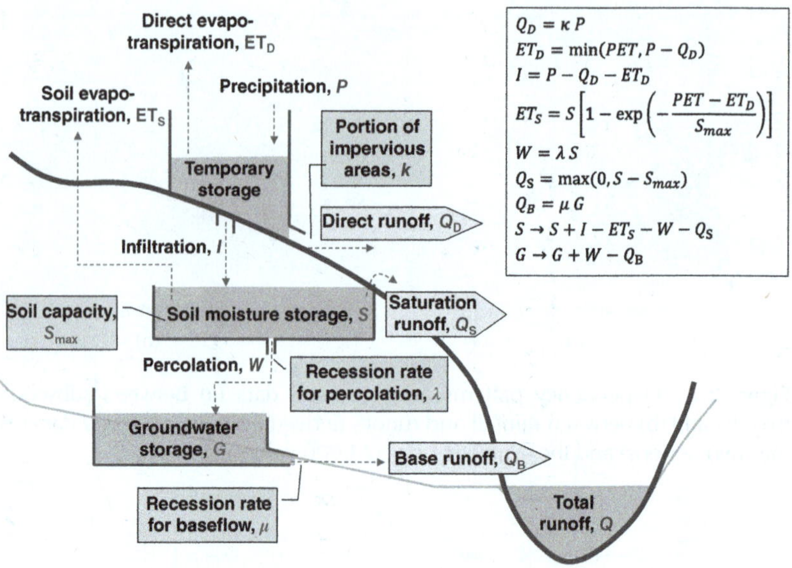

Figure 2.4 Schematic view of the conceptual model and the associated fluxes and parameters; governing equations are shown in the top right panel.

2016). NSE is quite large in calibration (0.834), but drops to 0.701 in validation, while the overall efficiency for the 50 yr of data is 0.765. The optimized parameter values are $S_{max} = 150\,mm$, $\kappa = 0.152$, $\lambda = 0.399$, and $\mu = 0.565$.

Next, we employ the stochastic calibration procedure to fit the model on the basis of synthetic series of 2,000 yr length by also employing SEEAS, and then we use the full sample of observations for validation. The generation of synthetic data is conducted for the rainfall and runoff, by using the recently introduced anySim package (Tsoukalas et al., 2020), which is in accordance with the specifications discussed in section 2.3. Regarding the PET inputs, we simply apply the mean monthly values of the historical sample. We remark that the use of a surrogate-based approach within the search procedure, which is subject to the computationally expensive hydrological model as a result of the long synthetic data, strongly facilitates the derivation of good solutions in reasonable time. The optimized parameter values against the synthetic rainfall-runoff data are quite different with respect to the ones derived by calibrating against the half of historical data, namely $S_{max} = 112\,mm$, $\kappa = 0.220$, $\lambda = 0.533$, and $\mu = 0.511$. Under this premise, the model exhibits slightly better performance over the full historical data, since the NSE value is 0.775.

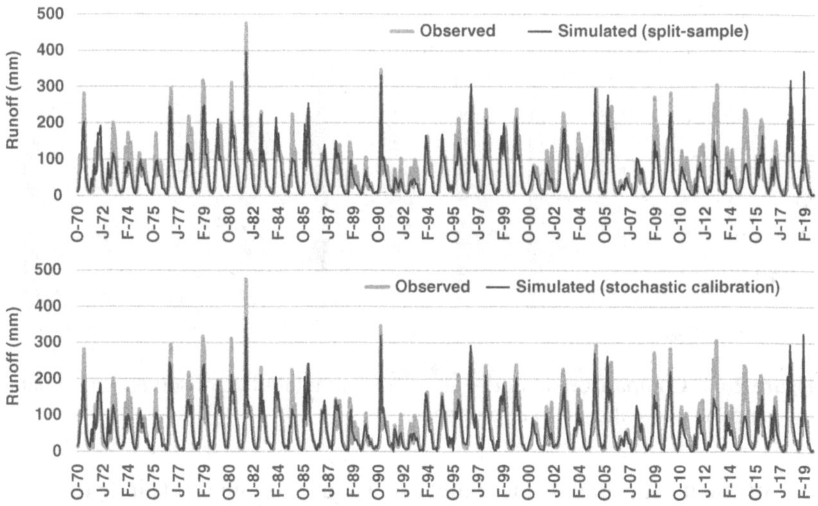

Figure 2.5 Comparison of model fitting (i.e., simulated vs. observed runoff data) by using the two calibration approaches.

Although the differences by the two calibration procedures are not fully detectable, when just contrasting simulated data with observations (Fig. 2.5), the stochastic approach actually ensures more stable model performance against different hydroclimatic periods. As shown in Figure 2.6, by splitting the observed data in 10 smaller samples of 5 yr length, the variability of NSE across subperiods is definitely larger with the parameters derived from the classical method, which indicates that the parameters that are inferred from the stochastic data are more robust. This outcome is reasonable, since the model is fitted against a wide range of statistically consistent hydroclimatic conditions, which capture all potential dependency patterns among the model inputs and outputs (i.e., rainfall and runoff), and they also include persistent clusters of wet and dry years, as a result of the Hurst-Kolmogorov dynamics.

2.4.3. Parameter Uncertainty Analysis

To further highlight the advantages of the stochastic calibration framework, we explore the parameter uncertainty through two alternative approaches that utilize the historical and synthetic time series, respectively. First, we employ the Generalized Likelihood Uncertainty Estimation (GLUE) method, initiated by Beven and Binley (1992), and still being

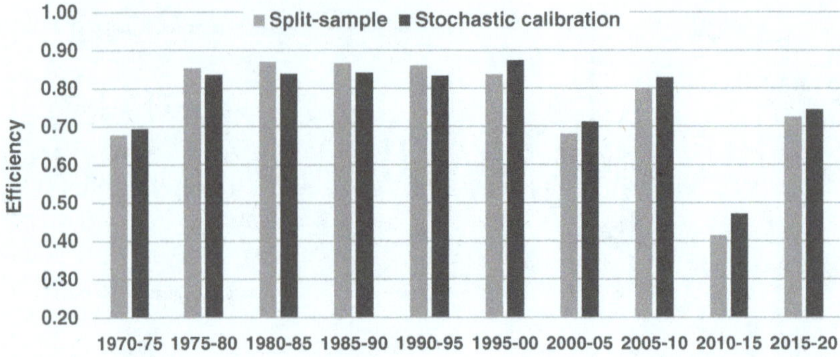

Figure 2.6 Comparison of model fitting (i.e., simulated vs. observed runoff data) by using the two calibration approaches.

one of the most popular uncertainty quantification methods in hydrology. In our case, we applied the GLUE method by sampling from 10,000 uniformly distributed parameter sets and then setting a NSE threshold equal to 0.75 in order to recognize behavioral from nonbehavioral solutions. Finally, we used the behavioral subset to extract its (a posteriori) empirical distributions, depicted by means of histograms (Fig. 2.7a–d).

The second analysis accounts for the synthetic data set, which is divided into 80 subsets, each of 25 yr length (i.e., equal to the data size

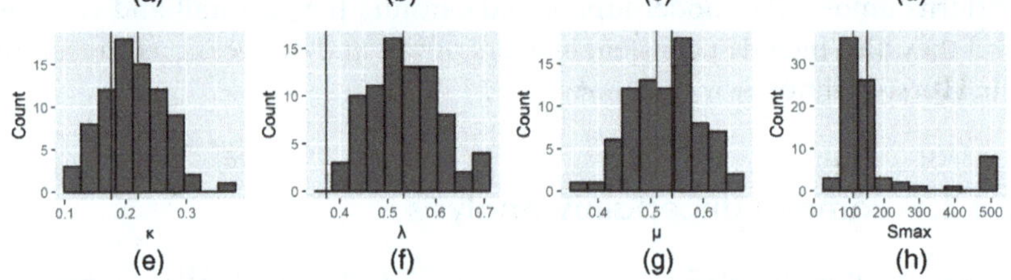

Figure 2.7 Empirical histograms of model parameters obtained with the GLUE method (a–d) and by calibrating the model across 80 subsets of the synthetic data (e–h). The dark and light lines indicate the optimized parameter values with the conventional and stochastic calibration approaches, respectively.

of conventional calibration). Next, we calibrate the model against each subset, thus obtaining a sample of 80 optimized parameter sets across the time-window of synthetic data. As made with the GLUE method, this sample is used to obtain the empirical distributions of the four model parameters (Fig. 2.7e–h).

While the distribution plots by the two uncertainty analysis methods are quite different, both of them clearly capture the parameter values derived via the stochastic calibration approach (light vertical lines). In contrast, the conventionally optimal (dark vertical lines) value of parameter λ (i.e., recession rate for percolation) lies far from the core of the uncertainty bounds, thus making questionable whether the entire calibration is robust or not. Another interesting remark is that the uncertainty bounds resulting from the second approach (i.e., the one that employs synthetic data and the notion of subperiod calibration) are narrower and sharper than those obtained by the GLUE method, thus indicating a more precise estimation of the model's parameter uncertainty. In our view, this behavior, which arguably is a promising topic of future research, may be attributed to the combined effect of using an optimization algorithm (instead of random sampling) along with the notion of subperiod calibration driven by synthetic data.

2.5. Concluding Remarks

Hydrological models that are calibrated against short records may be unable to reproduce the anticipated catchment behavior in the long run, while additional drawbacks arise due to the essential sacrifice of historical data in order to keep space for validation. In an attempt to address this issue, we highlight a promising strategy, aiming at inferring the model's parameters on the basis of synthetic rainfall-runoff data, and employ validation against the full historical sample of observations. The so-called stochastic calibration scheme is driven by a synthetic time series generation procedure, implemented within the R-package anySim, which takes advantage of the available data to represent the probabilistic properties and dependence structure of the rainfall-runoff processes across multiple scales, and the long-term fluctuations of the changing climate as well. Crucial requirements of the stochastic model are the faithful reproduction of the cross-dependencies across seasons and scales, which determines the causality between rainfall and runoff, and the fitting of suitable distribution functions to the processes of interest in order to reproduce extreme conditions that are impossible to be contained in historical data. Under this viewpoint, the proposed method actually allows for extracting more information from short time periods.

The advantages of stochastic calibration over conventional split-sample approaches are demonstrated by means of a proof of concept, involving the calibration of a lumped conceptual model with four parameters. It is interesting to remark that while the differences in terms of efficiency are actually not so much notable among the two approaches, the optimized parameter values are. Therefore, it is expected that the two model versions will respond quite differently under combinations of inputs and states that are outside of the ones found in the historical data. Given that the stochastically calibrated model is trained over many extended hydroclimatic conditions, it is expected to ensure stronger identifiability. This is also indicated by the fact that this version exhibits more stable predictive capacity across varying time windows. On the other hand, the obvious increase of computational effort, due to the substantially larger length of calibration data, can be addressed through recent advances in global optimization, where machine learning and surrogate modeling techniques are embedded within robust evolutionary algorithms (e.g., Tsoukalas et al., 2016).

Preliminary analyses of the proposed method against a large set of catchments, and by employing models of varying level of complexity, both at the monthly and daily scales, have already provided encouraging results regarding the generality of the stochastic calibration paradigm (Kourakos et al., 2021). Future efforts will focus to the uncertainty aspect, aiming to provide a holistic framework for model calibration and uncertainty quantification based on stochastically generated data.

Acknowledgments

The authors thank the editor, Dimitri Solomatine, for the kind invitation and his encouraging comments, and the two referees, Shervan Gharari and an anonymous referee, for their fruitful remarks and suggestions, which helped improve our chapter.

References

Andréassian, V., Le Moine, N., Perrin, C., Ramos, M. H., Oudin, L., Mathevet, T., et al. (2012). All that glitters is not gold: The case of calibrating hydrological models. *Hydrological Processes*, *26*, 2206–2210. https://doi.org/10.1002/hyp.9264.

Andréassian, V., Perrin, C., Berthet, L., Le Moine, N., Lerat, J., Loumagne, C., et al. (2009). Crash tests for a standardized evaluation of hydrological models.

Hydrology and Earth System Sciences, 13, 1757–1764. https://doi.org/10.5194/hess-13-1757-2009

Arsenault, R., Brissette, F., & Martel, J.-L. (2018). The hazards of split-sample validation in hydrological model calibration. *Journal of Hydrology, 566*, 346–362. https://doi.org/10.1016/j.jhydrol.2018.09.027

Bennett, N. D., Croke, B. F. W., Guariso, G., Guillaume, J. H. A., Hamilton, S. H., Jakeman, A. J., et al. (2013). Characterizing performance of environmental models. *Environmental Modeling & Software, 40*, 1–20. https://doi.org/10.1016/j.envsoft.2012.09.011.

Beven, K. J., & Binley, A. M., (1992). The future of distributed models: model calibration and uncertainty prediction. *Hydrological Processes, 6*, 279–298. https://doi.org/10.1002/hyp.3360060305

Blöschl, G., & Montanari, A. (2010). Climate change impacts: Throwing the dice? *Hydrological Processes, 24*, 374–381. https://doi.org/10.1002/hyp.7574

Brigode, P., Oudin, L., & Perrin, C. (2013). Hydrological model parameter instability: A source of additional uncertainty in estimating the hydrological impacts of climate change? *Journal of Hydrology, 476*, 410–425. https://doi.org/10.1016/j.jhydrol.2012.11.012

Broderick, C., Matthews, T., Wilby, R. L., Bastola, S., & Murphy, C. (2016.). Transferability of hydrological models and ensemble averaging methods between contrasting climatic periods. *Water Resources Research, 52*, 8343–8373. https://doi.org/10.1002/2016Wc028850

Coron, L., Andréassian, V., Perrin, C., Bourqui, M., & Hendrickx, F. (2014). On the lack of robustness of hydrologic models regarding water balance simulation: A diagnostic approach applied to three models of increasing complexity on 20 mountainous catchments. *Hydrology and Earth System Sciences, 18*(2), 727–746. https://doi.org/10.5194/hess-18-727-2014

Coron, L., Andréassian, V., Perrin, C., Lerat, J., Vaze, J., Bourqui, M., et al. (2012). Crash testing hydrological models in contrasted climate conditions: An experiment on 216 Australian catchments. *Water Resources Research, 48*, W05552. https://doi.org/10.1029/2011Wc021721

Dakhlaoui, H., Ruelland, D., & Tramblay, Y. (2019). A bootstrap-based differential split-sample test to assess the transferability of conceptual rainfall-runoff models under past and future climate variability. *Journal of Hydrology, 575*, 470–486. https://doi.org/10.1016/j.jhydrol.2019.05.056

Dakhlaoui, H., Ruelland, D., Tramblay, Y., & Bargaoui, Z. (2017). Evaluating the robustness of conceptual rainfall-runoff model under climate variability in northern Tunisia. *Journal of Hydrology, 550*, 201–217. https://doi.org/10.1016/j.jhydrol.2017.04.032.

Duethmann, D., Blöschlv, G., & Parajka, J. (2020). Why does a conceptual hydrological model fail to correctly predict discharge changes in response to climate change? *Hydrology and Earth System Sciences, 24*, 3493–3511. https://doi.org/10.5194/hess-24-3493-2020

Efstratiadis, A. (2011). New insights on model evaluation inspired by the stochastic simulation paradigm. EGU General Assembly 2011. *Geophysical Research*

Abstracts, Vol. 13, Vienna, 1852, European Geosciences Union. http://www.itia.ntua.gr/1116/

Efstratiadis, A., & Koutsoyiannis, D. (2010). One decade of multiobjective calibration approaches in hydrological modeling: A review. *Hydrological Sciences Journal*, *55*(1), 58–78. https://doi.org/10.1080/02626660903526292

Efstratiadis, A., Dialynas, Y., Kozanis, S., & Koutsoyiannis, D. (2014). A multivariate stochastic model for the generation of synthetic time series at multiple time scales reproducing long-term persistence. *Environmental Modeling & Software, 62,* 139–152. https://doi.org/10.1016/j.envsoft.2014.08.017

Efstratiadis, A., Nalbantis, I., & Koutsoyiannis, D. (2015). Hydrological modeling of temporally-varying catchments: Facets of change and the value of information. *Hydrological Sciences Journal*, *60*(7–8), 1438–1461. https://doi.org/10.1080/02626667.2014.982123

Finger, D., Vis, M., Huss, M., & Seibert, J. (2015). The value of multiple data set calibration versus model complexity for improving the performance of hydrological models in mountain catchments. *Water Resources Research*, *51*, 1939–1958. https://doi.org/10.1002/2014wc025712

Fowler, K., Coxon, G., Freer, J., Peel, M., Wagener, T., Western, A., et al. (2018). Simulating runoff under changing climatic conditions: A framework for model improvement. *Water Resources Research*, *54*, 9812–9832. https://doi.org/10.1029/2018WR023989

Fowler, K. J. A., Peel, M. C., Western, A. W., Zhang, L., & Peterson, T. J. (2016). Simulating runoff under changing climatic conditions: Revisiting an apparent deficiency of conceptual rainfall-runoff models. *Water Resources Research*, *52*, 1820–1846. https://doi.org/10.1002/2015Wc028068

Gharari, S., Gupta, H. V., Clark, M. P., Hrachowitz, M., Fenicia, F., Matgen, P., et al. (2021). Understanding the information content in the hierarchy of model development decisions: Learning from data. *Water Resources Research*, *57*(6), e2020WR027948. https://doi.org/10.1029/2020WR02794

Gharari, S., Hrachowitz, M., Fenicia, F., & Savenije, H. H. G. (2013.). An approach to identify time consistent model parameters: sub-period calibration. *Hydrology and Earth System Sciences*, *17*, 149–161. https://doi.org/10.5194/hess-17-149-2013

Gowda, P. H., Mulla, D. J., Desmond, E. D., Ward, A. D., & Moriasi, D. N. (2012). ADAPT: Model use, calibration, and validation. *Transactions of the ASABE*, *55*, 1345. https://doi.org/10.13031/2013.42246

Gupta, V. K., & Sorooshian, S. (1985). The relationship between data and the precision of parameter estimates of hydrological models. *Journal of Hydrology*, *81*(1–2), 57–77. https://doi.org/10.1016/0022-1694(85)90167-2

Jaafar, A. B., & Bargaoui, Z. (2020). Generalized split-sample test interpretation using rainfall-runoff information gain. *Journal of Hydrological Engineering*, *25*(1), 04019057. https://doi.org/10.1061/(ASCE)HE.1943-5584.0001868

Juston, J., Seibert, J., & Johansson, P. O. (2009). Temporal sampling strategies and uncertainty in calibrating a conceptual hydrological model for a small boreal catchment. *Hydrological Processes, 23*, 3093–3109. https://doi.org/10.1002/hyp.7421

Klemeš, V. (1986). Operational testing of hydrological simulation models. *Hydrological Sciences Journal*, *31*(1), 13–24. https://doi.org/10.1080/02626668609491024

Kling, H., Stanzel, P., & Preishuber, M. (2014). Impact modeling of water resources development and climate scenarios on Zambezi River discharge. *Journal of Hydrology: Regional Studies, 1,* 17–43. https://doi.org/10.1016/j.ejrh.2014.05.0022214-0025818

Kossieris, P., Tsoukalas, I., Makropoulos, C., & Savic, D. (2019). Simulating marginal and dependence behaviour of water demand processes at any fine time scale. *Water*, *11*(5), 885. https://doi.org/10.3390/w11050885

Kourakos, V., Efstratiadis, A., & Tsoukalas, I. (2021). Can hydrological model identifiability be improved? Stress-testing the concept of stochastic calibration. *EGU General Assembly 2021, online*, EGU21-11704. https://doi.org/10.5194/egusphere-egu21-11704, European Geosciences Union.

Koutsoyiannis, D. (2013). Hydrology and change. *Hydrological Sciences Journal*, *58*(6), 1177–1197. https://doi.org/10.1080/02626667.2013.804626

Koutsoyiannis, D. (2020). Simple stochastic simulation of time irreversible and reversible processes. *Hydrological Sciences Journal*, *65*(4), 536–551. https://doi.org/10.1080/02626667.2019.1705302

Koutsoyiannis, D., & Economou, A. (2003). Evaluation of the parameterization-simulation-optimization approach for the control of reservoir systems. *Water Resources Research, 39*(6), 1170. https://doi.org/10.1029/2003WR002148

Koutsoyiannis, D., Karavokiros, G., Efstratiadis, A., Mamassis, N., Koukouvinos, A., & Christofides, A. (2003). A decision support system for the management of the water resource system of Athens. *Physics and Chemistry of the Earth*, *28*(14–15), 599–609. https://doi.org/10.1016/S1474-7065(03)00106-2

Matalas, N. C., & Wallis, J. R. (1976). Generation of synthetic flow sequences. In A. K. Biswas (Ed.), *Systems approach to water management*. New York: McGraw-Hill.

Merz, R., Parajka, J., & Blöschl, G. (2013). Time stability of catchment model parameters: implications for climate impact analyses. *Water Resources Research*, *47*(2), W02531. https://doi.org/10.1029/2010WR009505

Mohammed, S. A., Solomatine, D. P., Hrachowitz, M., & Hamouda, M. A. (2021). Impact of dataset size on the signature-based calibration of a hydrological model. *Water*, *13*(7), 970. https://doi.org/10.3390/w13070970

Myers, D. T., Ficklin, D. L., Robeson, S. M., Neupane, R. P., Botero-Acosta, A., & Avellaneda, P. M. (2021). Choosing an arbitrary calibration period for hydrologic models: How much does it influence water balance simulations? *Hydrological Processes 35*(2), e14045. https://doi.org/10.1002/hyp.14045

Nalbantis, I., Efstratiadis, A., Rozos, E., Kopsiafti, M., & Koutsoyiannis, D. (2011). Holistic versus monomeric strategies for hydrological modeling of human-modified hydrosystems. *Hydrology and Earth System Sciences*, *15*, 743–758. https://doi.org/10.5194/hess-15-743-2011

Nicolle, P., Andréassian, V., Royer-Gaspard, P., Perrin, C.,Thirel, G., Coron, L., et al. (2021). Technical Note. RAT: a Robustness Assessment Test for calibrated and uncalibrated hydrological models. *Hydrology and Earth System Sciences*, *25*, 5013–5027. https://doi.org/10.5194/hess-25-5013-2021

Perrin, C., Andréassian, V., Mathevet, T., & Le Moine, N. (2008). Discrete parameterization of hydrological models: evaluating the use of parameter sets libraries over 900 catchments. *Water Resources Research*, *44*, W08447. https://doi.org/10.1029/2007WR006579

Perrin, C., Oudin, L., Andreassian, V., Rojas-Serna, C., Michel, C., & Mathevet, T. (2007). Impact of limited streamflow data on the efficiency and the parameters of rainfall-runoff models. *Hydrological Sciences Journal*, *52*, 131–151. https://doi.org/10.1623/hysj.52.1.131

Razavi, S., & Tolson, B. A. (2013). An efficient framework for hydrologic model calibration on long data periods. *Water Resources Research*, *49*, 8418–8431. https://doi.org/10.1002/2012Wc023442

Seibert, J. (2003). Reliability of model predictions outside calibration conditions. *Hydrology Research*, *34*(5), 477–492. https://doi.org/10.2166/nh.2003.0019

Seibert, J., & McDonnell, J. J. (2002). On the dialog between experimentalist and modeler in catchment hydrology: Use of soft data for multicriteria model calibration. *Water Resources Research*, *38*(11), 23-1-23-14. https://doi.org/10.1029/2001WR000978

Singh, S. L., & Bárdossy, A. (2012). Calibration of hydrological models on hydrologically unusual events. *Advances in Water Resources*, *38*, 81–91. https://doi.org/10.1016/j.advwatres.2011.12.006

Stephens, C. M., Marshall, L. A., Johnson, F. M., Lin, L., Band, L. E., & Ajami, H. (2020). Is past variability a suitable proxy for future change? A virtual catchment experiment. *Water Resources Research*, *56*, e2019WR026275. https://doi.org/1029/2019WR026275

Thirel, G., Andréassian, V., & Perrin, C. (2015a). On the need to test hydrological models under changing conditions. *Hydrological Sciences Journal*, *60*(78), 1165–1173. https://doi.org/10.1080/02626667.2015.10500272015

Thirel, G., Andréassian, V., Perrin, C., Audouy, J.-N., Berthet, L., Edwards, P., et al. (2015b). Hydrology under change: An evaluation protocol to investigate how hydrological models deal with changing catchments. *Hydrological Sciences Journal*, *60*(7–8), 1184–1199. https://doi.org/10.1080/02626667.2014.967248

Tsoukalas, I., Efstratiadis, A., & Makropoulos, C. (2018a). Stochastic periodic autoregressive to anything (SPARTA): Modeling and simulation of cyclostationary processes with arbitrary marginal distributions. Water Resources Research, 54 (1), 161–185, WRCR23047. https://doi.org/10.1002/2017WR021394

Tsoukalas, I., Efstratiadis, A., & Makropoulos, C. (2019). Building a puzzle to solve a riddle: A multi-scale disaggregation approach for multivariate stochastic processes with any marginal distribution and correlation structure. *Journal of Hydrology*, *575*, 354–380. https://doi.org/10.1016/j.jhydrol.2019.05.017

Tsoukalas, I., Kossieris, P., & Makropoulos, C. (2020). Simulation of non-Gaussian correlated random variables, stochastic processes and random fields: Introducing the anySim R-Package for environmental applications and beyond. *Water*, *12*(6), 1645. https://doi.org/10.3390/w12061645

Tsoukalas, I., Kossieris, P., Efstratiadis, A., & Makropoulos, C. (2016). Surrogate-enhanced evolutionary annealing simplex algorithm for effective and efficient optimization of water resources problems on a budget. *Environmental*

Modeling & Software, 77, 122–142. https://doi.org/10.1016/j.envsoft.2015.12.008

Tsoukalas, I., Papalexiou, S. M., Efstratiadis, A., & Makropoulos, C. (2018b). A cautionary note on the reproduction of dependencies through linear stochastic models with non-Gaussian white noise. *Water*, *10*(6), 771. https://doi.org/10.3390/w10060771

Vormoor, K., Heistermann, M., Bronstert, A., & Lawrence, D. (2018). Hydrological model parameter (in)stability: "Crash testing" the HBV model under contrasting flood seasonality conditions. *Hydrological Sciences Journal*, *63*(7), 991–1007. https://doi.org/10.1080/02626667.2018.1466056

Zhao, B., Mao, J., Dai, Q., Han, D., Daiand, H., & Rong, G. (2020). Exploration on hydrological model calibration by considering the hydro-meteorological variability. *Hydrology Research*, *51*(1), 30–46. https://doi.org/10.2166/nh.2019.047

3

A TWO-STAGE SURROGATE-BASED PARAMETER CALIBRATION FRAMEWORK FOR A COMPLEX DISTRIBUTED HYDROLOGICAL MODEL

Haiting Gu, Yue-Ping Xu, Li Liu, Di Ma, Suli Pan, and Jingkai Xie

Institute of Hydrology and Water Resources, Zhejiang University, Hangzhou, China

Distributed hydrologic models have been widely used for their functional diversity and rationality in theory. However, calibration of distributed models is computationally expensive since it requires a large number of model runs, even if an efficient multi-objective optimization algorithm is employed. To alleviate the burden of computation, we develop a two-stage surrogate-based model calibration framework by coupling standard backpropagation neural network with AdaBoost (ADBP) to efficiently calibrate the parameters of a large and complex distributed hydrologic model, that is, the Variable Infiltration Capacity (VIC) model. The first-stage model (SM-I) picks out the parameter sets whose simulated outputs are in the crucial range and the second-stage model (SM-II) estimates the values of outputs accurately with the parameter sets picked out by SM-I. The proposed surrogate model is tested in the Lanjiang River basin and the Xiangjiang River basin for parameter calibration. With the similar calibrated parameter sets for VIC, the surrogate model gains up to 20 times speedup

Advanced Hydroinformatics: Machine Learning and Optimization for Water Resources, Special Publications 78, First Edition. Edited by Gerald A. Corzo Perez and Dimitri P. Solomantine.

DOI: 10.1002/9781119639268.ch3

compared with the original VIC model in the basin with a large area and complex physical conditions.

3.1. Introduction

The distributed hydrologic model had its first application at the watershed scale by Freeze and Harlan (1969). The models that take full consideration into the effect of spatial nonuniformity of precipitation and land surface conditions on hydrologic processes can be preferable to lumped models theoretically, although in practical applications, the lumped models may perform better (Carpenter & Georgakakos, 2006; Liu et al., 2018). Nowadays, with the rapid development of computer technology, the geographic information system (GIS), digital elevation models (DEM), and remote sensing (RS), distributed hydrologic models with different configurations have been developed and used widely in practice (Abbaspour et al., 2015; Kovalets et al., 2015; Marek et al., 2016; Patil & Ramsankaran, 2017; Sun et al., 2016; Xu et al., 2015; Zhang et al., 2017; Zhao et al., 2009). They can even be combined with big data techniques (Guo et al., 2017; Zhang et al., 2016a,b), such as the MapReduce framework and Hadoop distribution file system.

The Variable Infiltration Capacity Macroscale Hydrologic model (VIC), which we used in this study, is a macroscale model that the areas above a critical scale are taken to be about 10 km × 10 km typically and the simulated runoff is shown to match observations quite well over large river basins (Maurer et al., 2002). Numerous studies have applied VIC successfully and obtained good performances in ensemble streamflow forecasting (Liu et al., 2017; Mo & Lettenmaier, 2014; Schumann et al., 2013). It is also used to study the hydrology under nonstationary conditions (Chawla & Mujumdar, 2018; Leng et al., 2015). The distributed models are widely used in water resources assessment, hydrologic forecast, and sediment transport modeling (Pérez et al., 2017; Xu et al., 2015). However, these kinds of models also have a limitation. It requires a lot of computational power and costs substantial time to solve partial derivative equations with numerical computation methods at each grid cell. Especially, tens of thousands of simulations are required when parameters of a distributed model (Razavi et al., 2012) are calibrated.

One way to accelerate parameter calibration is by employing an efficient auto-calibration method. Multi-objective calibration strategy for hydrological models can reduce the model equifinality and parameter uncertainty and acquire a valuable parameter set (Tuo et al., 2018).

In recent years, many multi-objective optimization algorithms (MOAs) for the hydrologic model calibration have been developed and applied in the hydrologic model parameter calibration (Gupta et al., 2009), such as the Multi-Objective Complex Evolution method (MOCOM-UA) (Yapo et al., 1998), the Strength Pareto Evolutionary Algorithm II (SPEA2) (Zitzler et al., 2001), Nondominated Sorting Genetic Algorithm II (NSGA-II) (Deb et al., 2002), and so on. Numerous studies (De Vos & Rientjes, 2007; Gharari et al., 2013; Guo et al., 2014; Huo et al., 2016; Kollat et al., 2012; Lee et al., 2013; Tang et al., 2005; Yassin et al., 2017) have shown that MOAs are efficient in the hydrologic model parameter calibration automatically.

Although an efficient optimization algorithm can reduce the number of simulations of model calibration for complex models, it may still need a large number. The surrogate model (SM) is also an important method to improve the efficiency of the optimization algorithms (Maier et al., 2014). The data-driven models, such as Artificial Neural Network (ANN) (Zhang et al., 2009), Radial Basis Function (RBF) (Christelis et al., 2016), Support Vector Machine (SVM) (Asefa et al., 2004; Zhang et al., 2009), and Gaussian Process Regression (GPR or Kriging) (Zhang et al., 2016a,b) are usually used to construct the response surface surrogates of physical models (Razavi et al., 2012). Several studies have been done through using surrogate models to accelerate the optimizations efficiently (Behzadian et al., 2009; Gong et al., 2016; Kourakos & Mantoglou, 2009; Yan & Minsker, 2006; Yan & Minsker, 2011; Yazdi et al., 2019). This kind of algorithm is called *meta model-enabled optimizers* or *SM-based optimization algorithms* in which the best method for building a SM with data generated by the physical model is first chosen (Santana-Quintero et al., 2010) and then a MOA scheme is developed to use SM instead of the physical model (Maier et al., 2014). However, most of the studies mentioned above are planned for groundwater or water allocation since they generally require finer resolution and lateral communication. In our study, we focus mainly on runoff calibration for a large and complex hydrologic model and investigate the performance of the SM-based optimization algorithm.

In our study, we aim to overcome the computational limitation in the calibration process of VIC. For this purpose, we propose a new modeling calibration framework. ANN is used to construct our surrogate model, which often shows better accuracy for spatial interpolation (Chowdhury et al., 2010; Sivapragasam et al., 2010). To further improve the accuracy of our surrogate model, two measures are adopted. The first one is using AdaBoost algorithm (Kearns & Valiant, 1994), which can improve the accuracy of ANN as well as make it work more stably. The second is building a two-stage model that tailors the surrogate model upon the parameter range near optimal solutions. In addition, the more efficient MOA, ε-NSGA II, is also applied to accelerate the parameter calibration. The two-stage surrogate model that we build here for VIC aims to accelerate the process of calibration to guarantee its accuracy. The method also provides an alternative idea to calibrate parameters and should be useful to alleviate the computational burden for other large and complex hydrologic models.

3.2. Method

3.2.1. Hydrologic Model and Its Calibration

VIC was originally developed by Liang et al. (1994, 1996). First, the user provides the model with grids of the study area, which includes meteorological data, digital elevation data, soil data, land cover data, and so on. Then the model simulates the hydrologic processes respectively on the basis of full water and energy balances in each grid cell. As a distributed hydrologic model in macroscale, the VIC model itself does not include the routing model. Therefore, the routing model based on a source-to-sink model (Lohmann et al., 1998; Lohmann et al., 1996) is utilized for runoff and base flow routing. For this application, the VIC model of version 4.2.d is used.

Multi-objective optimization algorithms involve more than one objective function to be optimized simultaneously. Multi-objective optimization algorithms produce a set of Pareto solutions, instead of a single optimal solution. The ε-NSGA II (Kollat & Reed, 2006) is used as a multi-objective optimization algorithm to calibrate parameters of VIC, which is updated from NSGA-II (Deb et al., 2002). Compared with NSGA-II, ε-NSGA II owns three new features, they are the ε-dominance archiving (Laumanns et al., 2002), the adaptive population sizing (Harik & Lobo, 1999), and the spontaneous termination (Reed et al., 2003). The ε-NSGA II is an effective and robust optimization algorithm (Fu et al., 2012; Hadka & Reed, 2012; Hurford et al., 2014; Kollat & Reed, 2007; Wang et al., 2014) and has been successfully applied in some previous studies (Liu et al., 2017; Liu et al., 2019; Pan et al., 2018). Here we operate the ε-NSGA II algorithm in parallel with a high-performance and widely portable implementation of the Message Passing Interface (MPI) standard named MPICH.

The Nash-Sutcliffe efficiency (*NSE*) (Nash & Sutcliffe, 1970) and relative bias (*Bias*) are used to mathematically quantify the performance between observed streamflow and streamflow simulated by VIC, whose detailed formulas are defined as

$$NSE = 1 - \frac{\sum_{i=1}^{n}(Q_o(i) - Q_s(i))^2}{\sum_{i=1}^{n}(Q_o(i) - \overline{Q}_0)^2} \tag{3.1}$$

and

$$Bias = \left| \frac{\sum_{i=1}^{n}(Q_s(i) - Q_o(i))}{\sum_{i=1}^{n} Q_o(i)} \right|, \tag{3.2}$$

where n is the length of the daily runoff series; $Q_o(i)$ and $Q_s(i)$ are the observed and simulated streamflow of the ith day respectively; and $\overline{Q_o}$ is the mean value of the observed streamflow. The value of *NSE* closer to 1 means a less difference between the observed and simulated flows, while that of *Bias* prefers to be 0.

The values of objective functions coverage in a range around the optimization objectives with the increasing iterations of optimization. We define the range that the values of objective functions converge in as crucial range and define the remaining range outside the crucial range as trivial range.

3.2.2. Algorithm of Response Surface Surrogates

Adaptive boost (AdaBoost) is a boosting algorithm (Freund et al., 1996; Kearns & Valiant, 1994). This algorithm assembles several weak predictors into a strong predictor to reduce deviation and variance brought by the random generation of initial sample weights and thresholds of ANN (Schapire, 1990). In our study, the standard backpropagation neural network (BPNN) is employed as a weak learner in AdaBoost. The algorithm with AdaBoost and BPNN is called ADBP. The detailed information about ADBP (Fries & Kerkez, 2018; Zhao et al., 2014) is described as follows:

1. Initialize training data and network. The data of training are normalized first. N sets of training data are given, where x_i ($i = 1,2, \ldots, N$) is the input parameters and y_i is the training objective. The initial weight of training sample x_i is set as $D_1(i) = 1/N$.
2. Train the weak learner and calculate the weight of weak learner. Here, BPNN is used as the weak learner to predict the outputs of training data. For $t = 1,\ldots,T$, where T is the size of weak learners, the BPNN

prediction error and the weight of the weak learner are denoted as the following:

$$e_t(i) = h_t(\mathbf{x}_i) - y_i$$

$$a_t = \frac{1}{2exp(|Err_t|)}, \ where\ Err_t = \sum_{i=1}^{N} D_t(i), \tag{3.3}$$

where $e_t(i)$ is the BPNN prediction error; $h_t(x_i)$ is the prediction value of y_i with BPNN; and a_t is the weight of the weak learner.

3. Update the weight of the training sample. If $|e_t(i)|>0.1$, the weight of the training sample would be updated as equation (3.2) or else $D_{t+1}(i) = D_t(i)$,

$$D_{t+1}(i) = 1.1 \times D_t(i). \tag{3.4}$$

4. Output the final hypothesis. Repeat the steps above and T weak learners with their weights are acquired. Then the weights are normalized and the T weak learners are combined into a strong learner as the following:

$$w_t = \frac{a_t}{\sum_{j=1}^{T} a_j}$$

$$h_{final} = \sum_{t=1}^{T} w_t h_t, \tag{3.5}$$

where h_{final} is the ensemble prediction value of the T weak learners; w_t is the normalized weight of the tth weak learner. The inputs of SM are the adjustable parameters of VIC and the outputs are the objective functions.

3.2.3. Surrogate Model for Model Calibration

Methodological Framework of SM-Based Parameter Calibration Algorithms

Figure 3.1 shows the model calibration framework for a complex distributed hydrologic model. The main steps in the model calibration framework are summarized here.

1. Training the initial two-stage surrogate model of VIC. The AdaBoost and BPNN are used to build SM. The detailed process to initialize SM is described in section 3.2.3.2.
2. Searching the Pareto solutions of SM. The optimization of SM uses ε-NSGA II. In the optimization, SM-I seeks out the input parameter

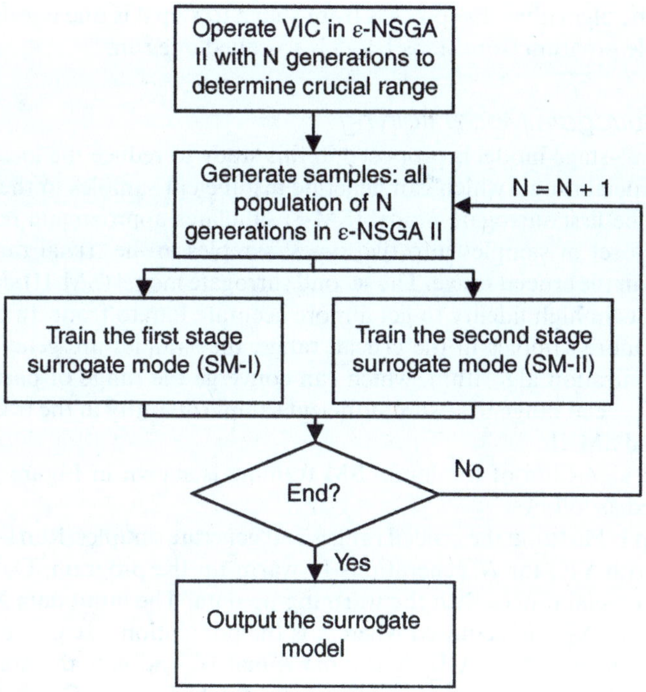

Figure 3.1 Framework of the multi-objective surrogate-based parameter calibration for VIC.

sets whose outputs are in the crucial range and then SM-II predicts the values of outputs with high accuracy.

3. Evaluating the Pareto solutions of SM. Here, the objective values of these Pareto solutions are calculated in SM and VIC, respectively. The performance metric of the optimal results is described in section 3.2.4.
4. Updating SM. If the results of SM do not meet the termination condition, new samples are generated whose input is the Pareto solutions of SM from step 2 and output is the values of objective functions calculated with these solutions in VIC. The samples are expanded with these new samples and the network of SM is updated. Then return to step 2.
5. Outputting the approximate parameter set of VIC. If the termination condition is satisfied, the optimal solutions of SM are considered to be the optimal solutions of VIC.

In the algorithm, the process from step 2 to step 4 is one iteration and the whole program from steps 1 to 5 is so-called one run.

Initial Surrogate Model Training

A two-stage model is proposed in this study to reduce the local errors in the crucial range, which can generate insufficient samples in the crucial range. The first surrogate model (SM-I) simulates approximate results to classify a set of samples into two types: samples in the trivial range and samples in the crucial range. The second surrogate model (SM-II) simulates the results in high fidelity to get a more accurate Pareto front. In order to acquire more samples in the crucial range, the samples are generated by the optimization algorithm, which can converge the range of parameters into the crucial range to ensure sufficient sample quantity in the training of SM-I and SM-II.

The algorithm of the initial SM training is shown in Figure 3.2 and described as follows:

- Step 1: Partition the crucial ranges and generate samples. Run ε-NSGA II with VIC for N generations to warm up the program. Determine the crucial ranges with the warming-up data. The input data $X = [X_1, X_2, \ldots, X_k]$ are acquired where k is the population size in the N generations of ε-NSGA II, X is a $m \times k$ matrix, and m is the number of the calibration parameters. The output data $E = [E_1, E_2, \ldots, E_k]$ are acquired where k is the initial population size of the warming-up data, E is a $n \times k$ matrix, and n is the number of the objective functions.
- Step 2: Train SM-I. After eliminating duplicates of the samples acquired in Step 1, SM-I is trained. Here, we use the hold-out cross validation to validate the training results in which generally 70% of samples are extracted randomly for the model training and the remaining 30% for the model testing. In this stage, SM was used to screen the parameter sets in the crucial range. The rate of correct classification is more important than the simulation accuracy. As a result, we adopted the rate of misclassification (*ER*) to show the performance of the model as the following:

$$ER = \frac{e_1 + e_2}{N_e} \times 100\%, \tag{3.6}$$

where e_1 is the number of parameter sets misinterpreted as those in the crucial range; e_2 is the number of parameter sets in the crucial range assigned to the trivial range; N_e is the number of parameter sets

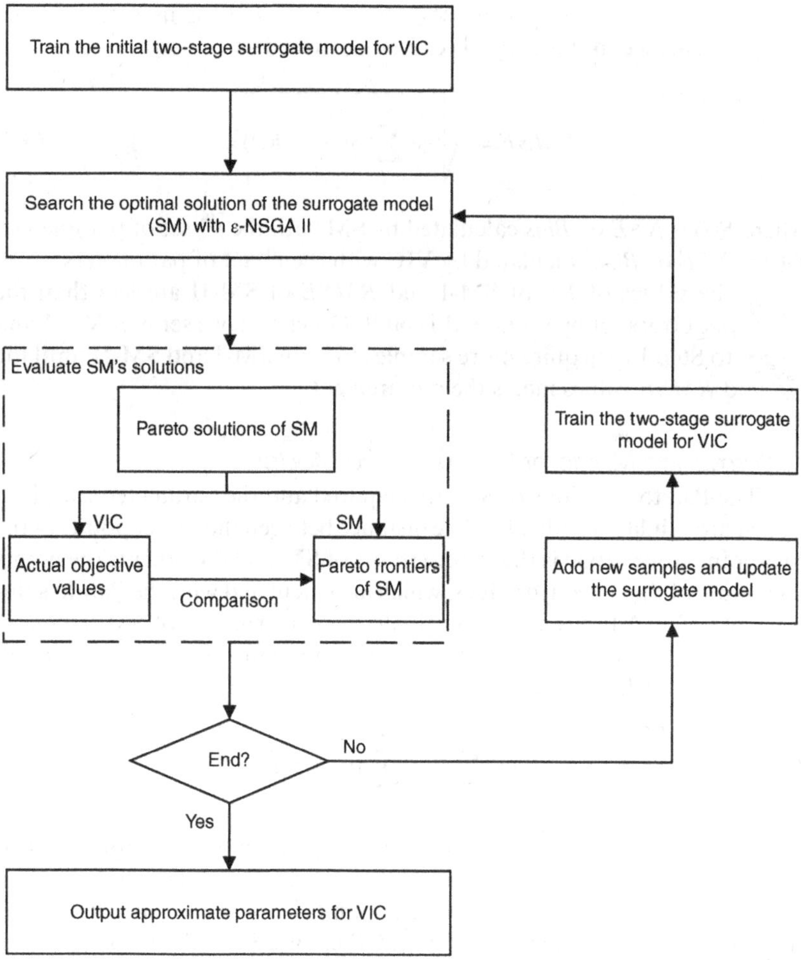

Figure 3.2 Flowchart of the initial surrogate model training.

actually in the crucial range. SM-I is said to misclassify the points when points in the crucial/trivial range are divided into the trivial/crucial range by SM-I.

- Step 3: Train SM-II. Samples within the critical range are first screened from the total samples. The method of the hold-out cross validation is still used in this stage. The difference from Step 2 is the performance evaluation. The root mean square error (*RMSE*) is used to evaluate performance in this stage. It can be described as

$$RMSE = \sqrt{\frac{1}{N}\sum_{i=1}^{N}(S(i) - O(i))^2}, \tag{3.7}$$

where $S(i)$ is *NSE* or *Bias* calculated by SM with the ith set of parameters; $O(i)$ is *NSE* or *Bias* calculated by VIC with the ith set of parameters.

If the values of *ER* of SM-I and *RMSE* of SM-II are less than the allowance errors set by users, SM is built. Otherwise, we set $N = N + 1$ and return to Step 1 to acquire more samples to train SM-I and SM-II until the *ER* and *RMSE* values meets the requirement.

Performance Metric of the Surrogate Model

The Pareto solutions of SM are acquired and the parameter sets given by SM are validated with VIC. The distance between the center points of the two-point groups (one is the Pareto fronts of SM and the other is the points consisting of the objective values, which are calculated with the Pareto solutions of SM in VIC) is used to evaluate the performance of SM. Here, we define a center point as a point with the minimum sum of distance to all the points, which is shown as

$$\min \sum d = \sum_{i}^{n} \left\| \mathrm{x}_0 - \mathrm{x}_i \right\|, \tag{3.8}$$

where the vector x_0 is the position of center point and the vectors x_i ($i = 1, 2, 3, \ldots, n$) are the positions of n Pareto optimal points; $\|\cdot\|$ is the norm of vector. If the distance of the two center points is less than a specified value, the parameter calibration is complete. Otherwise, SM will be updated with the data mixing the original samples, which is used in the previous iteration with the validated values and then the next iteration will be conducted.

3.3. Study Area and Data

To show the performance of the proposed surrogate model, we choose the Lanjiang River (LJR) basin and the Xiangjiang River (XJR) basin, which are located in different parts of China. They have different features in land cover, dominating processes, and meteorological conditions in case studies to investigate the role and efficiency of a large and complex hydrologic model. The locations of the two basins are shown in Figure 3.3.

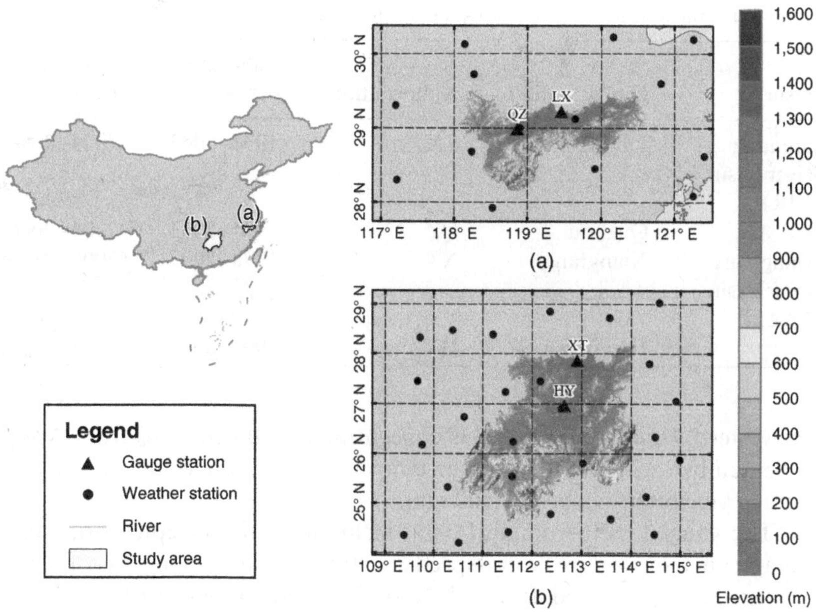

Figure 3.3 Location of (a) the Lanjiang River basin and (b) the Xiangjiang River basin.

The LJR is located on the upper reach of the Qiantang River basin in Zhejiang Province in the southeastern China on the west coast of the Pacific Ocean. The LJR basin covers an area of about 1.8×10^4 km^2 controlled by the Lanxi station. The basin is a humid region with annual average precipitation of about 1,600 mm and annual average temperature about 17°C, dominated by the maritime air masses in summer and the continental air masses in winter. Rainfall in the basin is concentrated at the turn of spring and summer and the high flow occurs during the plum rains from May to July. Besides the plum rains, the typhoon in summer also influences the hydrologic regime in the LJR basin, sometimes resulting in catastrophic floods.

The XJR is located in a humid region in southern China, about 856 km long, which is one of the largest tributaries of the Yangtze River. The river flows into the Dongting Lake, the second largest freshwater lake in middle China. The area of the basin above the Xiangtan station is about 8.2×10^4 km^2. The basin elevation ranges from 1 m above sea level to over 2,000 m in the upstream mountainous areas, but most regions in the Xiang River basin are plains. Dominated by a subtropical high over

Table 3.1 Basic information about the study area.

Basin	Gauge station	Abbreviation	Calibration period	Validation period
Lanjiang River basin (LJR)	Lanxi	LX	1994–2003	2004–2006
	Quzhou	QZ	1994–2003	2004–2006
Xiangjiang River basin (XJR)	Xiangtan	XT	1998–2007	2008–2013
	Hengyang	HY	1998–2007	2008–2013

the western Pacific, the rainfall is concentrated in the months from April to September, with annual average temperature around 17°C and annual average precipitation about 1,500 mm.

The gauged meteorological data, including daily precipitation, daily average air temperature, wind speed, relative humidity, and atmospheric pressure, were collected from 41 National Meteorological Observatory stations located in the LJR basin and the XJR basin, as shown in Figure 3.3 (black points). The daily runoff series of two gauge stations (black triangles) in each basin (i.e., Quzhou station (QZ) and Lanxi station (LX) in the LJR basin, Xiangtan station (XT) and Hengyang station (HY) in the XJR basin) are utilized for calibration and validation shown in Table 3.1. The digital elevation model data with 90 m resolution is downloaded from the National Aeronautics and Space Administration Shuttle Radar Topographic Mission; the soil data set is Harmonized World Soil Database version 1.1 (HWSD v1.1) (Fischer et al., 2008) in a resolution of 1 km; the land cover data set is WESTDC land cover data set version 2.0 (Ran et al., 2010) in a resolution of 1 km.

3.4. Experimental Setup

In this study, the VIC model (version 4.2.d) is run at a 6 hr time step in water and energy balance mode. The study areas are divided into $0.25° \times 0.25°$ grid cells, in which surface characteristics such as elevation, soil, and vegetation are specified (Nijssen et al., 2001). The gauged meteorological data are also interpolated into $0.25° \times 0.25°$ with the Inverse Distance Weighted method (IDW).

According to previous work (Isenstein et al., 2015; Kang et al., 2014; Liu et al., 2017; Lohmann et al., 1996; Mishra et al., 2010;

Table 3.2 Ranges of VIC calibration parameters.

Type	Parameter	Description	Unit	Range
Soil	b_{inf}	Variable infiltration curve parameter	–	0–0.4
	Ds_{max}	Maximum velocity of baseflow	mm/day	0–30
	Ds	Fraction of Ds_{max} where nonlinear baseflow begins	Fraction	0–1
	Ws	Fraction of maximum soil moisture where nonlinear baseflow occurs	Fraction	0–1
	d_2	Thickness of the second soil moisture layer	m	0.1–1.5
	d_3	Thickness of the third soil moisture layer	m	0.1–1.5
Routing	v	Mean flow velocity	m/s	0.5–5

Parada et al., 2003; Voisin et al., 2011; Xie et al., 2007), seven parameters in VIC are selected for calibration and their feasible ranges are shown in Table 3.2. Generally, the parameters selected to be calibrated are the thickness of the second and third soil moisture layer (d_2, d_3), the fraction of maximum soil moisture where nonlinear baseflow occurs (Ws), the maximum velocity of baseflow (Ds_{max}), fraction of Ds_{max} where nonlinear baseflow begins (Ds), and the variable infiltration curve parameter (b_{inf}).

Both the surrogate model and VIC parameters are optimized via ε-NSGA II in parallel. The initial sample size of ε-NSGA II is set as 24. The maximum generations per run and maximum number of populations were specified as 250 and 100,000, respectively. To reduce the runtime of VIC parameters optimization, the program would be terminated when the solutions were not improved in 30 generations.

BPNN is employed as a weak learner in AdaBoost. Since there is no need to train a weak predictor with high accuracy, a three-layer BPNN model was chosen according to the empirical formula proposed by Li and Liu (2009) as shown in equation (3.9),

$$m = \sqrt{n + l} + a, \tag{3.9}$$

where m is the number of nodes in hidden layer, n is the number of neural cells in network input layer, l is the number of neural cells in output layer, a is a number between 0 and 10. The structure was determined as 7-10-1. The numbers indicated that there are 7 nodes in the input layer, 10 nodes in the hidden layer, and one output in this BPNN model.

The maximum number of iterations and the training rate are set as 100 and 0.01, respectively, here. The sigmoid activation function is selected to compute the output of each neuron. The default values suggested by Beale et al. (2010) are employed in the training of weak learners for other parameters. In our study, 10 weak predictors were trained and then assembled into a strong predictor. In order to make sure that SM meets the demand of the accuracy in our study, we use the hold-out cross validation to validate the training results. In this study, when the *ER* value in SM-I is less than 10% and the *RMSE* value in SM-II is less than 5% of the crucial range, the training of the two-stage surrogate model is completed.

3.5. Results

3.5.1. Benchmark Test of the Surrogate Model

To show that the algorithm based on the SM is able to infer the initially used parameter set reliably, we selected the LJR basin and created a parameter set as an artificial benchmark, which is shown in Table 3.3. The artificial parameters were set as the middle points of the flexible ranges of the parameters in Table 3.2. As *NSE* and *Bias* of LX and QZ are able to reach 1 and 0, respectively, in this case, the crucial ranges of NSE_{LX}, $BIAS_{LX}$, NSE_{QZ}, and $BIAS_{QZ}$ were set as ≥0.99, ≤0.01, ≥0.99, and ≤0.01. The 50 generations of population in ε-NSGA II were used as the initial samples. We set the maximum update times for SM as 15.

Figure 3.4 shows the Pareto fronts in the fifth, tenth, and fifteenth iterations, respectively. In Figure 3.4, the final populations (FPs) and the final nondominated populations (FNPs) are the individuals and the nondominated individuals of the final generation in the optimization, respectively. Then the FPs and FNPs are validated by VIC, the values of which are also shown in Figure 3.4. The FNPs in the fifth iteration are in the crucial ranges

Table 3.3 Benchmark values and optimal solutions of the surrogate model.

Parameters	D_2	D_3	b_{inf}	Ds	Ws_{max}	Ws	v
Benchmark	0.75	0.75	0.2	0.5	20	0.8	2.75
Optimal solutions	0.7477	0.7854	0.2029	0.4749	22.2631	0.8457	2.6006
Relative errors (%)	−0.31	4.72	1.45	−5.02	11.32	5.71	−5.43

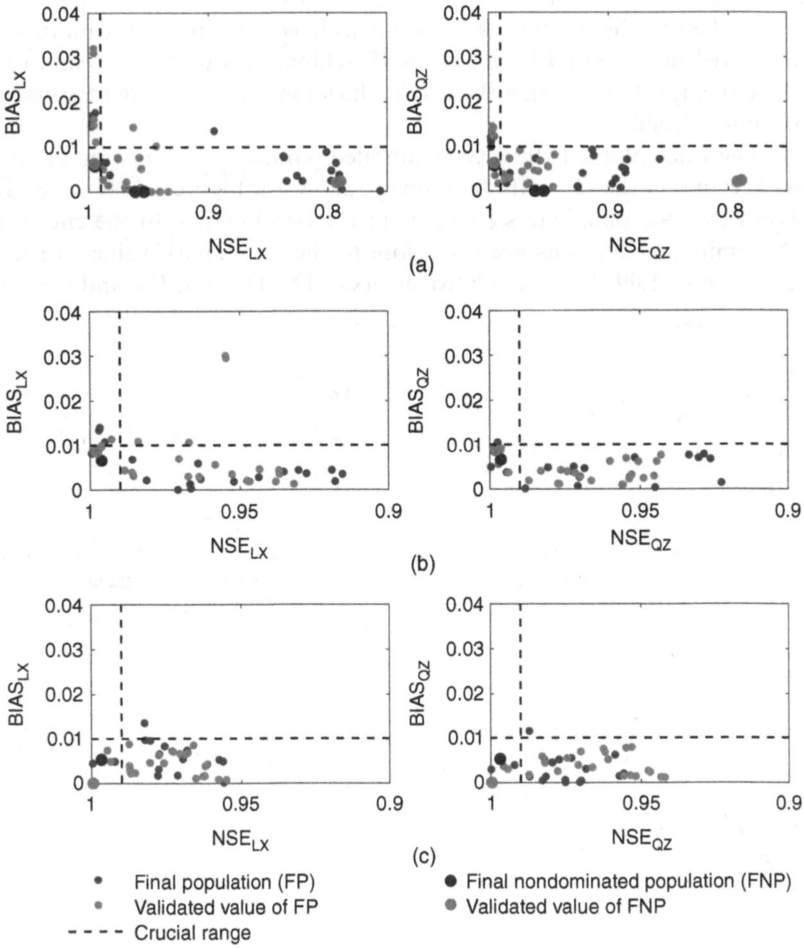

Figure 3.4 Multi-objective optimization procedure of the benchmark test in the (a) fifth, (b) tenth, and (c) fifteenth iterations.

but there are large deviations between the FNPs and their validated values. In the tenth and fifteenth iterations, the deviations between the FNPs and their validated values decrease. As shown in Figure 3.4b and c, the FNPs and their validated values are all in the crucial ranges but the validated values of FNPs in the tenth iterations are closer to the split line of $BIAS_{LX}$ and $BIAS_{QZ}$. It is interesting that the deviation in the fifteenth iteration is slightly larger than the one in the tenth iteration. This phenomenon is caused by the marginal error of ANN when the simulated objectives are reached to the bounds. It does not usually happen when calibrating the model in an actual basin, for *NSE* seldom gets to 1 and *Bias* to 0. Therefore, it is thought that the results shown in Figure 3.4 are reasonable and interpretable.

The trace plots of the seven calibrated parameters versus the benchmark values in the fifteenth iteration are shown in Figure 3.5. It is clearly shown that the parameters converge to the benchmarks. In the end, the nondominated solutions are very close to the benchmark values we set. As shown in Table 3.3, the relative errors of D_2, D_3, b_{inf}, Ds, and v given

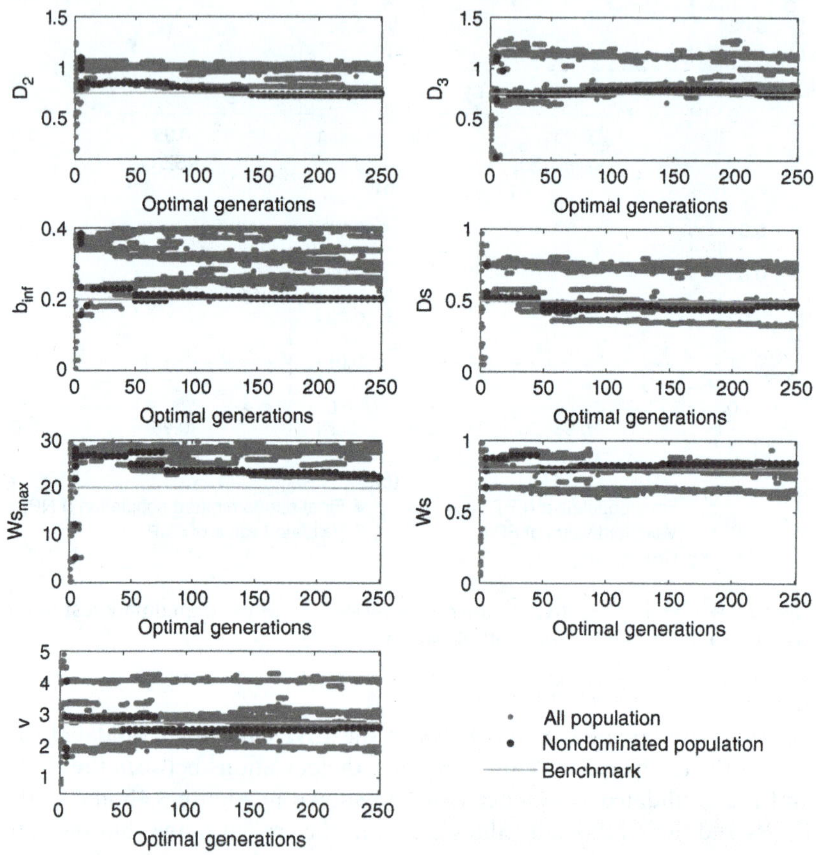

Figure 3.5 Trace plots of the seven calibrated parameters. The All population dots are the variable parameters for the nondominated population and the Nondominated population dots are the variable parameters for all the population in the fifteenth optimization. The Benchmark lines are the artificial benchmarks.

by SM are less than 10%, where the value of v has the largest relative error of 5.71%. Meanwhile, the relative error of Ws_{max} is 11.13%, which is larger than the other parameters but less than the 10% of the optimal scope (0 to 30). This parameter set is also validated in the VIC model whose objective functions are 0.9996, 0.0001, 0.9995, and 0.0009 for NSE_{LX}, $BIAS_{LX}$, NSE_{QZ}, and $BIAS_{QZ}$, respectively.

3.5.2. Pareto Solutions of the Surrogate Model in Two Study Areas

In order to choose the crucial ranges of *NSE* and *Bias*, the multi-objective optimization program was operated for 20 generations and then the crucial ranges were selected according to the ranges of the population. The ε-NSGA II was operated to select crucial ranges for the four optimization objectives. In each generation, the values of population were counted. In Figure 3.6a, the ranges of *NSE* are decreasing with the

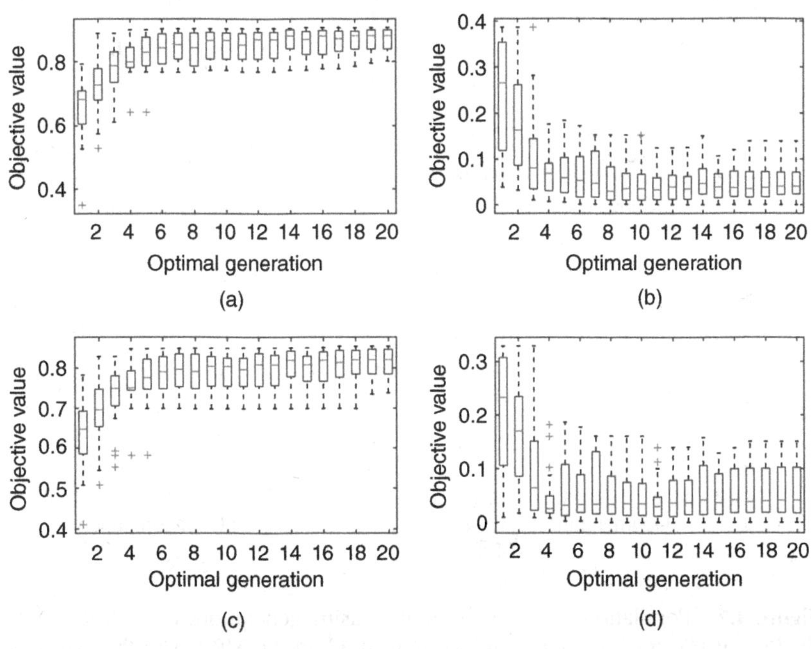

Figure 3.6 Population values with the increasing generations of (a) *NSE* at LX, (b) *Bias* at LX, (c) *NSE* at QZ, and (d) *Bias* at QZ in the LJR basin; the boxplots contain the medians (middle lines), the outliners (plus signs), the 25% and 75% quantiles (the edges of the box), and most extreme data points (whispers).

increasing iterations. At first, the range of *NSE* is from around 0.6 to around 0.8; then the ranges decrease gradually and the medians increase gradually; after the number of iterations reaches the value of 8, the ranges of population become stable in a relatively small range from about 0.8 to 0.9. The similar results are shown in Figure 3.6b–d and Figure 3.7.

With the results of Figure 3.6 and Figure 3.7, the crucial ranges of the objective function at each station in each basin are defined. The crucial ranges in the LJR basin and the XJR basin are shown in Table 3.4.

Since a stochastic algorithm is used, the calibration results (objective value and parameter set) vary when multiple calibration runs are performed. It is necessary that independent runs of calibration using SM be repeated. In this study, the calibration using SM was run 10 times in each study area. The obtained optimal points given by SM, the corresponding validating points, and the Pareto fronts estimated with ε-NSGA II in the LJR basin and the XJR basin are shown in Figure 3.8

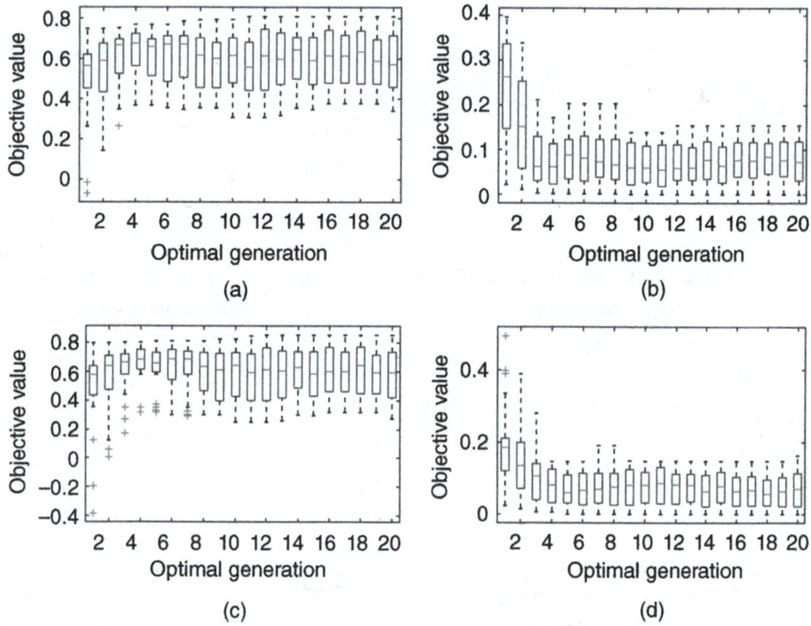

Figure 3.7 Population values with the increasing generations of (a) *NSE* at HY, (b) *Bias* at HY, (c) *NSE* at XT, and (d) *Bias* at XT in the XJR basin; the boxplots contain the medians (middle lines), the outliners (plus signs), the 25% and 75% quantiles (the edges of the box) and most extreme data points (whispers).

Table 3.4 Crucial ranges of NSE and Bias in each gauge station.

Gauge station	Crucial ranges	
	NSE	*Bias*
Lanxi	0.85–1	0–0.1
Quzhou	0.8–1	0–0.1
Xiangtan	0.7–1	0–0.15
Hengyang	0.7–1	0–0.15

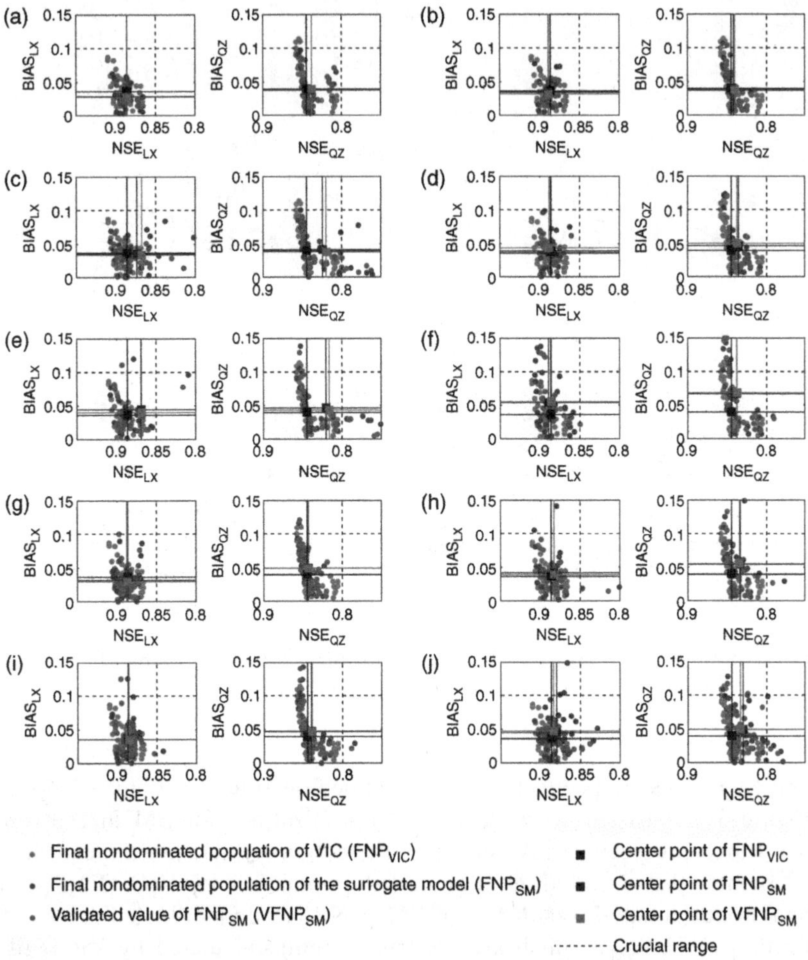

Figure 3.8 Two-dimensional Pareto plots for *Bias* versus *NSE* in the LJR basin at the last iteration in 10 parameter optimizations (a–j).

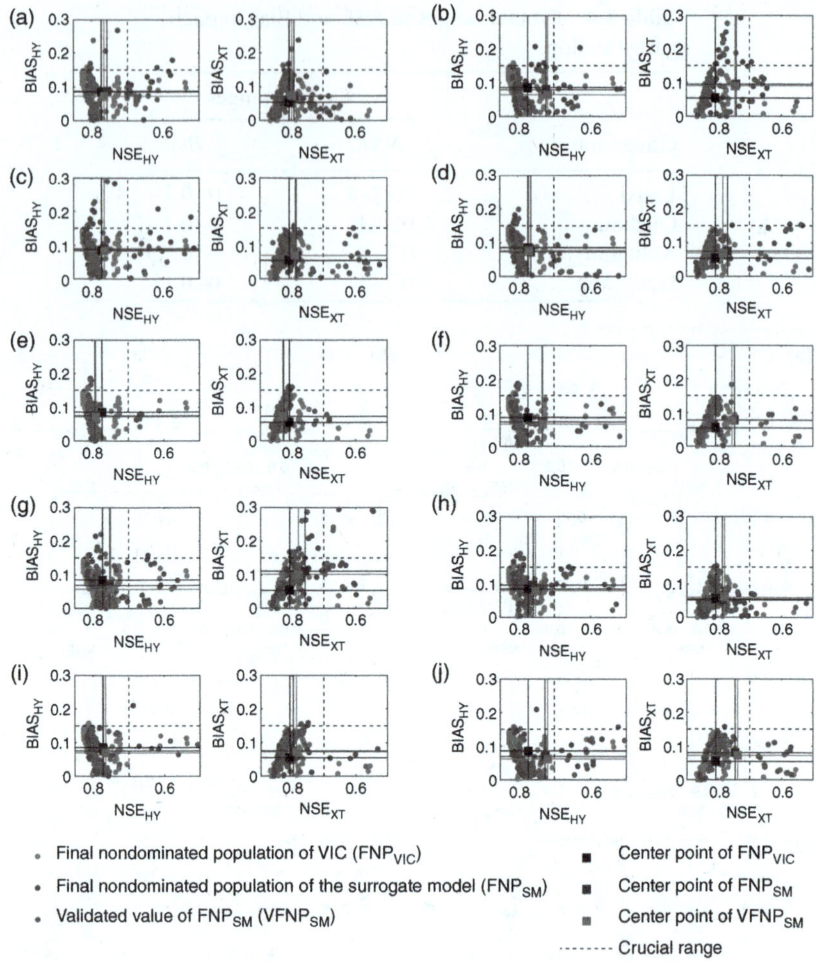

Figure 3.9 Two-dimensional Pareto plots for *Bias* versus *NSE* in the XJR basin at the last iteration in 10 parameter optimizations (a–j).

and Figure 3.9, respectively. The results demonstrated in Figure 3.8 and Figure 3.9 consist of 10 independent optimizations for SM in the two study areas, respectively. As shown in these figures, there are three groups of points, that is, final nondominated population of VIC (FNP_{VIC}), final nondominated population of SM (FNP_{SM}), and validated values of FNP_{SM} ($VFNP_{SM}$), which are the true outputs calculated by VIC with the parameter sets given by SM. FNP_{VIC} is the baseline. The center points of the FNP_{VIC}, FNP_{SM}, and $VFNP_{SM}$ are also shown in each subplot denoted as CP_{VIC}, CP_{SM}, and CP_{VSM}. It is seen that there are errors between FNP_{SM} and $VFNP_{SM}$, but CP_{SM} and CP_{VSM} are close in each optimization in both study areas. Furthermore, CP_{SM} and CP_{VIC} are also compared in these scatter plots. Since a stochastic algorithm is used in this study, the results are slightly different in each optimization for SM. As seen in two figures, CP_{SM} are very close to CP_{VIC} in most cases. In Figure 3.8c and e, Figure 3.9b, f, and j, the distances between CP_{SM} and CP_{VIC} are relatively large but acceptable (less than 0.1 in all the cases).

The compromise parameter sets were selected when the four objective functions reached a good tradeoff. The compromise parameter values (normalized) of VIC and SM in the two study areas are presented in Figure 3.10. The black lines in Figure 3.10 are the results of VIC, which is

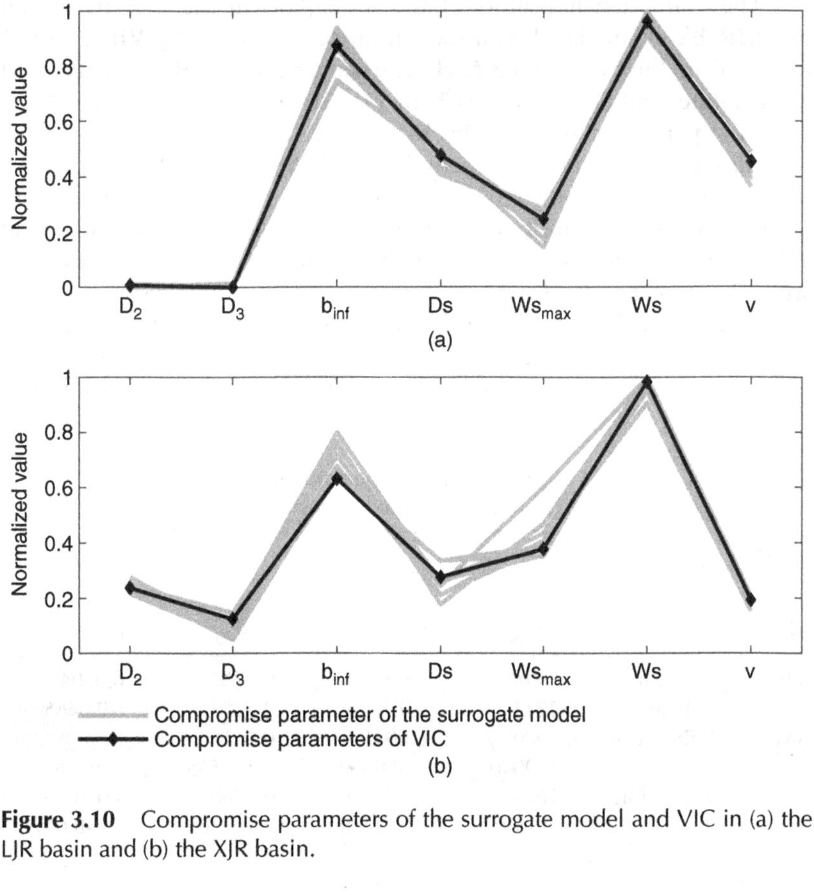

Figure 3.10 Compromise parameters of the surrogate model and VIC in (a) the LJR basin and (b) the XJR basin.

the baseline here. The grey lines are the results of SM. The 10 compromise parameter sets of SM are around the parameter set of VIC as shown in Figure 3.10. The optimal parameters calibrated with SM could be a very accurate approximation of VIC. In Figure 3.10a and b, the parameter set of the LJR basin is very similar to that of the XJR basin since the two study areas are dominated by similar climatic conditions. The large values of Ws in these two study areas, which is a fraction ranged from 0 to 1, mean that the base flows are both small and stable (Ws near 1 in VIC show the base flow generation is linear and small). The phenomenon accords with the feature that the runoff in the LJR basin and the XJR basin is dominated by rainfall.

3.5.3. Evaluation of the Surrogate Model in Two Study Areas

The evaluation indicators of the streamflow in the LJR basin and the XJR basin simulated with the parameter sets given by VIC and SM are listed in Tables 3.5 and 3.6. Here, when calculating *Bias*, we removed the absolute symbol to evaluate whether the streamflow was overestimated. With the parameter set given by VIC, the values of *NSE* at LX during the calibration and validation periods are 0.8998 and 0.7744, respectively, while those of *Bias* are 0.0032 and −0.1025, respectively. The ranges of *NSE* for 10 optimizations of SM at LX during the calibration and validation periods are 0.8990–0.9040 and 0.7616–0.7995, respectively, while those of *Bias* are 0.0001–0.0560 and −0.1659––0.0965, respectively. Both *NSE* and *Bias* of VIC in the calibration and validation period are in the range of its corresponding parameter of SM at the LX station. Similarly, the results also can be seen at all the stations in the two study

Table 3.5 Evaluations of simulation with the parameter sets given by VIC and SM during calibration and validation in the LJR basin.

		Objective			
Model	Period	NSE_{LX}	$BIAS_{LX}$ (+/−)	NSE_{QZ}	$BIAS_{QZ}$ (+/−)
VIC	Calibration	0.8998	0.0032	0.8465	0.0504
	Validation	0.7744	−0.1025	0.7788	−0.0465
SM	Calibration	0.8980–0.9040	0.0001–0.0560	0.8416–0.8555	0.0474–0.0909
	Validation	0.7616–0.7995	−0.1659–−0.0965	0.7760–0.7860	−0.1030–−0.0415

Table 3.6 Evaluations of simulation with the parameter sets given by VIC and SM during calibration and validation in the XJR basin.

		Objective			
Model	Period	NSE_{HY}	$BIAS_{HY}$ (+/–)	NSE_{XT}	$BIAS_{XT}$ (+/–)
VIC	Calibration	0.8075	−0.1059	0.8587	0.0194
	Validation	0.8094	−0.0542	0.8685	0.0265
SM	Calibration	0.8035–0.8156	−0.1172–−0.0580	0.8324–0.8624	0.0080–0.0682
	Validation	0.7917–0.8147	−0.0731–−0.0008	0.8152–0.8652	0.0070–−0.0854

areas. For the XJR basin, the model performance during the validation period is more satisfactory than that during the calibration period. It is probably caused by the existence of considerable extraordinary floods during the calibration period.

The parameter set with the best performance in the 10 sets mentioned above is denoted as P_{SM} and the compromise parameter set given by VIC is denoted as P_{VIC}. Simulated daily streamflow compared with observation in the LJR basin and the XJR basin during the calibration and validation periods are displayed in Figure 3.11 and Figure 3.12, respectively. The streamflow simulated with P_{SM} and P_{VIC} is compared with the observation, respectively. The left panels of Figure 3.11 and Figure 3.12 are the hydrographs of the study station in the two study areas, and the right panels show the empirical CDFs of the simulated streamflow and the observation in each station. The simulated streamflow with P_{VIC} was used as the baseline. It illustrates that the empirical CDF distributions of the simulation of P_{SM} coincides well with those of P_{VIC}. Compared with the empirical CDF distributions of the observations, the empirical CDF distributions of the simulations of P_{SM} also show good performances in the two study areas.

Figure 3.13 and Figure 3.14 show the annual average runoff depth calculated with P_{SM} and P_{VIC} (the left map and the middle map) and the difference of them (P_{SM} minus P_{VIC}, the right map), respectively. As shown in Figure 3.13a, the runoff depths of P_{SM} and P_{VIC} are almost the same and most grid cells in the map of P_{SM}-P_{VIC} are around 0. Figure 3.13b and Figure 3.14 are similar to Figure 3.13a. The figures indicate that the results of the parameter sets acquired by SM agree well with those of the parameter sets acquired by the VIC model.

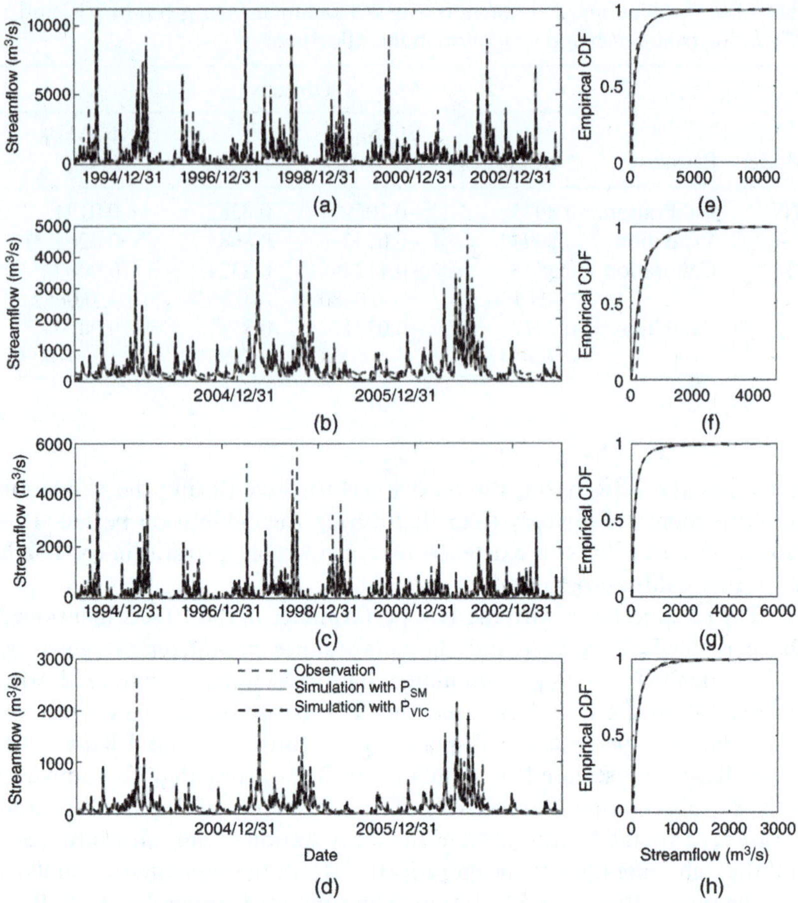

Figure 3.11 Simulated daily streamflow series and observed streamflow (a) at the LX during the validation period (1994 to 2003); (b) at the LX during the calibration period (2004 to 2006); (c) at the QZ during the validation period; (d) at the QZ during the calibration period. The right panels show the eCDFs of the observed and simulated streamflow (e) at the LX during the validation period; (f) at the LX during the calibration period; (g) at the QZ during the validation period; and (h) at the QZ during the calibration period. Here P_{SM} is the best parameter set in the 10 compromise parameter sets and P_{VIC} is the compromise parameter set given by VIC.

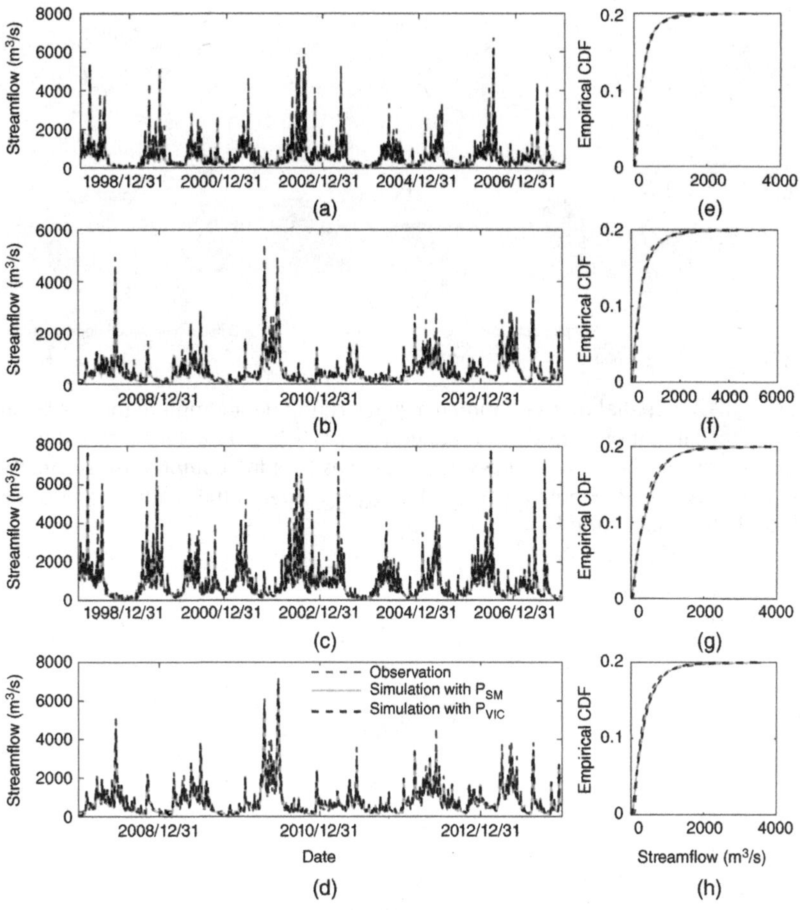

Figure 3.12 Simulated daily streamflow series and observed streamflow (a) at the XT during the validation period (1998 to 2007); (b) at the XT during the calibration period (2008 to 2013); (c) at the HY during the validation period; and (d) at the HY during the calibration period. The right panels show the eCDFs of the observed and simulated streamflow (e) at the XT during the validation period; (f) at the XT during the calibration period; (g) at the HY during the validation period; and (h) at the HY during the calibration period. Here P_{SM} is the best parameter set in the 10 compromise parameter sets and P_{VIC} is the compromise parameter set given by VIC.

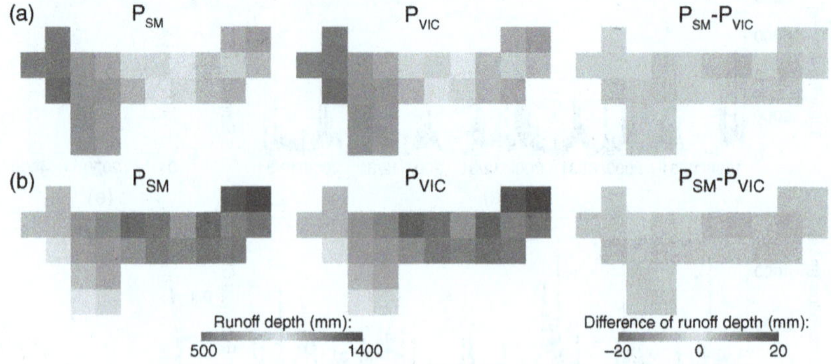

Figure 3.13 Spatial maps of annual average runoff depth (mm) in the LJR basin, during (a) calibration period; (b) validation period. The parameter set with the best performance in the 10 runs of SM is denoted as P_{SM}; the compromise parameter set given by VIC is denoted as P_{VIC}; the P_{SM}-P_{VIC} means the difference of runoff depth between P_{SM} and P_{VIC}.

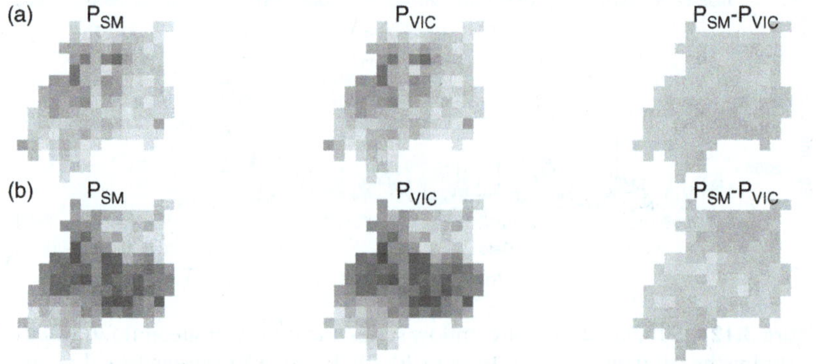

Figure 3.14 Spatial maps of annual average runoff depth (mm) in the XJR basin, during (a) calibration period; (b) validation period. The parameter set with the best performance in the 10 runs of SM is denoted as P_{SM}; the compromise parameter set given by VIC is denoted as P_{VIC}; the P_{SM}-P_{VIC} means the difference of runoff depth between P_{SM} and P_{VIC}.

3.5.4. Efficiency of the Surrogate Model in Two Study Areas

The efficiency of SM is also a key point as our original intention is to reduce the computation time in hydrologic model calibration. The computation time of one execution of VIC is decided by catchment area, grid cell size, length of the simulated series, and number of modules used in simulations such as snow melting and lake. In our study, the VIC model was run in parallel on 24 processors of Intel Xeon CPU E5-4620 v2 2.60 GHz (CentOS 6 x86_64). One execution of VIC required 3.6 s and 22.3 s in the LJR basin and the XJR basin, respectively. The computation time of one execution of SM is less affected by the physical conditions of the study area. The time is just very slightly increased along with the increase of population per generation, which is very small and can be negligible. In our study, the average time consumed by SM (to generate a generation of population with SM) required 7.3 s in both study areas.

The average time consumed for parameter calibrations with SM of the 10 runs in the LJR basin and XJR basin were 16,908 s and 73,154 s, respectively, and the time with VIC were 31,363 s and 876,390 s, respectively. The detailed information is listed in Tables 3.7 and 3.8. It is found that the initial samples and the times of network update (the number of the execution of SM) are different in different basins. In the LJR basin, the size of initial samples and the number of SM execution are 574 and 7 in

Table 3.7 Time consumed by multi-objective optimization algorithms in the LJR basin when calibrating the surrogate model and VIC.

Model	No.	Initial samples	Iteration	Update samples	Total runs of VIC	Time (s)
Surrogate model	1	528	9*250	269	797	19,294.2
	2	552	5*250	319	871	12,260.6
	3	576	7*250	255	831	15,766.6
	4	504	4*250	137	641	9,607.6
	5	504	10*250	323	827	21,227.2
	6	480	6*250	195	675	13,380.0
	7	648	10*250	358	1,006	21,871.6
	8	672	9*250	347	1,019	20,093.4
	9	648	8*250	273	921	17,915.6
	10	624	5*250	226	850	17,660.0
VIC			250+111		22,428	81,889.0

Note: Initial samples = the samples for the surrogate model training; update samples = samples for updating the network in the surrogate; iteration = the number of the executions of the surrogate model.

Table 3.8 Time consumed by multi-objective optimization algorithms in the XJR basin when calibrating the surrogate model and VIC.

Model	No.	Initial samples	Iteration	Update samples	Total runs of VIC	Time (s)
Surrogate model	1	1,440	19*250	1,148	2,588	92,387.4
	2	1,224	12*250	891	2,115	69,064.5
	3	1,392	18*250	1,059	2,451	87,507.3
	4	1,128	19*250	1,259	2,387	87,905.1
	5	1,368	19*250	1,362	2,730	95,554.0
	6	1,344	12*250	720	2,064	67,927.2
	7	1,368	8*250	479	1,847	55,788.1
	8	1,368	8*250	505	1,873	56,367.9
	9	1,200	14*250	867	2,067	71,644.1
	10	1,368	5*250	348	1,716	47,391.8
VIC			250 + 208		43,440	960,545.0

Note: Initial samples = the samples for the surrogate model training; update samples = samples for updating the network in the surrogate; iteration = the number of the executions of the surrogate model.

average, respectively. Meanwhile, in the XJR basin, they are 1,320 and 13, respectively. The total runs of VIC (including runs for initial samples and update samples) are related with the convergence rates of model, where the ε-NSGA II is terminated in the 22,428 runs of VIC in the LJR basin and in the 43,440 runs of VIC in the XJR basin. SM reduces the time consumed for the parameter calibration in the LJR basin but not so outstandingly. However, the time consumed shrinks sharply by using SM instead of VIC in the XJR basin (speeding up about 20 times).

3.6. Discussion

The two-stage surrogate model was a black box model constructed from the VIC model. The ADBP mined the information of the samples generated by the VIC model and then simulated the output according to the mined information. Therefore, the results from SM cannot be better than those from the original model. Although the optimal parameter sets acquired from SM and VIC have little difference and the Pareto optimal points of the two models are also similar, there are still some differences between the Pareto fronts of SM and VIC and errors existing between the real values and the values predicted by SM. In addition, it should be noted that this study set the crucial ranges according to a priori optimization.

If parameters converge slowly in other cases, it would take a longer time to derive the crucial ranges. In these cases, the crucial ranges can be set by obeying the rule of thumb or referring to previous work. Furthermore, the crucial ranges can also be determined or updated by the results of the previous parameter optimization. However, in this study, we have not studied the relationship of the ADBP's performance to the model improving times.

In our work, we used the distance between the center point of FNP_{SM} and the center point $VFNP_{SM}$ as the termination condition of the optimization, based on the assumption that FNP_{SM} can be used as FNP_{VIC} when FNP_{SM} is the same as $VFNP_{SM}$. To acquire FNP_{SM} in each iteration, ε-NSGA II was used, which can adapt population size and terminate automatically. In order to get FNP_{SM} close to $VFNP_{SM}$ more quickly and reduce the number of iterations, the two-stage surrogate model was designed and a better-performance surrogate method, ADBP, was used. Furthermore, the sample sets in this study were chosen from the sample space generated by ε-NSGA II to add the samples in the crucial range. From the results shown in Figures 3.8 and 3.9, ADBP performs stably with the different sample sets chosen from the sample space generated by ε-NSGA II. This indicates that the nonuniformity of initial training sample for ADBP is acceptable, which may be due to the feature of the optimal process where not all the samples in the sample space were used and the samples in the crucial range were more important than others.

In the meantime, the time consumed for calibration with SM is heavily curtailed compared with VIC. In order to accelerate the parameter calibration for VIC, the parallel computations were applied in a 24 processor platform, while SM constructed by ADBP was operated in single thread. As a result, there will be a remarkable increase in the saved time by SM if SM and the original model are both operated in single thread. Besides, the time SM saved is related with the VIC runtime. When the study area is larger and the physical conditions are more complex, the efficiency improvement with SM is more remarkable.

3.7. Conclusions

In this study, we developed a surrogate-based parameter calibration framework for a large complex distributed hydrologic model to accelerate the parallel automatic parameter calibration of VIC. The two-stage surrogate model in the framework is based on AdaBoost and BPNN. SM-I predicted outputs roughly and sought out the parameter sets in the crucial range. SM-II simulated the outputs in the crucial range accurately. The method was successfully applied in the LJR basin and the XJR basin.

RMSE and *ER* were chosen as the evaluation indices for SM-I and SM-II, respectively. Via this study, we find the following:

1. In the benchmark test, a parameter set with a very small error from the artificial parameter set can be acquired through using our surrogate-based parameter calibration framework.
2. The parameter sets of VIC acquired from the surrogate-based parameter calibration framework have no large difference from those from ε-NSGA II. The streamflow simulated with parameter sets from SM accords with that of VIC very well.
3. The proposed method can accelerate the calibration process markedly but the time saved is related to the study area. When the study area is larger and the physical conditions are more complex, the efficiency improvement with SM is more remarkable.

SM can save much time for calibration of complex distributed hydrologic models and improve in particular the state of the art in parameter calibration. Future work may focus on the extension of SM to study hydrologic model calibrations under changing environments and, in addition, the role of different optimization algorithms can be investigated.

Acknowledgments

The authors thank Zhejiang Key Research and Development Plan (2021C03017), the Key Project of Zhejiang Natural Science Foundation (LZ20E090001), and National Key Research and Development Plan "Inter-governmental Cooperation in International Scientific and Technological Innovation" (2016YFE0122100) for financial support. We thank the Environment and Ecological Science Data Center for West China (WestDC) for providing the soil data and the land cover data (https://westdc.westgis.ac.cn). We are also grateful to China Meteorological Administration for providing meteorological data used in this study. It can be downloaded from the website https://data.cma.cn.

References

Abbaspour, K. C., Rouholahnejad, E., Vaghefi, S., Srinivasan, R., Yang, H., & Kløve, B. (2015). A continental-scale hydrology and water quality model for Europe: Calibration and uncertainty of a high-resolution large-scale SWAT model. *Journal of Hydrology*, *524*, 733–752.

Asefa, T., Kemblowski, M. W., Urroz, G., Mckee, M., & Khalil, A. (2004). Support vectors-based groundwater head observation networks design. *Water Resources Research*, *40*(11), 151–175.

Beale, M. H., Hagan, M. T., & Demuth, H. B. (2010). Neural network toolbox™ user's guide. *The MathWorks, 2*, 77–81.

Behzadian, K., Kapelan, Z., Savic, D., & Ardeshir, A. (2009). Stochastic sampling design using a multi-objective genetic algorithm and adaptive neural networks. *Environmental Modeling & Software, 24*(4), 530–541.

Carpenter, T. M., & Georgakakos, K. P. (2006). Intercomparison of lumped versus distributed hydrologic model ensemble simulations on operational forecast scales. *Journal of Hydrology, 329*(1–2), 174–185.

Chawla, I., & Mujumdar, P. P. (2018). Partitioning uncertainty in streamflow projections under nonstationary model conditions. *Advances in Water Resources, 112*, 266–282.

Chowdhury, M., Alouani, A., & Hossain, F. (2010). Comparison of ordinary kriging and artificial neural network for spatial mapping of arsenic contamination of groundwater. *Stochastic Environmental Research and Risk Assessment, 24*(1), 1–7.

Christelis, V., Bellos, V., & Tsakiris, G. (2016). *Employing surrogate modeling for the calibration of a 2D flood simulation model*. Sustainable Hydraulics in the Era of Global Change, Proceedings of 4th IAHR Congress, Liege, Belgium.

Deb, K., Pratap, A., Agarwal, S., & Meyarivan, T. (2002). A fast and elitist multi-objective genetic algorithm: NSGA-II. *IEEE Transactions Evolutionary Computation, 6*(2), 182–197.

De Vos, N. J., & Rientjes, T. (2007). Multi-objective performance comparison of an artificial neural network and a conceptual rainfall-runoff model. *Hydrological Sciences Journal, 52*(3), 397–413.

Fischer, G., Nachtergaele, F., Prieler, S., van Velthuizen, H. T., Verelst, L., & Wiberg, D. (2008). Global agro-ecological zones assessment for agriculture (GAEZ 2008). IIASA, Laxenburg, Austria, and FAO, Rome, Italy.

Freeze, R. A., & Harlan, R. L. (1969). Blueprint for a physically-based, digitally-simulated hydrologic response model. *Journal of Hydrology, 9*(3), 237–258.

Freund, Y., Schapire, R. E., & Others (1996). *Experiments with a new boosting algorithm*. International Conference on Machine Learning, Bari, Italy.

Fries, K. J., & Kerkez, B. (2018). Using sensor data to dynamically map large-scale models to site-scale forecasts: A case study using the national water model. *Water Resources Research, 54*(8), 5636–5653.

Fu, G., Kapelan, Z., Kasprzyk, J. R., & Reed, P. (2012). Optimal design of water distribution systems using many-objective visual analytics. *Journal of Water Resources Planning and Management, 139*(6), 624–633.

Gharari, S., Hrachowitz, M., Fenicia, F., & Savenije, H. (2013). An approach to identify time consistent model parameters: Sub-period calibration. *Hydrology and Earth System Sciences, 17*(1), 149–161.

Gong, W., Duan, Q., Li, J., Wang, C., Di, Z., Ye, A., et al. (2016). Multiobjective adaptive surrogate modeling-based optimization for parameter estimation of large, complex geophysical models. *Water Resources Research, 52*(3), 1984–2008.

Guo, J., Zhou, J., Lu, J., Zou, Q., Zhang, H., & Bi, S. (2014). Multi-objective optimization of empirical hydrological model for streamflow prediction. *Journal of Hydrology, 511*, 242–253.

Guo, X., Wang, B., Xiong, W., & Jin, S. (2017). SWAT *hydrological model and big data techniques*. International Conference on 5G for Future Wireless Networks, Beijing, China.

Gupta, H. V., Kling, H., Yilmaz, K. K., & Martinez, G. F. (2009). Decomposition of the mean squared error and NSE performance criteria: Implications for improving hydrological modeling. *Journal of Hydrology*, *377*(1–2), 80–91.

Hadka, D., & Reed, P. (2012). Diagnostic assessment of search controls and failure modes in many-objective evolutionary optimization. *Evolutionary Computation*, *20*(3), 423–452.

Harik, G. R., & Lobo, F. G. (1999). *A parameterless genetic algorithm*. Proceedings of the Genetic and Evolutionary Computation Conference (GECCO 1999), Orlando, Florida.

Huo, J., Liu, L., & Zhang, Y. (2016). Comparative research of optimization algorithms for parameters calibration of watershed hydrological model. *Journal of Computational Methods in Sciences and Engineering*, *16*(3), 653–669.

Hurford, A. P., Huskova, I., & Harou, J. J. (2014). Using many-objective trade-off analysis to help dams promote economic development, protect the poor and enhance ecological health. *Environmental Modeling & Software*, *38*, 72–86.

Isenstein, E. M., Wi, S., & Ethan Yang, Y. C. (2015). *Calibration of a distributed hydrologic model using streamflow and remote sensing snow data*. World Environmental and Water Resources Congress 2015, Austin, Texas.

Kang, D. H., Shi, X., Gao, H., & Déry, S. J. (2014). On the changing contribution of snow to the hydrology of the Fraser River Basin, Canada. *Journal of Hydrometeorology*, *15*(4), 1344–1365.

Kearns, M., & Valiant, L. G. (1994). Crytographic limitations on learning Boolean formulae and finite automata. *Journal of the ACM*, *1*(41), 67–95.

Kollat, J. B., & Reed, P. M. (2006). Comparing state-of-the-art evolutionary multi-objective algorithms for long-term groundwater monitoring design. *Advances in Water Resources*, *29*(6), 792–807.

Kollat, J. B., & Reed, P. M. (2007). A computational scaling analysis of multiobjective evolutionary algorithms in long-term groundwater monitoring applications. *Advances in Water Resources*, *30*(3), 408–419.

Kollat, J. B., Reed, P. M., & Wagener, T. (2012). When are multiobjective calibration trade-offs in hydrologic models meaningful? *Water Resources Research*, *48*(3).

Kourakos, G., & Mantoglou, A. (2009). Pumping optimization of coastal aquifers based on evolutionary algorithms and surrogate modular neural network models. *Advances in Water Resources*, *32*(4), 507–521.

Kovalets, I. V., Kivva, S. L., & Udovenko, O. I. (2015). Usage of the WRF/DHSVM model chain for simulation of extreme floods in mountainous areas: A pilot study for the Uzh River Basin in the Ukrainian Carpathians. *Natural Hazards*, *75*(2), 2049–2063.

Laumanns, M., Thiele, L., Deb, K., & Zitzler, E. (2002). Combining convergence and diversity in evolutionary multiobjective optimization. *Evolutionary Computation*, *10*(3), 263–282.

Lee, G., Yu, W., Jung, K., et al. (2013). Catchment-scale soil erosion and sediment yield simulation using a spatially distributed erosion model. *Environmental Earth Sciences*, *70*(1), 33–47.

Leng, G., Tang, Q., & Rayburg, S. (2015). Climate change impacts on meteorological, agricultural and hydrological droughts in China. *Global and Planetary Change*, *126*, 23–34.

Li, F., & Liu, C. (2009). *Application study of BP neural network on stock market prediction*. 2009 Ninth International Conference on Hybrid Intelligent Systems, IEEE, Shenyang, China.

Liang, X., Lettenmaier, D. P., & Wood, E. F. (1996). One-dimensional statistical dynamic representation of subgrid spatial variability of precipitation in the two-layer variable infiltration capacity model. *Journal of Geophysical Research Atmospheres*, *101*(D16), 21403–21422.

Liang, X., Lettenmaier, D. P., Wood, E. F., & Burges, S. J. (1994). A simple hydrologically based model of land surface water and energy fluxes for general circulation models. *Journal of Geophysical Research Atmospheres*, *99*(D7), 14415–14428.

Liu, G., He, Z., Luan, Z., & Qi, S. (2018). Intercomparison of a lumped model and a distributed model for streamflow simulation in the Naoli River watershed, northeast China. *Water*, *10*(8), 1004.

Liu, L., Gao, C., Xuan, W., & Xu, Y. P. (2017). Evaluation of medium-range ensemble flood forecasting based on calibration strategies and ensemble methods in Lanjiang Basin, southeast China. *Journal of Hydrology*, *554*, 233–250.

Liu, L., Xu, Y. P., Pan, S. L., & Bai, Z. X. (2019). Potential application of hydrological ensemble prediction in forecasting floods and its components over the Yarlung Zangbo River Basin, *China. Hydrology and Earth System Sciences*, *23*(8), 3335–3352.

Lohmann, D. A. G., Nolte-Holube, R., & Raschke, E. (1996). A large-scale horizontal routing model to be coupled to land surface parameterization schemes. *Tellus Series A-Dynamic Meteorology and Oceanography*, *48*(5), 708–721.

Lohmann, D., Raschke, E., Nijssen, B., & Lettenmaier, D. P. (1998). Regional scale hydrology: I. Formulation of the VIC-2L model coupled to a routing model. *Hydrological Sciences Journal*, *43*(1), 131–141.

Maier, H. R., Kapelan, Z., Kasprzyk, J., Kollat, J., Matott, L. S., Cunha, M. C., et al. (2014). Evolutionary algorithms and other metaheuristics in water resources: Current status, research challenges and future directions. *Environmental Modeling & Software*, *62*, 271–299.

Marek, G. W., Gowda, P. H., Evett, S. R., Baumhardt, R. L., Brauer, D. K., Howell, T. A., et al. (2016). Calibration and validation of the SWAT model for predicting daily ET over irrigated crops in the Texas High Plains using lysimetric data. *Transactions of the Asabe*, *59*(2), 611–622.

Maurer, E. P., Wood, A. W., Adam, J. C., Lettenmaier, D. P., & Nijssen, B. (2002). A long-term hydrologically based data set of land surface fluxes and states for the conterminous United States. *Journal of Climate*, *15*(22), 3237–3251.

Mishra, V., Cherkauer, K. A., & Bowling, L. C. (2010). Parameterization of lakes and wetlands for energy and water balance studies in the great lakes region. *Journal of Hydrometeorology, 11*(5), 1057–1082.

Mo, K. C., & Lettenmaier, D. P. (2014). Hydrologic prediction over the conterminous United States using the national multi-model ensemble. *Journal of Hydrometeorology, 15*(4), 1457–1472.

Nash, J. E., & Sutcliffe, J. V. (1970). River flow forecasting through conceptual models part I: A discussion of principles. *Journal of Hydrology, 10*(3), 282–290.

Nijssen, B., Schnur, R., & Lettenmaier, D. P. (2001). Global retrospective estimation of soil moisture using the variable infiltration capacity land surface model, 1980–93. *Journal of Climate, 14*(8), 1790–1808.

Pan, S., Liu, L., Bai, Z., & Xu, Y. P. (2018). Integration of remote sensing evapotranspiration into multi-objective calibration of distributed hydrology-soil-vegetation model (DHSVM) in a humid region of China. *Water, 10*(12), 1841.

Parada, L. M., Fram, J. P., & Liang, X. (2003). Multi-resolution calibration methodology for hydrologic models: Application to a sub-humid catchment. *Calibration of Watershed Models, 6*, 197–212.

Patil, A., & Ramsankaran, R. (2017). Improving streamflow simulations and forecasting performance of SWAT model by assimilating remotely sensed soil moisture observations. *Journal of Hydrology, 555*, 683–696.

Pérez, F. F., Sweeck, L., Elskens, M., & Bauwens, W. (2017). A discontinuous finite element suspended sediment transport model for water quality assessments in river networks. *Hydrological Processes, 31*(9), 1804–1816.

Ran, Y., Li, X., & Lu, L. (2010). Evaluation of four remote sensing based land cover products over China. *International Journal of Remote Sensing, 31*(2), 391–401.

Razavi, S., Tolson, B. A., & Burn, D. H. (2012). Review of surrogate modeling in water resources. *Water Resources Research, 48*(7), W7401.

Reed, P., Minsker, B. S., & Goldberg, D. E. (2003). Simplifying multiobjective optimization: An automated design methodology for the nondominated sorted genetic algorithm-II. *Water Resources Research, 39*(7), 1196.

Santana-Quintero, L. V., Montano, A. A., & Coello, C. A. C. (2010). A review of techniques for handling expensive functions in evolutionary multi-objective optimization. In Y. Tenne & C. Goh (Eds.), *Computational intelligence in expensive optimization problems* (pp. 29–59). Berlin, Heidelberg, DE: Springer.

Schapire, R. E. (1990). The strength of weak learnability. *Machine Learning, 5*(2), 197–227.

Schumann, G., Neal, J. C., Voisin, N., Andreadis, K. M., Pappenberger, F., Phanthuwongpakdee, N., et al. (2013). A first large-scale flood inundation forecasting model. *Water Resources Research, 49*(10), 6248–6257.

Sivapragasam, C., Arun, V. M., & Giridhar, D. (2010). A simple approach for improving spatial interpolation of rainfall using ANN. *Meteorology and Atmospheric Physics, 109*(1–2), 1–7.

Sun, X., Bernard-Jannin, L. E. O., Garneau, C., Volk, M., Arnold, J. G., Srinivasan, R., et al. (2016). Improved simulation of river water and groundwater exchange in an alluvial plain using the SWAT model. *Hydrological Processes, 30*(2), 187–202.

Tang, Y., Reed, P., & Wagener, T. (2005). How effective and efficient are multiobjective evolutionary algorithms at hydrologic model calibration? *Hydrology and Earth System Sciences Discussions*, *2*(6), 2465–2520.

Tuo, Y., Marcolini, G., Disse, M., & Chiogna, G. (2018). A multi-objective approach to improve SWAT model calibration in alpine catchments. *Journal of Hydrology*, *559*, 347–360.

Voisin, N., Pappenberger, F., Lettenmaier, D. P., Buizza, R., & Schaake, J. C. (2011). Application of a medium-range global hydrologic probabilistic forecast scheme to the Ohio River basin. *Weather and Forecasting*, *26*(4), 425–446.

Wang, Q., Guidolin, M., Savic, D., & Kapelan, Z. (2014). Two-objective design of benchmark problems of a water distribution system via MOEAs: Toward the best-known approximation of the true Pareto front. *Journal of Water Resources Planning and Management*, *141*(3), 4014060.

Xie, Z., Yuan, F., Duan, Q., Zheng, J., Liang, M., & Chen, F. (2007). Regional parameter estimation of the VIC land surface model: Methodology and application to river basins in China. *Journal of Hydrometeorology*, *8*(3), 447–468.

Xu, Y. P., Gao, X., Zhu, Q., Zhang, Y., & Kang, L. (2015). Coupling a regional climate model and a distributed hydrological model to assess future water resources in Jinhua River Basin, east China. *Journal of Hydrologic Engineering*, *20*(4), 4014054.

Yan, S., & Minsker, B. (2006). Optimal groundwater remediation design using an adaptive neural network genetic algorithm. *Water Resources Research*, *42*(5).

Yan, S., & Minsker, B. (2011). Applying dynamic surrogate models in noisy genetic algorithms to optimize groundwater remediation designs. *Journal of Water Resources Planning and Management*, *137*(3), 284–292.

Yapo, P. O., Gupta, H. V., & Sorooshian, S. (1998). Multi-objective global optimization for hydrologic models. *Journal of Hydrology*, *204*(1–4), 83–97.

Yassin, F., Razavi, S., Wheater, H., Sapriza-Azuri, G., Davison, B., & Pietroniro, A. (2017). Enhanced identification of a hydrologic model using streamflow and satellite water storage data: A multicriteria sensitivity analysis and optimization approach. *Hydrological Processes*, *31*(19), 3320–3333.

Yazdi, J., Hokmabadi, A., & JaliliGhazizadeh, M. R. (2019). Optimal size and placement of water hammer protective devices in water conveyance pipelines. *Water Resources Management*, *33*(2), 569–590.

Zhang, D., Chen, X., Yao, H., & James, A. (2016a). Moving SWAT model calibration and uncertainty analysis to an enterprise Hadoop-based cloud. *Environmental Modeling & Software*, *84*, 140–148.

Zhang, H., Franssen, H. H., Han, X., Vrugt, J. A., & Vereecken, H. (2017). State and parameter estimation of two land surface models using the ensemble Kalman filter and the particle filter. *Hydrology and Earth System Sciences*, *21*(9), 4927.

Zhang, J., Li, W., Zeng, L., & Wu, L. (2016b). An adaptive Gaussian process-based method for efficient Bayesian experimental design in groundwater contaminant source identification problems. *Water Resources Research*, *52*(8), 5971–5984.

Zhang, X. S., Srinivasan, R., & Liew, M. V. (2009). Approximating SWAT model using Artificial Neural Network and support vector machine. *Jawra Journal of the American Water Resources Association*, *45*(2), 460–474.

Zhao, Q., Liu, Z., Ye, B., Qin, Y., Wei, Z., & Fang, S. (2009). A snowmelt runoff forecasting model coupling WRF and DHSVM. *Hydrology and Earth System Sciences*, *13*(10), 1897–1906.

Zhao, X., Ning, B., Liu, L., & Song, G. (2014). A prediction model of short-term ionospheric foF2 based on AdaBoost. *Advances in Space Research*, *53*(3), 387–394.

Zitzler, E., Laumanns, M., & Thiele, L. (2001). SPEA2: Improving the strength Pareto evolutionary algorithm. *TIK-Report*, *103*.

4

FUZZY COMMITTEES OF CONCEPTUAL DISTRIBUTED MODEL

Mostafa Farrag[1,2,3], Gerald A. Corzo Perez[4], and Dimitri P. Solomatine[4,5,6]

[1] Department of Water Quality and Ecology Software, Deltares, Delft, The Netherlands
[2] Hydrology Section, GFZ German Research Centre for Geosciences, Potsdam, Germany
[3] Institute for Environmental Sciences and Geography, University of Potsdam, Potsdam, Germany
[4] IHE Delft Institute for Water Education, Delft, The Netherlands
[5] Water Resources Section, Delft University of Technology, Delft, The Netherlands
[6] Water Problems Institute of the Russian Academy of Sciences, Moscow, Russia

Conceptual hydrological models imply a simplification of the complexity of the hydrological system; however, they lack the flexibility in reproducing a wide range of the catchment responses. Usually, a trade-off is done to sacrifice the accuracy of a specific aspect of the system behavior in favor of the accuracy of other aspects. This study evaluates the benefit of using a modular approach, "The fuzzy committee model" of building specialized models to reproduce specific responses of the catchment. We also assess the applicability of using predicted runoff from specialized models to form a fuzzy committee model. In this paper, weighting schemes with power parameter values are

Advanced Hydroinformatics: Machine Learning and Optimization for Water Resources, Special Publications 78, First Edition. Edited by Gerald A. Corzo Perez and Dimitri P. Solomantine.

DOI: 10.1002/9781119639268.ch4

investigated. A thorough study is conducted on the relation between the fuzzy committee variables (the membership functions and the weighting schemes), and their effect on the model performance. Furthermore, the Fuzzy committee concept is applied on a conceptual distributed model with two cases, the first with lumped catchment parameters and the latter with distributed parameters. A comparison between different combinations of the fuzzy committee variables showed the superiority of all Fuzzy Committee models over single models. Fuzzy committee of distributed models performed well, especially in capturing the highest peak in the calibration data set; however, it needs further study of the effect of model parameterization on the model performance and uncertainty.

4.1. Introduction

Conceptual models are a handy tool as they provide a simplified description of the complexity and heterogeneity of reality; consequently, they produce only a narrow range of catchment response, and the accuracy in reproducing a specific behavior may come against the accuracy of other aspects (Fenicia et al., 2007). Using hydrological models with the best set of parameter results from the calibration of a single objective function might not capture the complexity of the rainfall-runoff relationship in all ranges of flow.

The source of error in conceptual models as described by Agawa and Takeuchi (2016) is that the model structure is not correct to some extent even with optimization. Furthermore, parameter values obtained using a real measurement or estimated from equations differ from values that would achieve an optimal fit. Besides, the spatial variation in both climate variables and catchment characteristics are ignored (Beven, 1996).

The rainfall-runoff relationship is very complex and characterized by highly variable spatial and temporal properties. According to Fenicia et al. (2007), catchment behavior may display different states characterized by phenomena that can generally be identified qualitatively but are challenging to be quantified. Some of these phenomena are seasonality effects due to vegetation, land cover, changes in the topsoil, which affect processes like interception and infiltration. Therefore, catchment responses are season dependent.

Furthermore, catchment hydrological states are the main driving forces for the catchment responses. Changes in the hydrological state could change the catchment conditions from dry to wet or high to low flow. The changes also affect the contributing areas (saturated areas close to the river streams) to discharge, leading to a change in the domain of the hydrological processes. Last, most hydrological processes are characterized by nonlinear behavior like rainfall intensities, groundwater level, and soil moisture conditions. These processes, in return, control the surface runoff, fast and slow runoff response.

Calibration of rainfall-runoff models depends mainly on the objective function used during calibration, and, therefore, different objective function results in different parameter sets. Having more than one set of parameters for one catchment violates the model's physical justification, which ideally should have one unique parameter set (Oudin et al., 2006). The unavoidable dependency of calibration on objective function can be turned into a solution to the different states of the catchment responses. Hence, objective function can be adjusted to handle the practical issue for which the model is built.

By acknowledging the limitations of the conceptual models, Corzo and Solomatine (2007), Fenicia et al. (2007), Fernando et al. (2009), and Oudin et al. (2006) have proposed approaches for the representation of a catchment behavior. Instead of searching for the best model to represent the whole domain in rainfall-runoff relation, runoff can be categorized into different flow regimes, and different models can be calibrated to reproduce each regime (Fig. 4.1). Such specialized models then can be dynamically combined to establish a more comprehensive representation of the catchment processes. This approach is called the *modular* or *multimodel* approach and is considered one way of adding hydrological knowledge into models (Corzo & Solomatine, 2007). The multimodel approach can implicitly consider the variability in the behavior of the catchment response that is not explicitly taken into account in the conceptual model realization.

The modular approach considers switching between different models representing different aspects of system behavior. Therefore, these models have to address different ranges of the flow hydrograph (Abrahart & See, 2000) or different seasons. Based on Corzo and Solomatine (2007), the flow hydrograph was partitioned using three different methods: clustering-based classification, baseflow separation, and temporal segmentation. Jain and Srinivasulu (2006) have decomposed the hydrograph into different segments relying on the concept that different physical

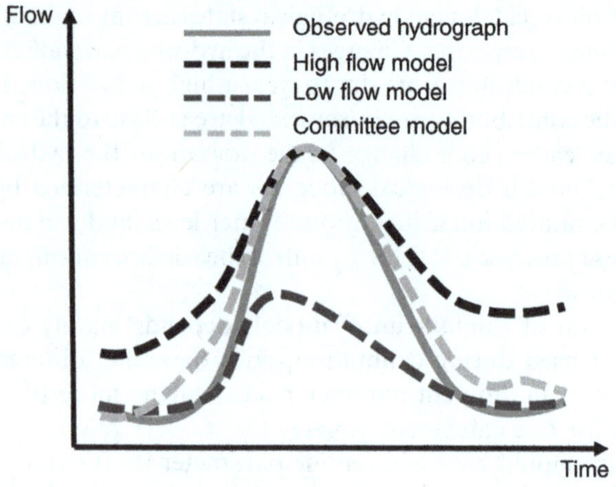

Figure 4.1 The concept of building local models represents different flow regimes and combining the flow simulated by these models using fuzzy membership function.

processes generate different hydrograph segments. Using multi-objective optimization to address different targeted hydrological responses (high and low flow) will result in a set of optimal solutions called *Pareto front*. The Pareto front represents a trade-off between the selected objectives. None of the optimization objectives can be further improved without the degradation of the other.

To incorporate the hydrological knowledge encapsulated in each of these specialized local models, Clemen (1989) has explored different architectures to combine models and forecasts; Jain and Srinivasulu (2006) have used a straightforward switching between models at different time steps. Various ways have been proposed to include a system of weighing, which can be very simple like the averaging method by Shamseldin et al. (1997) or a more complex technique (Ajami et al., 2006; Duan et al., 2007; Shamseldin et al., 1997; Xiong et al., 2001). The Bayesian model averaging scheme (BMA) (Duan et al., 2007) exploits predictions made by different hydrological models to form a probabilistic prediction. BMA weighs different models based on their probabilistic likelihood measures. Xiong et al. (2001) applies the Takagi-Sugeno fuzzy system (Takagi & Sugeno, 1985) to combine forecasted discharge from different models. Specialized models can also be combined based on a membership function. Membership function gives each model a different weight at every time step based on the runoff status (Kayastha et al., 2013). Models can also be combined using a time-varying seasonal index extracted from a state variable or climatic data to weigh the specialized models (Oudin et al., 2006).

The fuzzy committee approach by Solomatine (2006) uses a membership function to express the difference in the degree of believability of the outputs of two models (high and low flow models). The membership function allows for a soft combination of models' output and prevents unrealistic discontinuities in the simulated system behavior (Fig. 4.1).

This research explores the fuzzy committee models first presented by Solomatine (2006) and further developed by Fenicia et al. (2007) and Kayastha et al. (2013), by building and combining different specialized models across weighting schemes and membership functions. In this chapter, different fuzzy committee model setups are tested with different goodness of fit. As recommended by Kayastha et al. (2013), weighting schemes with a power parameter of a value of 2 and 3 are considered. A thorough study is performed on the relation between the fuzzy committee parameters (the membership function (gamma, delta), the weighting schemes' type, and power parameter) and their effect on the model performance. The fuzzy committee concept is applied to a conceptual distributed model. Two cases are considered. The first uses lumped catchment parameters where two specialized models represent the whole catchment. The second uses distributed parameters where each cell is represented by two specialized models.

4.2. Case Study

The study catchment is in the Jiboa River in the central part of El Salvador. The catchment spreads over 432 km^2 with the outlet stations Puente Viejo located at (13.52 °N, 88.98 °W) (Fig. 4.2). The catchment consists of four topographical areas: (1) San Vicente Volcano with an elevation of 2,171 m a.s.l (above sea level) and very steep slopes, (2) the Ilopango Lake (covers an area of 70 km^2), (3) the Balsamo mountain range area with an elevation of 1,000 m a.s.l, and (4) the coastal plain area with elevation ranges from 0 to 100 m a.s.l.

The basin is instrumented by a telemetry system of 15 min time step rainfall gauges and a water level gauge at the outlet of the catchment. For calibration of the model, rainfall and water level values were aggregated to the hourly time step, while the water level was further converted into discharge using a rating curve equation. Monthly potential evapotranspiration values were obtained for each station using a potential evapotranspiration-elevation relation derived by Hargreaves method and

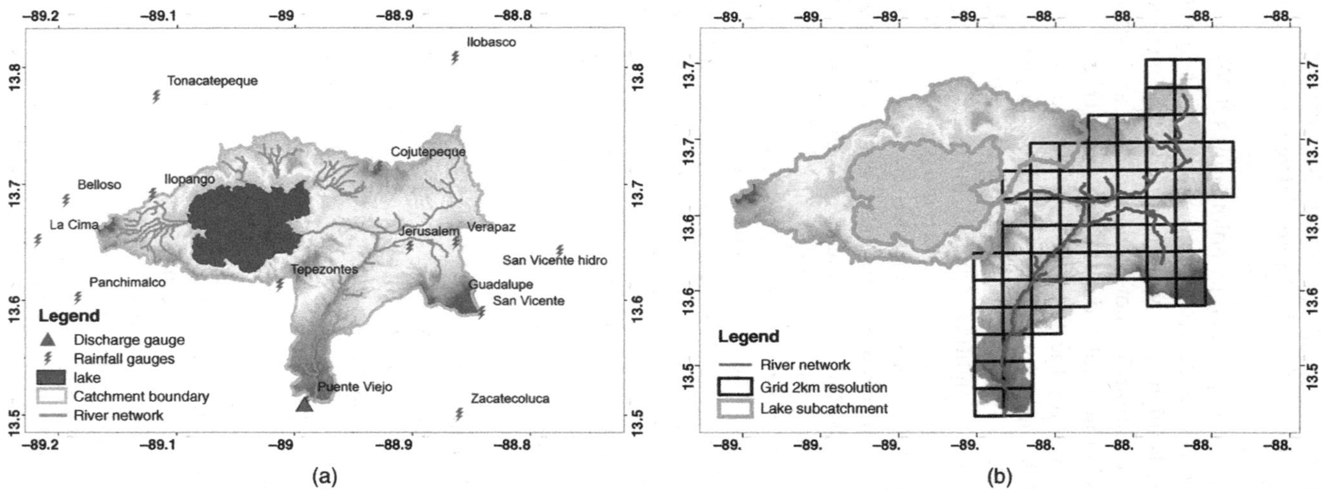

Figure 4.2 (a) Jiboa catchment stations and (b) model setup.

linear regression (SNET, 2005). Hourly temperature estimates were calculated from one available daily station using the sine function and were used for the whole catchment. The lake is simulated using a storage-outflow relation, where storage changes from inflow, direct rainfall, and evaporation are updated each time step. Outflow from the lake is calculated using the newly calculated storage. Inverse square distance weighting method was used to calculate distributed data for different spatial resolution for all data required for the hydrological simulation (precipitation, temperature, potential evapotranspiration).

4.3. Methodology

The first step in building a fuzzy committee is hard partitioning, where a pareto-optimal set of specialized nondominated models is created via multi-objective optimization. Out of many possible models generated, those better representing high and low regimes are selected. The first step results in two specialized high and low flow models addressing the catchment response they are built to reproduce. The second step is combining the simulated discharge from each model developed in the first step, using a fuzzy membership function. The fuzzy membership function consists of two parameters γ and σ whose values are obtained using optimization; the result of this step will be one simulated discharge output. (see Fig. 4.3). The term *committee model* will refer to a committee of two specialized models for low and high flow combined using a weighting scheme with optimized values of γ and δ to minimize the value of RMSE.

4.3.1. Model Structure Description

HBV-96 lumped conceptual model (Lindström et al., 1997) is used in this study to simulate the vertical movement of water for each cell/subcatchment. The raster-based distributed model is built where runoff from each cell is routed directly to the outlet using the Maxbas function (Fig. 4.4); each cell has a MAXBAS value depending on its flow path length. Discharge at the outlet is obtained by summing the routed runoff from all cells at each time step. Only superficial discharge (surface runoff and interflow generated from upper zone UZ reservoir in the HBV model structure) is routed, while flow generated from lower zone LZ is assumed to be lumped for the whole catchment (Fig. 4.4).

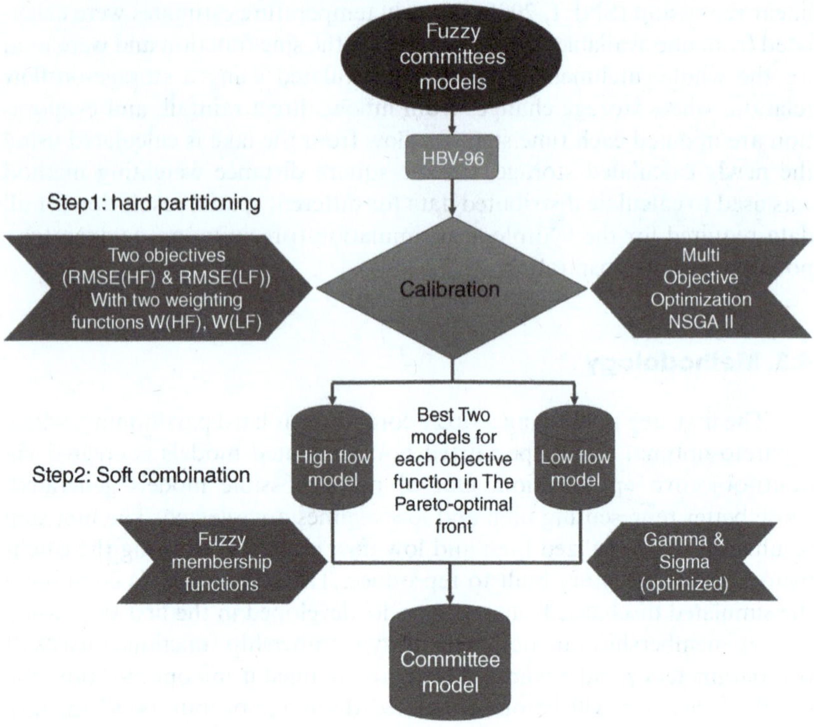

Figure 4.3 Proposed steps and approaches for improving committee models using a dynamic weighting scheme.

Several trials have been made to adapt the model structure to the spatial separation of the hydrological process, which happens when using the distributed model, as stated by Dehotin and Braud (2008). The beta parameter is responsible for passing water to the upper zone (equation (4.1)). To increase the sensitivity of the model structure to catchment fast response, beta is considered in the calibration with a value greater than 1. Additionally, a direct runoff rule is added to move water from the soil moisture tank to the upper zone whenever soil moisture exceeds the soil's field capacity (Schumann, 1993).

$$\frac{R}{IN} = \left(\frac{SM}{FC}\right)^{\beta}. \tag{4.1}$$

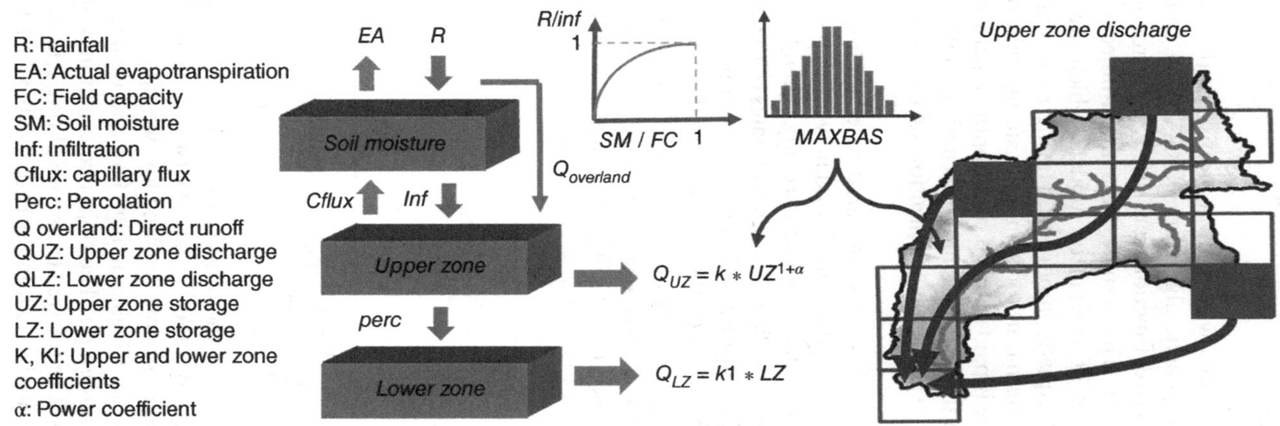

Figure 4.4 Model structure.

4.3.2. Model Calibration and Building Specialized Local Models

Model calibration follows a multi-objective optimization using the nondominated sorted genetic algorithm NSGA-II by Deb et al. (2002), where different objective functions are used as introduced by Gupta and Sorooshian (1998). The result of the multi-objective optimization is a pareto-optimal set of specialized models, which are nondominated solutions, and all are equally important. Having more than one solution indicates the conflicting nature of solutions, stating that optimal performance in one objective is paid in terms of suboptimal performance in other objectives. Every solution has its limitations and strengths in reproducing different catchment responses as a reflection of the associated objective function.

Model performance is always measured based on how close the simulated flow to the observed hydrograph is and how much the model succeeds in reproducing the streamflow volume correctly. Therefore, high and low flow were selected as distinct flow regimes and states of the system behavior; if the model managed to capture both states, the overall performance would be improved. The objectives of the modular model are to reproduce the system behavior during both regimes as accurately as possible. Root mean square error (RMSE) with W_{hf} and W_{lf} as weighting schemes (equations (4.2) and (4.3)) were used to put a higher emphasis on high and low flow values, respectively.

$$\mathrm{RMSE_{hf}} = \sqrt{\frac{1}{n}\left(\sum_{i=1}^{n}((\mathrm{Qobs_i} - \mathrm{Qsim_i}))^2 * \mathrm{W_{hf,i}}\right)} \tag{4.2}$$

and

$$\mathrm{RMSE_{lf}} = \sqrt{\frac{1}{n}\left(\sum_{i=1}^{n}((\mathrm{Qobs_i} - \mathrm{Qsim_i}))^2 * \mathrm{W_{lf,i}}\right)}, \tag{4.3}$$

where $Qobs_i$ and $Qsim_i$ are the observed flow and simulated flow respectively at each time step, W_{lf} and W_{hf} are the two weighting functions, with W_{lf} to place high weights on low flow and low weights on high flow values and W_{hf} to place stronger weights on high flow and weaker weights on low flow values on the hydrograph. The objective function, together with the weighting function, will be referred to as a weighting scheme (WStype), the corresponding figures and equation for each WStype are below (Fig. 4.5, Table 4.1)

Table 4.1 Weighting schemes for root mean square error for high ($RMSE_{hf}$) and low flow ($RMSE_{lf}$).

WStype	W_{lf}	W_{hf}	Eq	Figure
I	$=(l)^N$	$=(h)^N$	4.4	Figure 4.5a
II	$\begin{cases} 0, if\, l > \alpha \\ \left(\left(\frac{1}{\alpha^2}\right) * (1-l)^2 - \left(\frac{2}{\alpha}\right) * (1-l) + 1\right), if\, l \leq \alpha \end{cases}$	$\begin{cases} 1, if\, h > \alpha \\ (h/\alpha)^N, if\, h \leq \alpha \end{cases}$	4.5	Figure 4.5b
III	$\begin{cases} 0, if\, l > \alpha \\ \left(\left(\frac{1}{\alpha^2}\right) * (1-l)^2 - \left(\frac{2}{\alpha}\right) * (1-l) + 1\right), if\, l \leq \alpha \end{cases}$	$\begin{cases} 1, if\, h > \alpha \\ 0, if \leq \alpha \end{cases}$	4.6	Figure 4.5c
IV	$\begin{cases} 0, if\, l > \alpha \\ \left(1 - \left(\frac{1+1}{\alpha}\right)\right), if\, l \leq \alpha \end{cases}$	$\begin{cases} 1, if\, h > \alpha \\ 0, if \leq \alpha \end{cases}$	4.7	Figure 4.5d

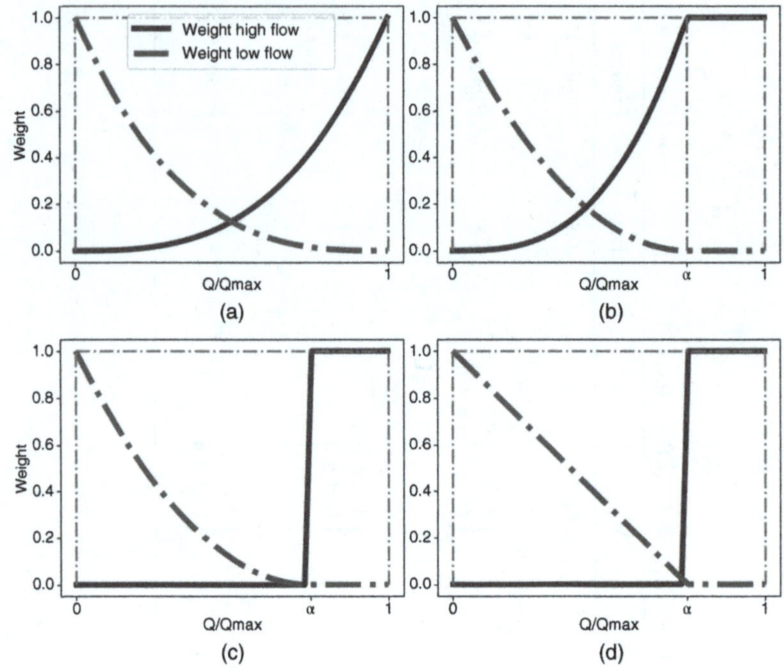

Figure 4.5 Weighting schemes for weighted root mean squared error as an objective function. W_{lf} and W_{hf} are obtained from equations (4.4)–(4.7): (a) WS-I, (b) WS-II, (c) WS-III, and (d) WS-IV.

l and h in the previous equations are calculated as equation (4.8):

$$l = \frac{(Q_{obs,max} - Q_{sim,i})}{Q_{obs,max}}, \quad h = \frac{Q_{obs,i}}{Q_{obs,max}}, \tag{4.8}$$

where $Q_{obs,\, max}$ is max observed flow value, α is a threshold for selecting weights, N is the power (see Table 4.1). N was chosen to be 2 ans 3, the value of α is a ratio between calculated discharge at a particular time step and the maximum discharge value in the calibration data set ($Q_i/Q_{obs,\, max}$). α, determines the range of values where both models are working together, and the range where discharge value from the high flow model is only considered, α was chosen to be 0.75.

It is important to mention that the values of RMSE_{lf}, RMSE_{hf}, and *RMSE* are calculated differently and cannot be compared with each other, as they have been multiplied by different weights. RMSE_{lf} evaluates the error in low flow and RMSE_{hf} evaluates the error in high flow, while RMSE evaluates the error in the general flow. Data used in the calibration are almost 2 yr and a half-hourly time step, two wet seasons were used in the calibration data set and one in the validation.

4.3.3. Combining Specialized Models

The main objective is to produce a flow simulation efficient in both low and high flow domains using time-varying weights, as mentioned earlier. Obtaining a competent solution in all flow regimes is very difficult, and using a specialized model in both high and low flow is a better way to achieve a general efficient model.

The challenge in combining different specialized models built under different flow regimes is to determine which model will work and how to switch smoothly from one model to another carrying the water balance information; and in case more than one model is working at a time, how to handle the compatibility at the boundaries.

In order to avoid such complication, using a weighting scheme, which guarantees a smooth transition between models, can be the solution. The share of each specialized model will be determined using a fuzzy membership function described by Fenicia et al. (2007), Kayastha et al. (2013), and Solomatine (2006).

In the weighting scheme, a trapezoidal function with two parameters γ and δ is used, wherein the membership function of the low flow model a weight of 1 will be given to low flow values if the relative flow is below parameter γ (Fig. 4.6c), then starts to decrease when relative flow values are between γ and δ (both models are working together), until it reaches zero beyond the boundary of parameter δ (Fig. 4.6c). Similarly, the membership function of a high flow model is working but in opposite logic. The membership functions are shown here (equations (4.9) and (4.10)).

$$M_{lf} = \begin{cases} 1, if\, h < \gamma \\ 1 - \left(\frac{(h-\gamma)}{(\delta-\gamma)}\right)^N, if\, \gamma \le h \le \delta \\ 0, if\, h \ge \delta \end{cases} \tag{4.9}$$

and

$$M_{hf} = \begin{cases} 0, if\, h < \gamma \\ \left(\frac{(h-\gamma)}{(\delta-\gamma)}\right)^{1/N}, if\, \gamma \le h \le \delta \\ 1, if\, h \ge \delta \end{cases} \tag{4.10}$$

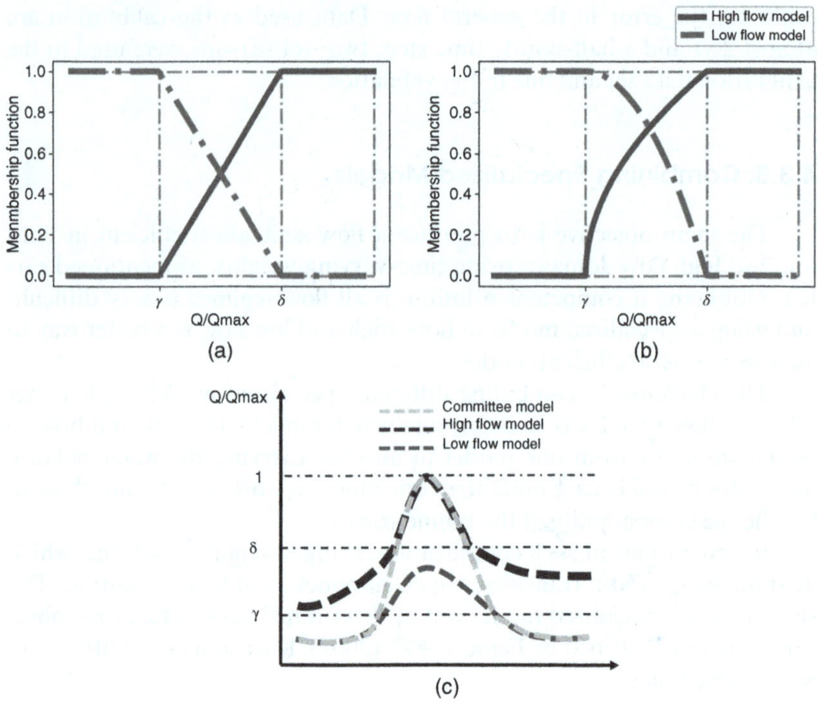

Figure 4.6 Fuzzy membership functions to combine specialized models: (a) MF-A uses a linear transition between γ and σ threshold (N equals 1); (b) MF-B uses nonlinear transition with N equals 2; (c) segments of the hydrograph where each part of the membership function is applied.

The output of the two specialized models is multiplied by the previous weights and then normalized, so the final calculated flow from the committee models will be as follows (equation (4.11)). The combining function of high and low flow specialized models

$$Q_{C,i} = \frac{M_{lf,i} * Q_{lf,i} + M_{hf,i} * Q_{hf,i}}{M_{lf,i} + M_{hf,i}}, \tag{4.11}$$

where $Q_{LF,i}$ and $Q_{hF,i}$ are simulated high and low flows, respectively; M_{LF} and M_{hF} are membership functions for low and high flow models, respectively; i is the time step; and γ and δ are two thresholds for low and high flows, respectively. N is the power value to smooth the transition between models; for Type A, N is equal to one, for Type B, N is equal to two (see Fig. 4.6).

The γ and σ are the membership function thresholds that represent switches in the catchment behavior; these switches could be a reflection of change in contributing area that feeds the outflow or change in the channel properties (Fenicia et al., 2007).When there is no way to quantify the change in the catchment behavior into thresholds, γ and σ could be determined using optimization to minimize the root mean squared error of the committee model.

4.3.4. Model Setup

Different spatial setups were used to simulate the hydrological response of the catchment. The presence of the lake prevented the representation of the whole catchment using one lumped model. Therefore, the lake and all the upstream subcatchments to the lake were represented as one lumped model (Fig. 4.2), which will be referred to as lake subcatchment. Different spatial resolutions are used to simulate the rest of the catchment (which will be referred to as the Jiboa subcatchment) in order to capture the spatial variability of the precipitation. Jiboa subcatchment is simulated using different model setups as follows:

1. Lumped model (two models in total)
2. Committee of specialized lumped models (Lumped-MF-A and Lumped-MF-B) where two models are combined using two different membership functions (four models in total, two specialized models for each subcatchment).
3. Conceptual distributed model built with lumped catchment parameters (Dist-L) using different spatial resolutions (4 km, 2 km, 1 km, and 500 m) (two models in total, one model for each subcatchment)
4. Committee of conceptual distributed models (Dist-L-C two specialized models for the entire subcatchment) built with lumped catchment parameters using a spatial resolution of 4 km^2 (four models in total, two specialized models for each subcatchment)
5. Conceptual distributed model built with distributed catchment parameters (Dist-D different parameter for each cell) using 4 km^2 spatial resolutions (one model for each cell, and one model for the lake subcatchment)
6. Committee of conceptual distributed models (Dist-D-C two specialized models for each cell) built with distributed catchment parameters using a spatial resolution of 4 km^2 (two models for each cell, and two specialized models for the lake subcatchment)

The lake subcatchment is represented the same way in all models, and the only difference between models is the spatial representation of the Jiboa subcatchment.

4.3.5. Evaluation Criteria

Different performance criteria were used to compare and assess the performance of each model: Nash-Sutcliffe efficiency coefficient (NSE) (equation (4.12)), Nash-Sutcliffe with logarithmic values (equation (4.13)), root mean square error (RMSE) (equation (4.15)).

$$\mathrm{NSE(HF)} = 1 - \frac{\sum_t (\mathrm{Qobs} - \mathrm{Qsim})^2}{\sum_t (\mathrm{Qobs} - \mathrm{Qavg})^2}. \tag{4.12}$$

As the standard formula of Nash-Sutcliffe overestimates high values, so changing the values of flow by logarithmic values (equation (4.13)) will flatten the high flow; the peaks and low flow values will be the same. As a result, the effect of the error in low flow values will dominate, and the error value will be more sensitive to error in low flow values (Krause et al., 2005).

$$NSE(lf) = 1 - \frac{\sum_t (\ln(Qobs) - \ln(Qsim))^2}{\sum_t (\ln(Qobs) - \ln(Qavg))^2}. \tag{4.13}$$

The mean cumulative error measures how much the model succeeds in reproducing the streamflow volume correctly (equation (4.14)).

$$WB = 100 * \left(1 - \left|1 - \frac{\sum Qsim}{\sum Qobs}\right|\right). \tag{4.14}$$

$$RMSE = \sqrt{\frac{1}{n}\sum_{i=1}^{n}(Qobs - Qsim)^2}. \tag{4.15}$$

Gupta et al. (2009) have shown the limitation of using a single error function to measure the efficiency of calculated flow and showed that Nash-Sutcliff efficiency (NSE) or RMSE could be decomposed into three components: correlation, variability, and bias. The KGE index (equation (4.16)) has aggregated this information into one formula, which measures how far the values of these properties to the ideal point.

$$KGE = 1 - \sqrt{(c-1)^2 + (\alpha - 1)^2 + (\beta - 1)^2}, \tag{4.16}$$

where *Qobs* is the observed discharge, *Qsim* is the simulated discharge, and *Qavg* is the mean of observed discharges, c = *cross correlation between*

Qobs & *Qsim*, α measures the variability in the data $\alpha = \frac{\sigma_{Qsim}}{\sigma_{Qobs}}$, and $\beta = \frac{\text{mean(Qsim)}}{\text{mean(Qobs)}}$.

4.4. Results and Discussion

The conceptual representation of the catchment responses in the HBV model does not allow for simultaneous optimal representation of various behaviors of the catchment response. Therefore, the committee model approach is used to capture simultaneously different aspects of the catchment behavior

For the lake subcatchment, two specialized models have been built and used for the whole study without change, while for the Jiboa subcatchment, different spatial discretization (lumped and distributed) were modeled using four different cell resolutions (4 km, 2 km, 1 km, and 500 m)(see Fig. 4.7 and section 4.3.4 Model Setup).

For the lumped representation of the Jiboa subcatchment, two specialized models have been built for each subcatchment (two models for lake subcatchment and two models for Jiboa subcatchment). The Table 4.2 shows a summary of the performance of a different combination of committee models.

The values of both gamma and delta in Table 4.1 are further visualized in Figure 4.8 to analyze the interaction between the weighting scheme, the membership function, and the power value N. Gamma and delta are values of discharge threshold that determine how much both specialized models are working together. Figure 4.8 shows that MF-A makes specialized models work together more as it goes from WS-1 to WS-IV (the length of the lines increase from WS-I to WS-IV). On the other side, MF-B and N = 2 for both WS-I and II specialized models are almost working individually (models works together only in the range of Q/Qmax between γ of 0.24 to δ of 0.27), while MF-B with WS-I, WS-II (N = 3), WS-III and IV make models work together most of the time (in the range of Q/Qmax between $\gamma = 0.06$ to $\delta = 0.88$).

More comprehensive studying on the relation between membership function, weighting scheme, and power value (N) shows that MF-B tends to make specialized models either to work together all the time or to work almost individually, as shown in Figure 4.8, which gives higher values of errors as shown in Figure 4.9 (points with bigger size). MF-A makes specialized models work individually and together almost equal times (the difference between delta and gamma around 0.5), which gives better results for committee model performance.

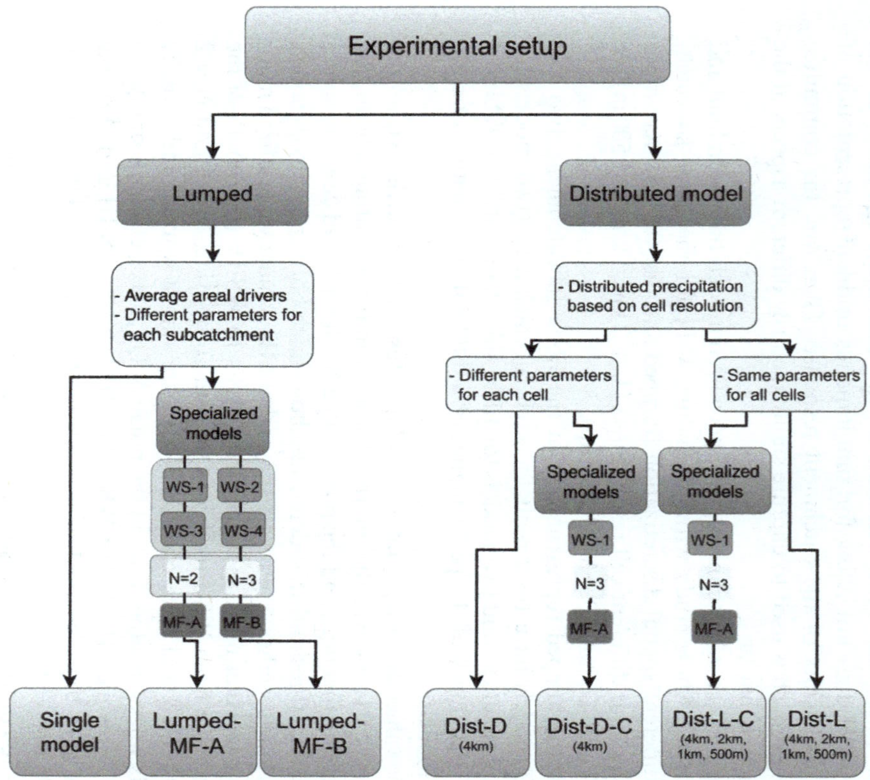

Figure 4.7 Experimental models setup.

Table 4.2 Performance of committee models with a different combination of weighting schemes and membership.

Model	WS type		Membership function			Low flow RMSE (lf)		High flow RMSE (Hf)		RMSE		KGE	
	WS type	N	MF	γ	Δ	Cal.	Val.	Cal.	Val.	Cal.	Val.	Cal.	Val.
	Single model					0.98	0.80	0.63	0.70	1.66	1.72	0.70	0.650
Committee model	I	2	A	0.19	0.36	0.95	0.84	0.58	0.71	1.36	1.45	0.84	0.84
			B	0.23	0.26	0.97	0.90	0.59	0.72	1.39	1.51	0.84	0.84
		3	A	0.17	0.43	0.80	0.68	0.42	0.47	1.28	1.34	0.88	0.87
			B	0.08	0.79	0.83	0.64	0.44	0.45	1.33	1.25	0.85	0.85
	II	2	A	0.15	0.35	0.80	0.65	0.75	0.88	1.33	1.37	0.85	0.83
			B	0.24	0.26	0.83	0.78	0.76	0.91	1.38	1.50	0.83	0.81
		3	A	0.18	0.41	0.81	0.70	0.60	0.73	1.28	1.41	0.87	0.86
			B	0.06	0.72	0.84	0.65	0.61	0.70	1.33	1.31	0.86	0.87
	III	-	A	0.16	0.57	0.82	0.67	0.35	0.29	1.31	1.28	0.87	0.87
			B	0.10	0.84	0.84	0.65	0.34	0.26	1.34	1.22	0.86	0.87
	IV	-	A	0.19	0.61	1.01	0.85	0.33	0.24	1.32	1.26	0.86	0.86
			B	0.13	0.87	1.03	0.79	0.34	0.24	1.35	1.18	0.86	0.90

Model	WS type		Membership function	Low flow NSE (ln)		High flow NSE		WB	
	WS type	N	MF	Cal.	Val.	Cal.	Val.	Cal.	Val.
Single model				0.6	0.821	0.615	0.691	99.48	85.08
Committee model	I	2	A	0.66	0.843	0.741	0.781	98.54	89.08
			B	0.66	0.82	0.731	0.762	99.02	89.09
		3	A	0.69	0.856	0.769	0.812	99.38	93.86
			B	0.68	0.864	0.750	0.836	99.89	87.57
	II	2	A	0.7	0.886	0.751	0.805	99.14	87.32
			B	0.69	0.84	0.734	0.764	97.35	85.08
		3	A	0.7	0.854	0.770	0.792	98.84	93.25
			B	0.69	0.862	0.752	0.820	97.05	91.25
	III	-	A	0.68	0.887	0.759	0.830	99.07	93.15
			B	0.68	0.885	0.749	0.844	99.42	88.93
	IV	-	A	0.66	0.893	0.755	0.833	96.23	96.48
			B	0.65	0.895	0.746	0.855	95.88	93.72

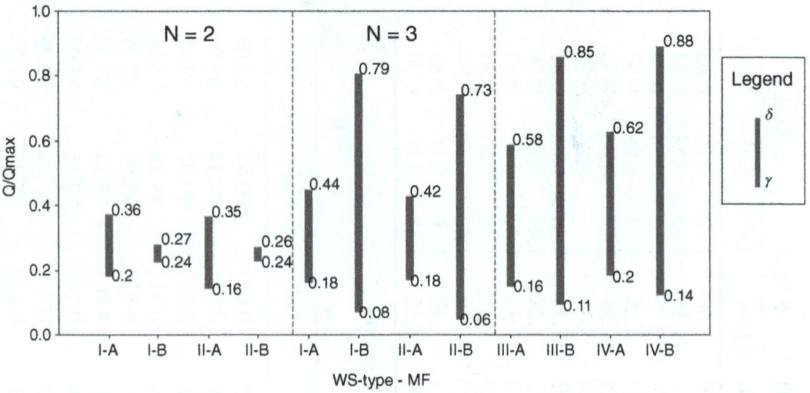

Figure 4.8 Weighting scheme parameters (gamma and delta). The first part of the x-axis notation represents the WS-type, and the second part represents the membership function. The figure is divided into parts based on the power N (WS-III and WS-IV do not have N). The line represents the range of discharge where both models (high and low flow models) are considered in the total calculated discharge for the committee model Qc (i.e., for WS I-A for Q/Qmax less than 0.2, only the low flow model is considered, and the opposite for Q/Qmax is greater than 0.36, while for Q/Qmax between 0.2 and 0.36, both models are considered in the Qc).

The weights W_{lf},W_{hf} are mainly affected by the proportional value of the calculated discharge to the maximum observed discharge during the calibration period ($Q_i/Q_{obs,max}$). Since $Q_{obs,max}$ depends on the chosen period of the calibration and can be surpassed in the validation data set or when the model is used in operation, then W_{lf} will have negative values. To avoid such a problem, a wider range of discharge could be used. Parameters obtained by the optimization algorithm are shown in Figure 4.10.

In Figure 4.10, the parameters' values are normalized with respect to the lower and upper bounds so that the feasible range of all parameters could be shown at the same figure between 0 and 1. Each line on the figure represents one set of parameters, and it indicates the relation between how the parameter range corresponds to the optimal solutions and the initial feasible parameter range. Any observations about the degree of variability of the parameters will not be in the right context as the figure represents only the values of the Pareto optimal set of solutions (see dark squares in Fig. 4.15), not all the solutions (other grey dots in Fig. 4.15).

The behavior of some parameters may follow a trend when moving from the best model for high flow to the best model for low flow; this behavior is evident in parameter K, which is the discharge coefficient of the upper

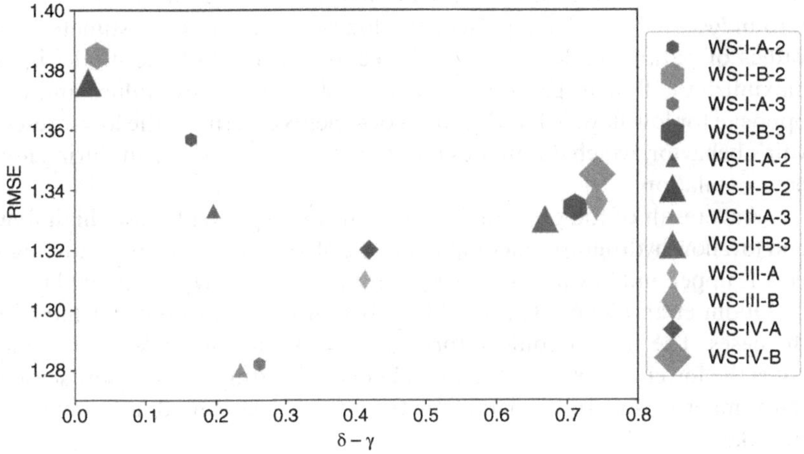

Figure 4.9 The relation between committee component and committee model performance, the first part of the points' notation, refers to the WS-type; the second part refers to the membership function (A or B); and the last part represents the power N (WS-III and WS-IV do not have N).

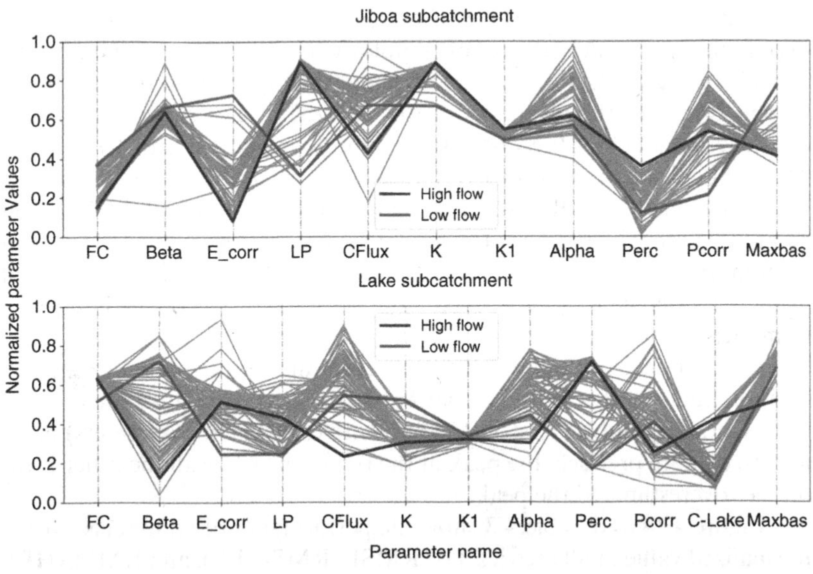

Figure 4.10 Normalized parameters plot (parameters corresponding to Pareto optimal solutions).

zone (subsurface flow) and K1 the discharge coefficient of the lower zone (groundwater flow). As expected, the high flow model overestimates the values of parameter K, K1, and C-lake as the trend of the model is to maximize the flow to get as close as possible to high flow values and the opposite for low flow. K1 and C-lake coefficients determine the lower reservoir's behavior, which dominates the response of the catchment during low flow simulation.

As a result of the previous behavior, it was expected to have high flow and low flow hydrograph encompassing the observed flow hydrograph (high flow is upper and low flow is lower), but this does not happen as explained by Oudin et al. (2006). This could happen in ideal cases but not most of the cases. The reason could return to the existence of a lake that dominates the lower response of the catchment, and during the low flow season (November to April), the total flow from the catchment comes only from the lake.

For conceptual distributed models, specialized models have been built using weighting scheme 1 (N = 3) for every single cell in the Jiboa subcatchment. Membership function A was used to combine specialized models obtained earlier using two parameters gamma and delta obtained by optimization using root mean square as an objective function. The committee of distributed models' performance shows very close results in both calibration and validation data sets (see Fig. 4.11, Table 4.2 and Table 4.3), which implies that the models' degree of complexity is close to system complexity. Models in Figure 4.11 are ordered from left to right based on how many models represent the responses of the catchment. The single model uses one general model for the whole Jiboa subcatchment, while the "Dist-D-C 4 km" model uses two models for each cell. Moreover, the performance has improved from a RMSE of 1.64 m^3/s in single models to 1.3 m^3/s in committee models; the performance of all committees is very close; and a committee model represent.ing Jiboa subcatchment spatially as lumped using MF-A is the best-performed model (see Fig. 4.12; for the indication of models' name see Fig. 4.7).

A closer check to the highest flood event during the period of the calibration data set, it was found that the distributed committee model with 4 km spatial resolution Dist-D-C 4 km (with distributed parameters) manages to capture precisely the peak at the right time while all the other committees overestimated the peak.

Figure 4.14 and Table 4.3 show the performance of all models, with a normalized value of all metrics. For RMSE, RMSE(LF), and RMSE(HF), the highest error is at the top of the graph with a value of 1.656, 1.05, and 0.64, respectively, and the lowest values at the bottom of the graph with values of 1.282, 0.8, and 0.43, respectively. While for NSE(HF) and

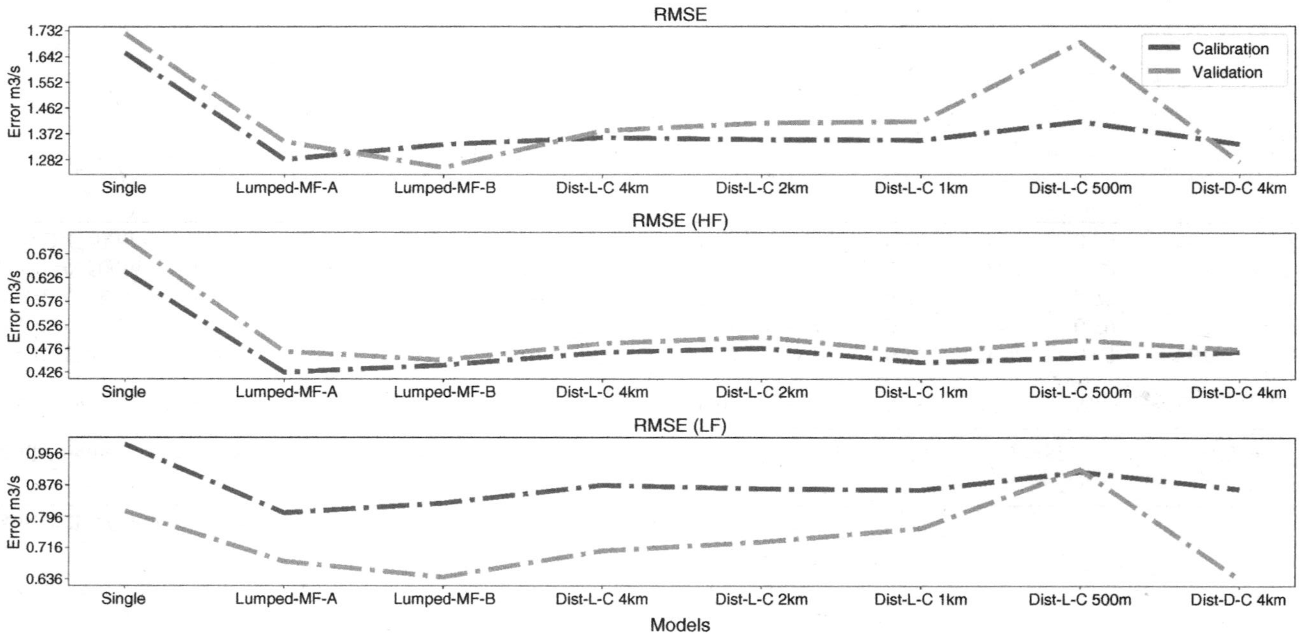

Figure 4.11 Performance graph of all committee models. Models are ordered from left to right based on how many models are used to represent the catchment.

Table 4.3 Committee of conceptual distributed models' performance.

Model name	Low flow RMSE (lf)		High flow RMSE (Hf)		RMSE		KGE		High flow NSE		Low flow NSE (ln)	
	Cal.	Val.	Cal.	Val.	Cal.	Val.	Cal.	Val.	Cal.	Val.	Cal.	Val.
Dist-L-C 4km	0.88	0.71	0.47	0.49	1.36	1.38	0.85	0.87	0.74	0.80	0.67	0.86
Dist-L-C 2km	0.87	0.73	0.48	0.50	1.35	1.41	0.84	0.86	0.74	0.79	0.68	0.85
Dist-L-C 1km	0.86	0.77	0.45	0.47	1.35	1.41	0.85	0.80	0.74	0.79	0.71	0.76
Dist-L-C 500m	0.91	0.92	0.46	0.49	1.41	1.69	0.79	0.79	0.72	0.70	0.66	0.78
Dist-D-C 4km	0.87	0.64	0.47	0.47	1.34	1.27	0.84	0.88	0.75	0.83	0.67	0.90

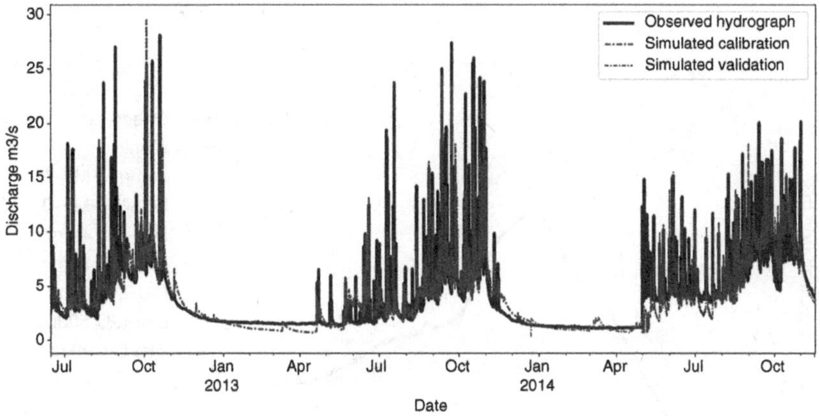

Figure 4.12 Flow hydrograph result from committee model Lumped-MF-A.

NSE(LF), the highest values are at the bottom of the graph with values of 0.71 and 0.77, respectively, and the lowest values at the top of the figure with values of 0.55 and 0.62, respectively. All single models (lumped and distributed) appear at the upper part of the figure (highest RMSE and lowest NSE) while committee models appear at the bottom of the graph (lowest RMSE and highest NSE), which indicates that committee models outperformed all single models in all performance criteria. Figure 4.13 with Figure 4.11 show that the effect of cells' size in distributed models is very slight, which indicates the need for more research regarding the conceptual distributed models.

The best 70 model parameterizations found by NSGA-II are presented in Figure 4.15 as a Pareto front of optimal nondominated solutions. It also illustrates the trade-off between selected objectives and indicates that the single models can not simultaneously match the full variability of the catchment response. The optimal set of parameters for the calibration data set does not have superiority over the validation data set, so the best-specialized models of high and low flow obtained for the calibration data set are not the best (among other Pareto optimal set) in the validation data set.

Figure 4.15 shows that committee models are closer to the ideal point than the single models and all members of the Pareto front. This ensures that committee models' performance is better than any other single model. The optimal models in calibration are not necessary to be the best in validation as well. However, the version in calibration and validation has to be very close. All combinations of committee models were better than all of Pareto members in both calibration and validation data sets.

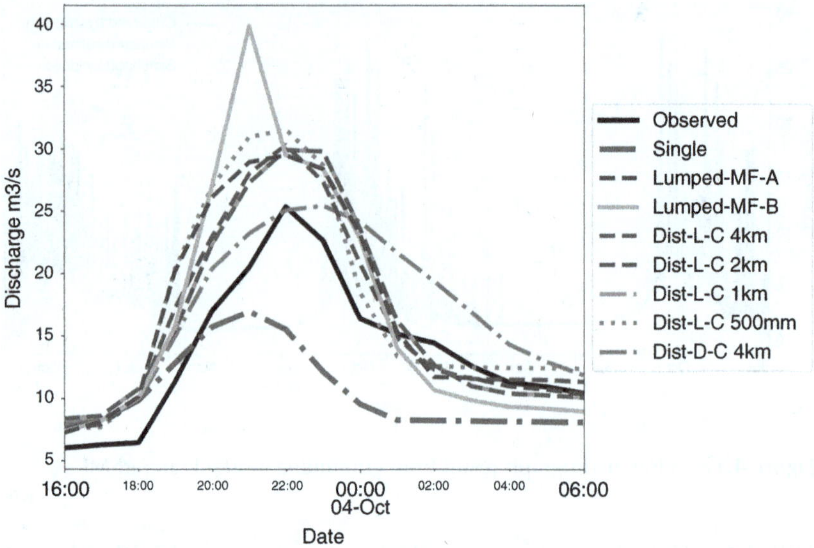

Figure 4.13 Performance of committee models at the highest flood event in the calibration period.

It is easily noticeable that the values of $RMSE_{LF}$ are always higher than the values of $RMSE_{HF}$. The reason is that the low flow period is much longer than the high flow period, so it is expected that longer periods would have higher error than shorter periods (the number of low flow values is more than high flow values) and the formula of weighted root mean squared error summation of the error is divided by the same total number of observations see (Fig. 4.15).

A committee of specialized models developed to reproduce a particular aspect of the hydrologic response (high and low flow models) combined using a fuzzy membership function managed to exploit the strength of each model to improve the performance of the committee. This gives the closest performance to the ideal point than all single models and all Pareto front members, which ensures that the committee models' performance is better than any other single model.

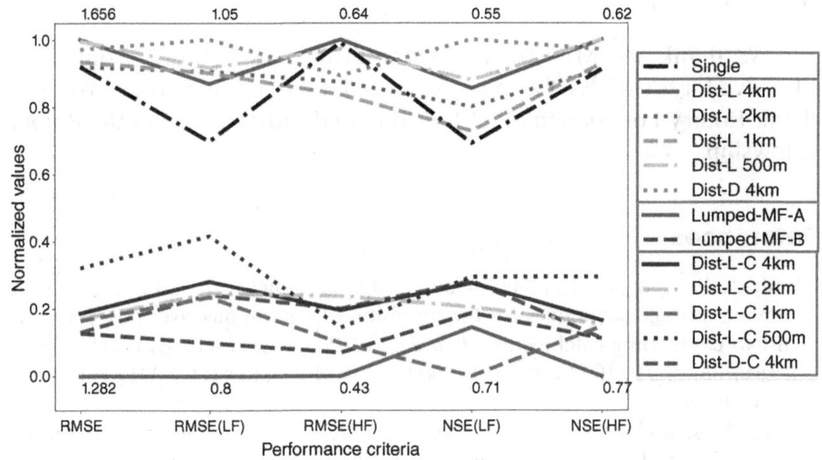

Figure 4.14 Normalized performance criteria of all models for the visual comparison, the maximum and minimum values of each error are presented. Committee models at the bottom have lower RMSE, RMSE(LF), and RMSE(HF), and higher NSE(LF) and NSE(HF); thus, the performance is better in all criteria.

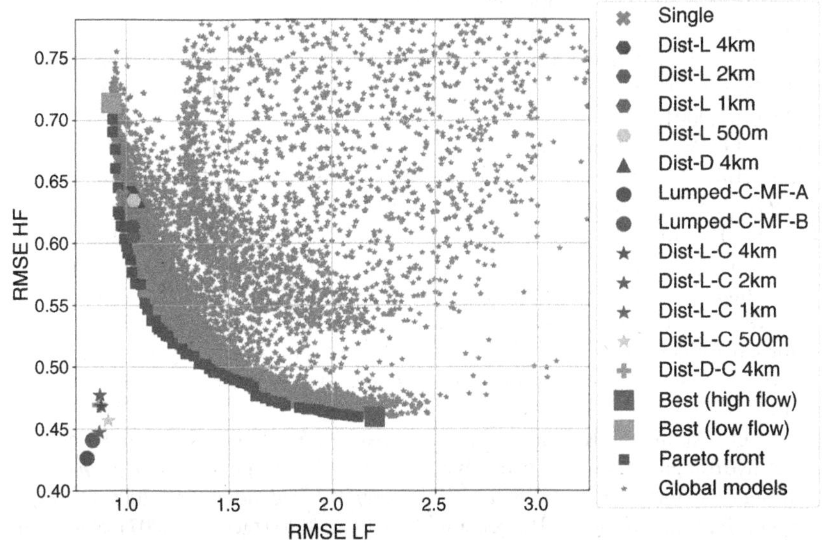

Figure 4.15 Pareto optimal sets of specialized models result from the multiobjective optimization and the different models.

Acknowledgments

We thank the Ministry of Environment and Natural resources of El Salvador (MSRN) for providing data used in this study. We also thank Rotary Foundation (TRF) for the fund provided within the Rotary fellowship.

References

Abrahart, R. J., & See, L. (2000). Comparing neural network and autoregressive moving average techniques for the provision of continuous river flow forecasts in two contrasting catchments. *Hydrological Processes*, *14*(11–12), 2157–2172. https://doi.org/10.1002/1099-1085(20000815/30)14:11/12<2157::AID-HYP57>3.0.CO;2-S

Agawa, T., & Takeuchi, O. (2016). Re-examination of psychological needs and L2 motivation of Japanese EFL learners: An interview study. *Asian EFL Journal*, *89*(89), 74–98.

Ajami, N. K., Duan, Q., Gao, X., & Sorooshian, S. (2006). Multimodel combination techniques for analysis of hydrological simulations: application to distributed model intercomparison project results. *Journal of Hydrometeorology*, *7*(4), 755–768. https://doi.org/10.1175/jhm519.1

Beven, K. (1996). *Equifinality and uncertainty in geomorphological modeling*. The Scientific Nature of Geomorphology. Proceeding of the 27th Binghamton Symposium in Geomorphology, *September 1996*, 289–314.

Clemen, R. T. (1989). Combining forecasts: A review and annotated bibliography. *International Journal of Forecasting*, *5*(4), 559–583. https://doi.org/10.1016/0169-2070(89)90012-5

Corzo, G., & Solomatine, D. (2007). Baseflow separation techniques for modular artificial neural network modeling in flow forecasting. *Hydrological Sciences Journal*, *52*(3), 491–507. https://doi.org/10.1623/hysj.52.3.491

Deb, K., Pratap, A., Agarwal, S., & Meyarivan, T. (2002). A fast and elitist multiobjective genetic algorithm: NSGA-II. *IEEE Transactions on Evolutionary Computation*, *6*(2), 182–197. https://doi.org/10.1109/4235.996017

Dehotin, J., & Braud, I. (2008). Which spatial discretization for distributed hydrological models? Proposition of a methodology and illustration for medium to large-scale catchments. *Hydrology and Earth System Sciences*, *12*(3), 769–796. https://doi.org/10.5194/hess-12-769-2008

Duan, Q., Ajami, N. K., Gao, X., & Sorooshian, S. (2007). Multimodel ensemble hydrologic prediction using Bayesian model averaging. *Advances in Water Resources*, *30*(5), 1371–1386. https://doi.org/10.1016/j.advwatres.2006.11.014

Fenicia, F., Solomatine, D. P., Savenije, H. H. G., & Matgen, P. (2007). Soft combination of local models in a multi-objective framework. *Hydrology and Earth System Sciences*, *11*(6), 1797–1809. https://doi.org/10.5194/hess-11-1797-2007

Fernando, D. A. K., Shamseldin, A. Y., & Abrahart, R. J. (2009). *Using gene expression programming to develop a combined runoff estimate model from conventional rainfall-runoff model outputs*. 18th World IMACS Congress and

MODSIM09 International Congress on Modeling and Simulation: Interfacing Modeling and Simulation with Mathematical and Computational Sciences, Proceedings, July, 748–754. https://www.scopus.com/inward/record.uri?eid=2-s2.0-80052983135&partnerID=40&md5=73dcd5eba92b4c026dcff2b6bf128f02

Gupta, H. V., & Sorooshian, S. (1998). Toward improved calibration of hydrologic models: Multiple and noncommensurable measures of information. *Water Resources Research*, *34*(4), 751–763.

Gupta, H. V., Kling, H., Yilmaz, K. K., & Martinez, G. F. (2009). Decomposition of the mean squared error and NSE performance criteria: Implications for improving hydrological modeling. *Journal of Hydrology*, *377*(1–2), 80–91. https://doi.org/10.1016/j.jhydrol.2009.08.003

Jain, A., & Srinivasulu, S. (2006). Integrated approach to model decomposed flow hydrograph using artificial neural network and conceptual techniques. *Journal of Hydrology*, *317*(3–4), 291–306. https://doi.org/10.1016/j.jhydrol.2005.05.022

Kayastha, N., Ye, J., Fenicia, F., Kuzmin, V., & Solomatine, D. P. (2013). Fuzzy committees of specialized rainfall-runoff models: Further enhancements and tests. *Hydrology and Earth System Sciences*, *17*(11), 4441–4451. https://doi.org/10.5194/hess-17-4441-2013

Krause, P., Boyle, D. P., & Bäse, F. (2005). Comparison of different efficiency criteria for hydrological model assessment. *Advances in Geosciences*, *5*(89), 89–97. https://doi.org/10.5194/adgeo-5-89-2005

Lindström, G., Johansson, B., Persson, M., Gardelin, M., & Bergström, S. (1997). Development and test of the distributed HBV-96 hydrological model. *Journal of Hydrology*, *201*(1–4), 272–288. https://doi.org/10.1016/S0022-1694(97)00041-3

Oudin, L., Andréassian, V., Mathevet, T., Perrin, C., & Michel, C. (2006). Dynamic averaging of rainfall-runoff model simulations from complementary model parameterizations. *Water Resources Research*, *42*(7), 1–10. https://doi.org/10.1029/2005WR004636

Schumann, A. H. (1993). Development of conceptual semi-distributed hydrological models and estimation of their parameters with the aid of GIS. *Hydrological Sciences Journal*, *38*(6), 519–528. https://doi.org/10.1080/02626669309492702

Shamseldin, A. Y., O'Connor, K. M., & Liang, G. C. (1997). Methods for combining the outputs of different rainfall-runoff models. *Journal of Hydrology*, *197*(1–4), 203–229. https://doi.org/10.1016/S0022-1694(96)03259-3

SNET (2005). *Balance hídrico integrado y dinámico en El Salvador*, 109. https://unesdoc.unesco.org/ark:/48223/pf0000228142

Solomatine, D. P. (2006). *Optimal modularization of learning models in forecasting environmental variables*. Proceedings of the IEMSs 3rd Biennial Meeting, Summit on Environmental Modeling and Software, July. http://www.scopus.com/inward/record.url?eid=2-s2.0-84858682542&partnerID=tZOtx3y1

Takagi, T., & Sugeno, M. (1985). Fuzzy identification of systems and its applications to modeling and control. *IEEE Transactions on Systems, Man, and Cybernetics, SMC-15* (1), 116–132. https://doi.org/10.1109/TSMC.1985.6313399

Xiong, L., Shamseldin, A. Y., & O'Connor, K. M. (2001). A non-linear combination of the forecasts of rainfall-runoff models by the first-order Takagi-Sugeno fuzzy system. *Journal of Hydrology*, *245*(1–4), 196–217. https://doi.org/10.1016/S0022-1694(01)00349-3

5

REGRESSION-BASED MACHINE LEARNING APPROACHES FOR DAILY STREAMFLOW MODELING

Vidya S. Samadi[1], Sadgeh Sadeghi Tabas[2], Catherine A. M. E. Wilson[3], and Daniel R. Hitchcock[1]

[1] *Department of Agricultural Sciences, Clemson University, Clemson, South Carolina, USA*
[2] *Department of Civil Engineering, Clemson University, South Carolina, USA*
[3] *Hydro-environmental Research Centre, School of Engineering, Cardiff University, Cardiff, United Kingdom*

The increasing rise of computational power has led to significant advances in the application of machine learning (ML) techniques in hydrological simulation. These methods are specifically developed to discover constitutive relations by implementing an unbiased implicit approach to capture unforeseen patterns in massive data sets. This study applied multiple ML approaches for daily rainfall-runoff simulation across a mixed urban-rural drainage system, the Northeast Cape Fear River Basin in North Carolina (NC), USA. Multiple ML algorithms such as Support Vector Machines (SVM), Bayesian Lasso, and Random Forest (RF) models, along with the Sacramento Soil Moisture Accounting (SAC-SMA) rainfall-runoff model, were applied to predict sequential daily streamflow records based on a set of collected data from climate and streamflow gauging

Advanced Hydroinformatics: Machine Learning and Optimization for Water Resources, Special Publications 78, First Edition. Edited by Gerald A. Corzo Perez and Dimitri P. Solomantine.

DOI: 10.1002/9781119639268.ch5

stations. Analysis suggests that the effects of input data on model performance, the error associated with forcing data, the amount of training data, and the correlation among different attributes of data series have strong influences on ML computation. Compared to the SAC-SMA, the Bayesian Lasso model was able to simulate the temporal dependencies among the observations and thereby was capable of accurately modeling the multivariate sequences of complex daily rainfall-runoff records across a mixed urban-rural catchment. The results provided an algorithmically informed simulation on the dynamics of daily streamflow simulation that may apply to other complex catchments and climate settings.

5.1. Introduction

Watershed modeling helps understand the past and current state of rainfall-runoff processes in a watershed and provides a way to explore the implications of management and planning decisions and imposed forcing function changes such as land use and climate. In complex watersheds (in this case, coastal plain watersheds), very significant modeling efforts have been conducted based on both lumped and physically based simulations (Amatya & Jha, 2011; Samadi & Meadows, 2017; Samadi et al. 2017, 2020; among others). However, rapid increase in the availability of measurement data has spurred the development of many data-driven methods or machine learning (ML) approaches for modeling and predicting rainfall-runoff dynamics. Indeed, in the realm of process-based hydrologic modeling, time-series ML algorithms have been recently utilized to improve short- and long-term streamflow predictions at various scales and domains (e.g., Solomatine & Dulal, 2003; Solomatine & Shrestha, 2009; Corzo et al., 2009; Shrestha et al., 2014; Fang et al., 2017; Kratzert et al., 2018; Shen, 2018; Shen et al., 2018; Wu et al., 2018). ML algorithms attempt to estimate the mapping function (f) from the input variables (x) to numerical or continuous output variables (y) and can learn patterns hidden in data. Different ML algorithms make different assumptions about the shape and structure of the function and how best to optimize a representation to approximate it. Several studies have demonstrated that ML models can learn complex nonlinear input-output relations (e.g., Solomatine & Dulal, 2003; Solomatine & Shrestha, 2009; Shen,

2018); thereby, they are capable of serving as a complementary alternative to other state-of-the-art techniques in rainfall-runoff applications (Yang et al., 2020).

Among various ML algorithms, regression-based models, such as Support Vector Machines (SVM), Random Forest (RF), Multi-layer Perceptron (MLP), and Decision Tree (DT), have been widely applied for rainfall-runoff simulations across various domains and scales. For example, Sivapragasam and Muttil (2005) applied SVM to compute rating curve values at three gauging stations in Washington, USA. Their results indicated that SVM is well suited for rating curves extrapolation compared with widely used logarithmic and higher order polynomial fitting methods. Bui et al. (2016) proposed a new ML approach based on a neural fuzzy inference system and metaheuristic optimization for flood susceptibility modeling (MONF) over the Tuong Duong district in central Vietnam. They found that MONF outperformed other ML algorithms, including MLP and DT for flood susceptibility mapping. Sadler et al. (2018) applied RF method for the coastal urban stormwater prediction in Virginia, USA. They used quality-controlled, crowd-sourced street flooding reports ranging from 1 to 159 per storm event for 45 storm events to train and evaluate the RF model. Their results showed that RF performed better than other ML algorithms, such as Poisson regression, at predicting the number of flood reports and had a lower false-negative error rate. Despite the expanded use of ML models in streamflow simulation, very few studies have used advanced ML methods to simulate coastal rainfall-runoff processes. The closest work may be the statistical analysis of streamflow records in the United States to compute the probabilities of high and low flow events in the past several decades along with the projected changes in the coming decades (Hidalgo et al. 2009; Campbell et al., 2011; Asadieh & Krakauer, 2017).

ML approaches can be divided into three broad categories: (1) supervised learning, (2) unsupervised learning, and (3) reinforcement learning. Supervised learning maps the observations or labeled data sets to known output and generates formula based on input and output values (e.g., Simeone, 2018). On the other hand, unsupervised learning discovers implicit relationships/patterns in a given unlabeled data set (unclassified data set) and finds an association between input values to group the data. Reinforcement learning follows a trial-and-error method for each predictive step or action with no precise label or error message (e.g., Shrestha & Mahmood, 2019). This study aims to use multiple state-of-the-art regression-based ML models to predict daily streamflow across a coastal plain basin in North Carolina, USA. Coastal plain modeling is a substantial challenge as this type of simulation is tempered by the heterogeneity in basin characteristics (low-lying elevation, evapotranspiration, wide floodplains, etc.) and complexities in soil storage capacity due to extensive alluvial and nonalluvial forested wetlands and riparian storage (e.g., Shirmohammadi et al., 1986; Samadi et al., 2017). This research simulated a coastal plain rainfall-runoff process using three different regression-based ML algorithms including SVM, Bayesian Lasso, and RF. While the SVM and RF models have been widely used for rainfall-runoff simulations (see Loos & Elsenbeer, 2011; Wang et al., 2015; Yang et al., 2016, among others), Bayesian Lasso has rarely been implemented for hydrologic modeling problems (Das Bhowmik et al., 2018) and have never performed for a coastal plain hydrologic simulation. This study is thus the first attempt to implement a Bayesian Lasso algorithm to exploit model inference via posterior distributions for a coastal plain rainfall-runoff simulation problem. This advanced algorithm, in addition to more widely used SVM and RF approaches, is employed to glean information from geospatial data and weather information and train proposed supervised learning models for daily streamflow simulation. Various ML simulations were also compared with the simulation result of a well-known conceptual rainfall-runoff model (i.e., Sacramento Soil Moisture Accounting (SAC-SMA)). It is important to note that the modeling approach presented here is general and applicable to any rainfall-runoff simulation problem, although the actual interpretations of the modeling diagnostics may vary from application to application.

This chapter is organized as follows. The theoretical aspects of ML models and their modeling structures, the SAC-SMA model, and the study area are presented in section 5.2. Results and findings are outlined in section 5.3. A detailed discussion of ML's potential use in rainfall-runoff applications is presented in section 5.4.

5.2. Materials and Methods

5.2.1. ML Models

ML algorithms are powerful models because they consider a learning style to adopt the procedures with a given data set in the training process. These approaches can be trained using time series data in which patterns are clearly defined. Once the ML models are well trained, they can output simulation results quickly without having to start from scratch, which enables the ML-centric methods to serve as a complementary alternative to any hydrologic simulation model. The conceptual formulations and mathematical concepts of each algorithm are briefly discussed in the following sections.

Support Vector Machines

The SVM algorithm is proposed by Vapnik, (2013) as a binary classification method. This algorithm uses the concept of Vapnik Chervonenkis (VC) dimension (see Vapnik, 2013), which is a measure of capacity or complexity of a machine to minimize structural risk to optimization and obtain learning and generalization ability. Given by a set of N samples of $\{x_k, y_k\}_{k=1}^{N}, x \in R^m, y \in R$, where x is an input vector of size m, and y is the corresponding output value. An SVM estimator f on regression can be expressed as

$$f(x) = \omega \cdot \emptyset(x) + b, \tag{5.1}$$

where ω is a weight vector and b denotes bias. The $\emptyset$ represents a nonlinear transfer function that maps the input vectors into a high dimensional feature space (e.g., Vapnik, 2013). A convex optimization problem with an ε-insensitivity loss function is used in this model to obtain optimization (equation (5.2)):

$$\begin{aligned} &\underset{\omega, b, \xi, \xi^*}{min} \ \frac{1}{2}\|\omega\|^2 + C\sum_{i=1}^{m}\left(\xi_i + \xi_i^*\right), \\ &\omega \cdot \emptyset(x_i) + \mathbf{b} - y_i \le \varepsilon + \xi_i, \\ &s \cdot t \cdot \left\{y_i - \omega \cdot \emptyset(x_i) - \mathrm{b} \le \varepsilon + \xi_i^*\right\}, \\ &\xi_i, \xi_i^* \ge 0; i = 1, 2, \ldots, m, \end{aligned} \tag{5.2}$$

where ξ and ξ^* are slack variables, which include soft margin to deal with infeasible constraints; C is a positive constant to penalize training errors by the loss function over the error tolerance ε and prevent overfitting. In this algorithm, the overfitting problem is typically solved using Duality Theory and Lagrangian multipliers and imposing Karush-Kuhn-Tucker (KKT) optimality condition (see Shevade et al., 2000; Smola, 2002).

Bayesian Lasso

In the classic linear regression setting, the original Lasso model, proposed by Tibshirani (1996), is an $L1$-penalized least squares estimate that aims to resolve the following optimization problem:

$$\min_{\beta} (y - X\beta)^T(y - X\beta) + \lambda\sum_{i=1}^{D}|\beta_i|, \tag{5.3}$$

where X is a $N \times D$ matrix of standardized predictors of dimension D, y is the mean-centered response vector, β denotes a vector that contains the D regression coefficient, and λ is the regularization parameter, which is a constant. The coefficient of parameters estimations are the solutions of the penalized regression, and sparsity in the Lasso model is obtained because some of the estimated parameters in β will be zero (see Efron et al., 2004; Helgøy & Li, 2019; Tibshirani, 1996). The Lasso solution can also be interpreted as a Bayesian posterior mode estimate, so-called Markov Chain Monte Carlo (MCMC) posterior estimation, when the prior for the coefficient parameters, $p()$, is a Laplace distribution or the double exponential distribution (Tibshirani, 1996). A fully Bayesian Lasso was introduced by Park and Casella (2008) with a conditional prior on β of the form $\rho(\beta|\sigma^2, \lambda)$. The posterior probability distribution for the coefficient parameters can be calculated using the Gibbs sampler. The Bayesian Lasso by Park and Casella (2008) has several attractive properties when it comes to parameter and hyperparameter estimation in the Bayesian setting, that is, that the joint posterior distribution for β and σ^2 generally has one mode instead of multiple posterior priors (Helgøy & Li, 2019). When conditioning Bayesian calculation on σ^2, it is important to guarantee a unimodal full posterior (see Park & Casella, 2008). Without this, the posterior may not be unimodal, and the convergence may slow down, making the true distribution less meaningful. Readers are referred to Tibshirani (1996) and Park and Casella (2008) for more information.

Random Forest

RF, proposed by Breiman (2001), is one of the most popular ML techniques for regression problems. These models consist of a large number of individual decision trees that operate as an ensemble. Each individual tree in the RF outputs a class (or group) prediction, and the class with the most votes becomes the model prediction. The prerequisites for the RF well performance include (1) the presence of some actual signals in time series data so that the decision tree models built using those signals perform better than random guessing, and (2) the prediction made by the individual trees need to have low correlations with each other.

Decision trees are very sensitive to the data sets and a slight change in the training data can result in significantly different tree structures. However, RF took advantage of this deficiency by allowing each individual tree to randomly sample from the data set with replacement, resulting in different tree structures. This process is known as bagging. Each tree in a random forest can be picked only from a random subset of features. This potential limitation makes more variation among the trees in the model, and eventually leads to a lower correlation among trees and more diversification. However, one of the main problems of regression trees is that they are prone to overfitting in the training period and thus perform poorly in validation when given unseen data (e.g., Murphy, 2012). Overfitting occurs when the model is flexible and has a high capacity to memorize and learn both the actual relationship in the training data and any noise present in the data set. The RF model can overcome this deficiency by learning the structure of the decision tree that varies considerably with the training data. Instead of searching for the most important feature while splitting a node, RF searches for the best feature among a random subset of features. This results in a wide diversity that generally leads to better performance, thus avoiding overfitting of the model to the training data set. Users have the ability to make trees more random by using random thresholds for each feature, rather than searching for the best possible thresholds.

Determination of Input Structure and Calibration Setup

Antecedent values of input variables could be considered to represent the catchment characteristics in the ML models (Solomatine & Xue, 2004; Tongal & Booij, 2018). In this study, a combination of precipitation (P), maximum and minimum temperatures (T^{max} and T^{min}), solar radiation ($Srad$), and humidity (H) was used in addition to the antecedent day streamflow (Q^s_{t-1}) to build the rainfall-runoff simulation. The antecedent day streamflow was calculated using a Feed-Forward Neural Network (FFNN). After a number of experiments on the training data set aimed at finding the most relevant inputs, the following common input structure was selected:

$$Q^s_t = f\left(P_t, \ldots, P_{t-3}, T^{max}_t, T^{min}_t, Srad_t, H_t, Q^s_{t-1}\right). \tag{5.4}$$

Prior to the model development, normalization was applied to all input variables as

$$X_{norm} = \frac{X_0 - \mu_x}{\sigma_x}, \tag{5.5}$$

where X_{norm} and X_0 indicate, respectively, the scaled and original data. The μ_x and σ_x represent the mean and standard deviation of the original data, correspondingly. For each ML method used in this study, the optimal model parameters were selected by maximizing the objective function

(Mean Squared Error; MSE) using training and calibration data sets. The models were developed on the training data set with a range of parameter values, and the objective function value was calculated for the calibration period. The optimal model parameters were then used to simulate streamflow for the validation period. In this study, the range of parameter values was determined by examining several hydrologic simulation studies, such as Shortridge et al. (2016), Deka (2014), Kumar et al. (2016), Naghibi et al. (2017), Worland et al. (2018), Tongal and Booij (2018), and Fotovatikhah et al. (2018).

SAC-SMA Model

The SAC-SMA model is developed by the U.S. National Weather Service (NWS) as a conceptual hydrologic simulation model. The tool utilizes a moisture accounting system to represent the catchment storage capacity. The SAC-SMA divides the basin into lower and upper soil zones at different depths, and defines the distribution of moisture, that is, tension water components (driven by evapotranspiration and diffusion) and free water components (driven by gravitational forces) in each of these two zones via a set of parameters (Burnash, 1995). The model uses climatic variables, such as precipitation and temperature, along with parameters on soil moisture states and the basin's relative permeability, to estimate the amount of water that enters, is stored in, and leaves the basin. Thus, the model computes several key hydrologic processes, such as evapotranspiration, percolation, interflow, and different forms of runoff from a basin. These are (1) direct runoff from permanent and temporary impervious areas, (2) surface runoff due to precipitation occurring at a rate faster than percolation and interflow can take place when both upper zone storages are full, (3) interflow resulting from the lateral drainage of a temporary free water storage, (4) supplemental base flow, and (5) primary base flow (Burnash, 1995); although, proper application of the SAC-SMA requires interpretations of the primary rainfall-runoff mechanisms (precipitation, evapotranspiration, and soil moisture) in a physically reasonable manner. A distortion in any of these characteristics may produce a compensating distortion in another element. Such distortion may appear to produce an effective streamflow simulation during the fitting period, but can reduce the model performance, particularly during high and low flow events.

Several performance metrics were used in this research for model benchmarking such as Nash-Sutcliffe efficiency (NSE; Nash & Sutcliffe, 1970), the Kling-Gupta efficiency (KGE; Gupta et al. 2009), and the Box-Cox transformed root-mean-square error (TRMSE; Box & Cox, 1964). In addition, three different signatures of the flow duration curve

(FDC) were used to evaluate the performance of specific ranges of discharge simulation. Following Yilmaz et al. (2008), we partitioned the FDC into three different segments (1) high-flow segment (0–0.02 flow exceedance probabilities) characterizing watershed response to significant precipitation events; (2) midflow segment (0.2–0.7 flow exceedance probabilities) that specifies by flows from moderate size precipitation events and also related to the intermediate-term primary and secondary base flow relaxation response of the watershed system; and (3) low-flow segment (0.7–1.0 flow exceedance probabilities), related to the long-term sustainability of flow and controlled by the interaction of baseflow with riparian evapotranspiration (ET) during extended dry periods. A steep slope in the midsegment FDC reflects a watershed having a quick or flashy response (due to small soil storage capacity and hence larger percentage of overland flow) while flatter midsegment slopes are associated with slower and more sustained watershed response to precipitation events.

5.2.2. Study Region and Data

The study region is the Northeast Cape Fear River Basin (hereafter NE-CFB), North Carolina, USA (HUC 03030007), which drains 1,714.70 mi^2 (4,441.1 km^2) of the Cape Fear River Basin (CFRB). The CFRB is one of the four river basins entirely contained within the North Carolina border. The river is formed by the Deep, Rocky, and Haw rivers, which converge in Chatham County just below the B. Everett Jordan Dam. The river ends in 32,000 acres of the estuary near Southport. The CFRB is the largest river in North Carolina, encompassing a 9,000 mi^2 basin that includes streams flowing within 29 of the state's 100 counties. With about 27% of the state's population and dozens of municipalities situated within its boundaries (including Greensboro, Burlington, Chapel Hill, Sanford, Fayetteville, Dunn, Clinton, Warsaw, Burgaw, and Wilmington), the CFRB is one of the most industrialized regions in North Carolina. These and other cities and towns rely on the river and its tributaries for freshwater, transportation, recreation, natural habitats, and other uses. The CFRB flows into the Atlantic Ocean near Cape Fear, from which it takes its name. This study used CAMELS (Catchment Attributes and Meteorology for Large-sample Studies) data set described in Newman et al. (2015) and Addor et al. (2017). CAMELS is a basin-scale hydrometeorological forcing data that covers 671 catchments in the contiguous United States (CONUS). The data set provides daily meteorological forcing from three data sets, Daymet (Thornton et al., 2012), Maurer (Maurer et al., 2002), and NLDAS (Xia et al., 2012), as well as a daily streamflow gauging data set from the U. S. Geological Survey (USGS). Meteorological forcing data used in this research include daily maximum and minimum temperatures, daily precipitation, solar radiation, and vapor pressure deficit. The NE-CFB data set was used to perform this research and predict daily streamflow records. A conceptual framework of proposed modeling research is presented in Figure 5.1. As illustrated, the NE-CFB streamflow records were calibrated using multiple ML approaches. The modeling results were then compared with the SAC-SMA simulation using various performance indices.

5.3. Results

This research used 10 yr of daily streamflow records, from 01 Jan. 2000 to 31 Dec. 2009, as a training data set, and then the next 4 yr (01 Jan. 2010 to 31 Dec. 2013) as an independent test period. The training period is critical to result in reasonable and reliable simulation because a small division of the data set from the training period would lead to overfitting. The ML results were also compared with the SAC-SMA simulation to determine the degree of modeling success. Various performance metrics were then calculated and presented in Table 5.1 for both calibration and validation performance assessments. The values of NSE and KGE for the SAC-SMA and multiple ML models suggest that the Bayesian Lasso method performed well in both calibration and validation periods. This revealed that a regression model with independent Laplace (i.e., double exponential) prior distribution could handle the complexities of high and low flow fluctuations across a coastal plain drainage system. As shown in Table 5.1, the Bayesian Lasso even outperformed the SAC-SMA hydrologic simulation. Furthermore, the SVM performed as well as the SAC-SMA while the RF performed better in rainfall-runoff simulation. Among different indices, KGE appears to be a better index for estimating the agreement between observed and simulated coastal plain streamflow records. KGE is a nonlinear combination of correlation, flow variability measure, and bias; thereby it appears to be an appropriate metric to evaluate the impacts of flow variabilities on simulation results. With respect to the FDC segments, Bayesian Lasso successfully captured high and low flow fluctuations while most of the ML models were less successful in calibrating moderate streamflow records or midsegment of the FDC (0.2–0.7 flow exceedance probabilities). This might reflect the fact that the response of watershed to moderate precipitation events varies largely over time depending on timing (e.g., month, season) and depth and duration of precipitation with space along the landscape.

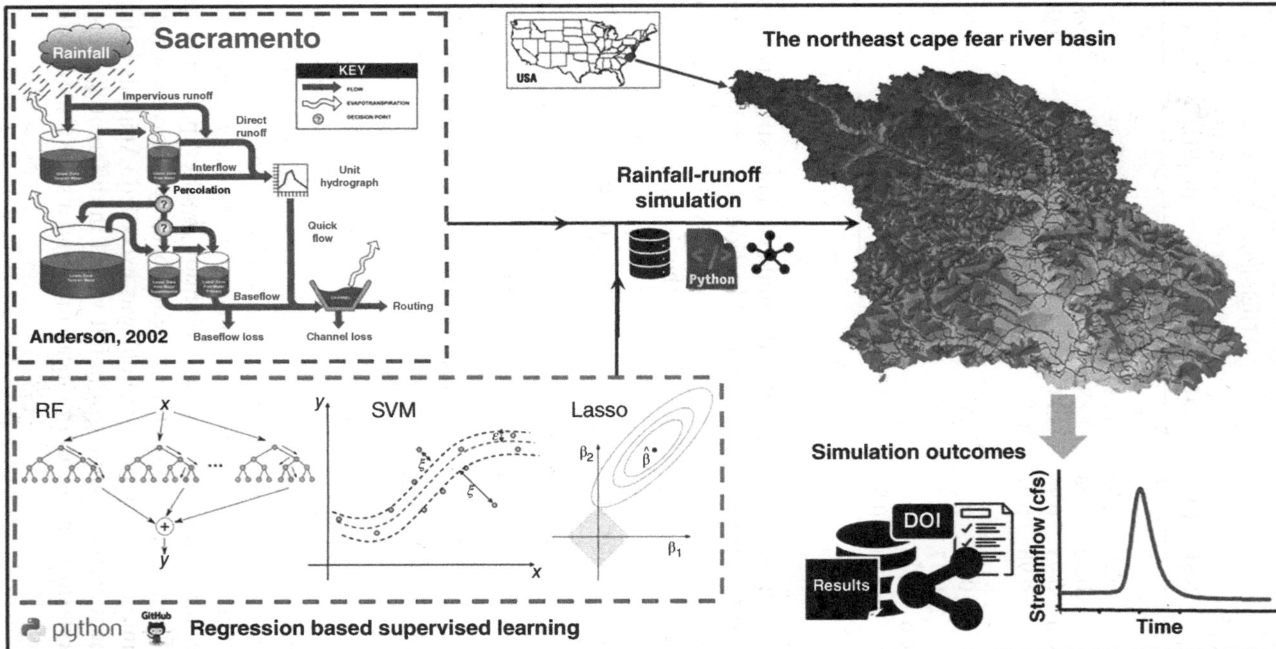

Figure 5.1 A conceptual framework of the proposed modeling workflow illustrated for the Northeast Cape Fear River Basin.

Table 5.1 Performance metrics of the ML and SAC-SMA models.

Model	NSE	KGE	TRMSE	High-flow segment of the FDC	Midflow segment of the FDC	Low-flow segment of the FDC
Calibration						
SAC-SMA	0.79	0.82	3.88	−13.14	−10.68	27.19
SVM	0.82	0.78	1.88	−24.31	−6.99	−208.41
Bayesian Lasso	0.95	0.96	3.21	−4.23	−8.62	−158.52
RF	0.98	0.97	0.7	−2.64	−2.47	19.21
Validation						
SAC-SMA	0.77	0.79	4.07	−17.88	−0.89	39.20
SVM	0.77	0.76	3.07	−26.99	−15.48	44.77
Bayesian Lasso	0.84	0.88	2.87	−8.43	−11.03	−26.31
RF	0.81	0.87	2.456	−8.69	−9.012	67.23

Figure 5.2 shows the observed and simulated daily streamflow using the ML and SAC-SMA methods during calibration and validation periods. As shown, the Bayesian Lasso model performed well particularly during high flow events (flood events). RF, on the other hand, performed satisfactorily while SVM and SAC-SMA showed similar performances. Although, all models were unable to simulate very high streamflow records caused by extraordinary precipitation events.

Figure 5.3 illustrates the FDCs for the ML, SAC-SMA, and observation during the calibration and validation. As shown, the SAC-SMA successfully calibrated intermediate flows while somewhat underestimated, and overestimated high- and low-flow events. Moreover, the SAC-SMA underestimated the peak rates considerably compared with the Bayesian Lasso model. The SAC-SMA model uses a lumped fashion to calibrate soil moisture states and the basin's relative permeability. These states are fed as several key basin's parameters into the model, such as interflow, evapotranspiration, and percolation. It seems hydrologic variables such as loss (due to dense vegetation) and soil moisture have strong control on rainfall-runoff processes and mechanisms in the study region. These factors determine catchment wetness conditions and evapotranspiration processes. Thus, runoff magnitude might abruptly increase when the catchment initial abstraction threshold was exceeded. Analysis indicates that catchment response time to rainfall varied in different seasons and years with a flashy response during peak rates and a slow response during

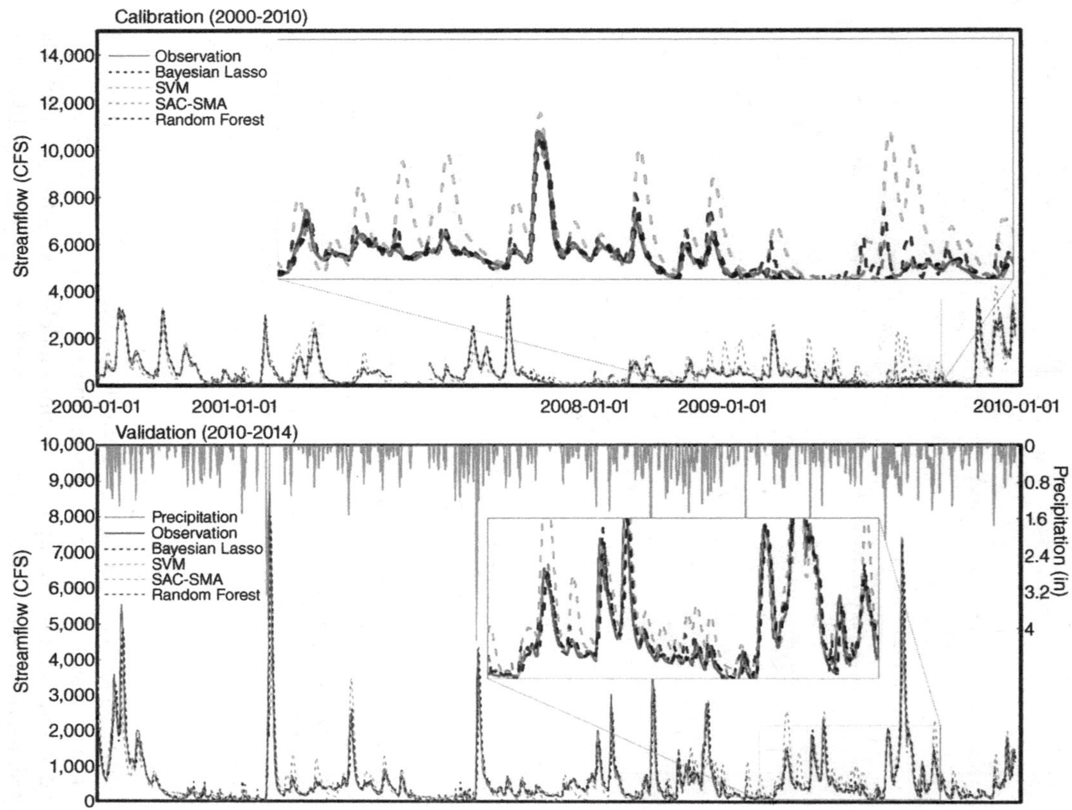

Figure 5.2 Observed versus simulated daily streamflow of the ML methods as well as the SAC-SMA for calibration and validation periods. Precipitation data are plotted as a secondary axis.

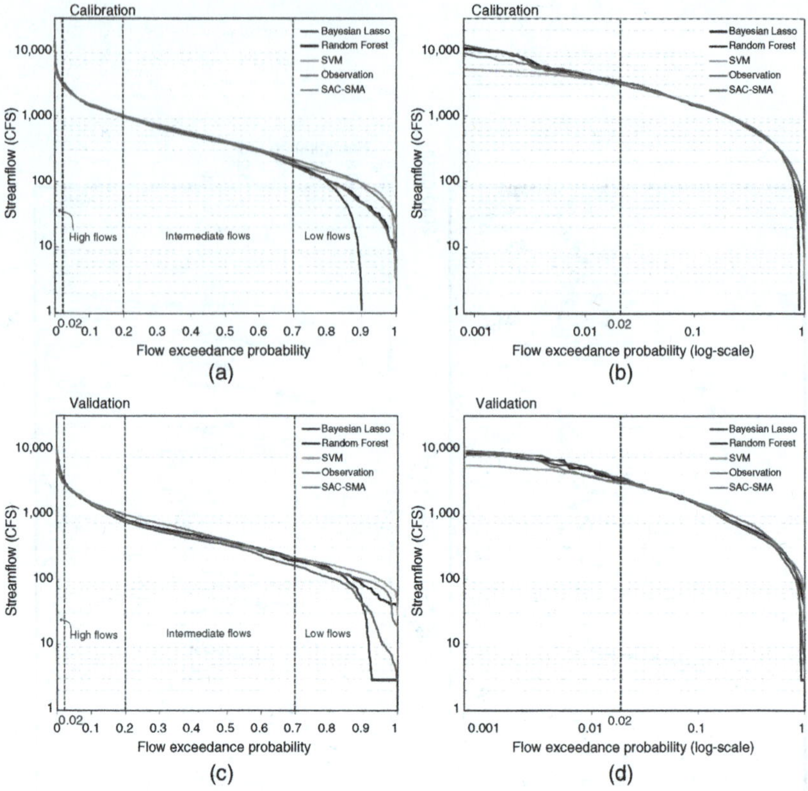

Figure 5.3 FDCs that display modeled and observed streamflow values against their relative exceedance time during calibration and validation period. FDC (a, c) represents the percentage of the time that a flow is exceeded. The zero value on the x-axis represents the highest observed flow and 1 represents the lowest observed flow that was exceeded over time. Flow exceedance probability is also presented using a log scale (b, d) to better visualize the low- and high-flow variabilities.

moderate to low rainfall events. Thus, the relationships between rainfall and runoff may link to the dynamics and timing of the rainfall amounts and distribution. Overall, among various models used in this research, the Bayesian Lasso offers promising results for the NE-CFB rainfall-runoff simulation. In a Bayesian setting, Lasso estimates can be interpreted as posterior mode estimates with independent and identical Laplace prior for coefficients that are well suited to model time series data occurring sequentially.

5.4. Conclusion and Future Works

Regression-based ML models can be used as a supplement to both conceptual and physics-based hydrologic simulation models. However, there continues to be a concern regarding their ability to provide skillful prediction performance, in part because of their need for much more training data to generate hydrological processes, varying degrees of noise in data, and overfitting, which can translate into a significant amount of error and uncertainty in the modeling results. We argue that the potential benefits of ML models for daily streamflow simulation of a coastal plain drainage system can be realized only via the formulation of robust and rigorous Bayesian Lasso interface as posterior mode estimate that is well suited to evaluate input-output behavior (behavioral consistency) for sequential streamflow time series data.

One obvious way to reduce the obstacles and improve ML fidelity is using preprocessing methods (e.g., singular spectrum and wavelet analyses; see Vautard et al., 1992), finding intricate patterns in large volumes of data, improving sequence-to-sequence learning, and calculating noise variance and then capturing the outliers through monitoring the noise component. Enhancing the ability of ML on modeling extreme observations within time-series records is also important and needs further attention. Furthermore, physical parameters (and process interactions) could be mathematically formulated and incorporated into the ML models to ensure proper consistency between the model input-output behavior and the available data and catchment properties. Our research group has sought, therefore, to develop appropriate ML models that are comparable with a catchment scale hydrologic simulation model. The main limitation of our proposed ML approaches, however, is their vast computational time and cost, and overfitting that make them somewhat poor predictors. Our ongoing work explores the development of more advanced and computational cost-effective ML algorithms such as Recurrent Neural Networks (RNNs) to integrate the information extractable from the recurrence/structure of the network and mathematical formulation with catchment physical attributes and observations. As always, we invite dialogue with others interested in these and related ML model simulation topics.

Acknowledgments

This work is supported in part by the U.S. National Science Foundation (NSF) Directorate for Engineering under grant CBET 1901646.

Any opinions, findings, and discussions expressed in this study are those of the authors and do not necessarily reflect the views of the NSF. Clemson University is acknowledged for generous allotment of computing time on the Palmetto cluster.

References

Addor, N., Newman, A. J., Mizukami, N., Clark, M. P. (2017). The CAMELS data set: catchment attributes and meteorology for large-sample studies. *Hydrology and Earth System Sciences*, *21*, 5293–5313.

Amatya K. M., & Jha, M. K. (2011). Evaluating the SWAT model for a lowgradient forested watershed in Coastal South Carolina. *Transactions of the ASABE*, *54*(6), 2151–2163.

Asadieh, B., & Krakauer, N.Y. (2017). Global change in streamflow extremes under climate change over the 21st century. *Hydrology and Earth System Sciences*, *21*, 5863.

Box, G.E.P., & Cox, D.R. (1964). An analysis of transformations. *Journal of the Royal Statistical Society, Series B*, *26*, 211–243.

Breiman, L. (2001). Random forests. *Machine Learning*, *45*, 5–32.

Bui, D.T., Pradhan, B., Nampak, H., Bui, Q.-T., Tran, Q.-A., & Nguyen, Q.-P. (2016). Hybrid artificial intelligence approach based on neural fuzzy inference model and metaheuristic optimization for flood susceptibilitgy modeling in a high-frequency tropical cyclone area using GIS. *Journal of Hydrology*, *540*, 317–330.

Burnash, R.J.C. (1995). The NWS River forecast system-catchment modeling. In V. Singh (Ed.), *Computer models of watershed hydrology*. Water Resources Publication, Colorado, 311–366.

Campbell, J. L., Driscoll, C. T., Pourmokhtarian, A., & Hayhoe, K. (2011). Streamflow responses to past and projected future changes in climate at the Hubbard Brook Experimental Forest, New Hampshire, United States. *Water Resources Research*, 47.

Corzo, G., Solomatine, D., Hidayat, M., Werner, M., Uhlenbrook, S., et al. (2009). Combining semi-distributed process-based and data-driven models in flow simulation: a case study of the Meuse river basin. *Hydrology and Earth System Sciences*, *13*, 1619–1634.

Das Bhowmik, R., Seo, S., & Sahoo, S. (2018). Streamflow simulation using Bayesian regression with multivariate linear spline to estimate future changes. *Water*, *10*(7), 875. https://doi.org/10.3390/w10070875

Deka, P.C. (2014). Support vector machine applications in the field of hydrology: A review. *Applied Soft Computing*, *19*, 372–386.

Efron, B., Hastie, T., Johnstone, I., & Tibshirani, R. (2004). Least angle regression. *The Annals of Statistics*, *32*, 407–499.

Fang, K., Shen, C., Kifer, D., & Yang, X. (2017). Prolongation of SMAP to spatiotemporally seamless coverage of continental US using a deep learning neural network. *Geophysical Research Letters*, *44*, 11–30.

Fotovatikhah, F., Herrera, M., Shamshirband, S., Chau, K., Faizollahzadeh Ardabili, S., & Piran, M. J. (2018). Survey of computational intelligence as basis to big flood management: Challenges, research directions and future work. *Engineering Applications of Computational Fluid Mechanics*, *12*, 411–437.

Gupta, H. V, Kling, H., Yilmaz, K. K., & Martinez, G. F. (2009). Decomposition of the mean squared error and NSE performance criteria: Implications for improving hydrological modelling. *Journal of Hydrology*, *377*, 80–91.

Helgøy, I.M., & Li, Y. (2019). A noise-robust fast sparse Bayesian learning model. *arXiv Prepr.*, arXiv1908.07220.

Hidalgo, H. G., Das, T., Dettinger, M. D., Cayan, D. R., Pierce, D. W., Barnett, T. P., et al. (2009). Detection and attribution of streamflow timing changes to climate change in the western United States. *Journal of Climate*, *22*, 3838–3855.

Kratzert, F., Klotz, D., Brenner, C., Schulz, K., & Herrnegger, M. (2018). Rainfall-runoff modeling using Long Short-Term Memory (LSTM) networks, 6005–6022.

Kumar, D., Pandey, A., Sharma, N., & Flügel, W.-A. (2016). Daily suspended sediment simulation using machine learning approach. *Catena*, *138*, 77–90.

Loos, M., & Elsenbeer, H. (2011). Topographic controls on overland flow generation in a forest: An ensemble tree approach. *Journal of Hydrology*, *409*, 94–103.

Maurer, E. P., Wood, A. W., Adam, J. C., Lettenmaier, D. P., & Nijssen, B. (2002). A long-term hydrologically based data set of land surface fluxes and states for the conterminous United States. *Journal of Climate*, *26*(23), 9384–9392. https://doi.org/10.1175/JCLI-D-12- 25 00508.1.

Murphy, K. P. (2012). *Machine learning: A probabilistic perspective*. MIT press.

Naghibi, S. A., Ahmadi, K., & Daneshi, A. (2017). Application of support vector machine, random forest, and genetic algorithm optimized random forest models in groundwater potential mapping. *Water Resource Management*, *31*, 2761–2775.

Nash, J. E., & Sutcliffe, J. V. (1970). River flow forecasting through conceptual models part I: A discussion of principles. *Journal of Hydrology*, *10*(3), 282–290. https://doi.org/10.1016/0022-1694(70) 90255-6

Newman, A. J., Clark, M. P., Sampson, K., Wood, A., Hay, L. E., Bock, A., et al. (2015). Development of a large-sample watershed-scale hydrometeorological data set for the contiguous USA: Data set characteristics and assessment of regional variability in hydrologic model performance. *Hydrology and Earth System Sciences*, *19*, 209–223. https://doi.org/10.5194/hess-19-209-2015

Park, T., & Casella, G. (2008). The Bayesian Lasso. *Journal of the American Statistical Associtation*, *103*, 681–686.

Sadler, J.M., Goodall, J.L., Morsy, M.M., Spencer, K. (2018). Modeling urban coastal flood severity from crowd-sourced flood reports using Poisson regression and Random Forest. *Journal of Hydrology*, *559*, 43–55.

Samadi, S., Pourreza-Bilondi, M., Wilson, C., & Hitchcock, D. B. (2020). Bayesian model averaging with fixed and flexible priors: Theory, concepts, and calibration experiments for rainfall-runoff modeling. *Journal of Advances in Modeling Earth Systems*, *12*, e2019MS001924.

Samadi, S., Tufford, D. L., & Carbone, G. J. (2017). Assessing parameter uncertainty of a semi-distributed hydrology model for a shallow aquifer dominated environmental system. *JAWRA Journal of the American Water Resources Association, 53*, 1368–1389.

Samadi, S. Z., & Meadows, M. E. (2017). The transferability of terrestrial water balance components under uncertainty and nonstationarity: A case study of the coastal plain watershed in the Southeastern USA. *River Research and Applications 33*, 796–808.

Shen, C. (2018). A transdisciplinary review of deep learning research and its relevance for water resources scientists. *Water Resources Research, 54*, 8558–8593. https://doi.org/10.1029/2018WR022643

Shen, C., Laloy, E., Elshorbagy, A., Albert, A., Bales, J., Chang, F.-J., et al. (2018). HESS Opinions: Incubating deep-learning-powered hydrologic science advances as a community. *Hydrology and Earth System Sciences, 22*.

Shevade, S. K., Keerthi, S. S., Bhattacharyya, C., & Murthy, K. R. K. (2000). Improvements to the SMO algorithm for SVM regression. *IEEE Transactions on Neural Networks and Learning Systems, 11*, 1188–1193.

Shirmohammadi, A., J. M. Sheridan, & L. E. Asmussen (1986). Hydrology of alluvial stream channels in southern coastal plain watersheds. *Transactions American Society of Agricultural Engineers, 29*(1), 135–142.

Shortridge, J. E., Guikema, S. D., & Zaitchik, B. F. (2016). Machine learning methods for empirical streamflow simulation: a comparison of model accuracy, interpretability, and uncertainty in seasonal watersheds. *Hydrology and Earth System Sciences, 20*.

Shrestha A., & Mahmood, A. (2019). Review of deep learning algorithms and architectures. *IEEE Access, 7*, 53040–53065. https://doi.org/10.1109/ACCESS.2019.2912200.

Shrestha, D.L., Kayastha, N., Solomatine, D., & Price, R. (2014). Encapsulation of parametric uncertainty statistics by various predictive machine learning models: MLUE method. *Journal of Hydroinformatics, 16* (1), 95–113.

Simeone, O. (2018). A very brief introduction to machine learning with applications to communication systems. *IEEE Transactions on Cognitive Communications and Networking, 4*(4), 648–664.

Sivapragasam, C., & Muttil, N. (2005). Discharge rating curve extension: A new approach. *Water Resource Management. 19*, 505–520.

Smola, A. (2002). Support vector machines, regularization, optimization, and beyond.

Solomatine, D. P., & Dulal, K. N. (2003). Model trees as an alternative to neural networks in rainfall: Runoff modeling. *Hydrological Sciences Journal, 48*(3), 399–411.

Solomatine D. P., & Shrestha, D. L. (2009). A novel method to estimate model uncertainty using machine learning techniques. *Water Resources Research*, 45, W00B11, https://doi.org/10.1029/2008WR006839.

Solomatine, D. P., & Xue, Y. (2004). M5 model trees and neural networks: application to flood forecasting in the upper reach of the Huai River in China. *Journal of Hydrologic Engineering, 9*, 491–501.

Thornton, P. E., Thornton, M. M., Mayer, B. W., Wilhelmi, N., Wei, Y., & Cook, R. B. (2012). Daymet: Daily surface weather data on a 1 km grid for North

America (1980–2012). Oak Ridge National Laboratory Distributed Active Archive Center.

Tibshirani, R. (1996). Regression shrinkage and selection via the lasso. *Journal of the Royal Statistical Society, Series B, 58*, 267–288.

Tongal, H., & Booij, M. J. (2018). Simulation and forecasting of streamflows using machine learning models coupled with base flow separation. *Journal of Hydrology, 564*, 266–282.

Vapnik, V. (2013). *The nature of statistical learning theory*. Springer Science & Business Media.

Vautard, R., Yiou, P., & Ghil M. (1992). Singular-spectrum analysis: A toolkit for short, noisy and chaotic signals. *Physica D, 58*, 95–126.

Wang, Z., Lai, C., Chen, X., Yang, B., Zhao, S., & Bai, X. (2015). Flood hazard risk assessment model based on random forest. *Journal of Hydrology, 527*, 1130–1141.

Worland, S. C., Farmer, W. H., & Kiang, J. E. (2018). Improving predictions of hydrological low-flow indices in ungaged basins using machine learning. *Environmental Modeling & Software, 101*, 169–182.

Wu, Y., Liu, Z., Xu, W., Feng, J., Palaiahnakote, S., & Lu, T. (2018). *Context-aware attention LSTM network for flood prediction*. 24th International Conference on Pattern Recognition (ICPR), Beijing, China, 2018. https://doi.org/10.1109/ICPR.2018.8545385

Xia, Y., Mitchell, K., Ek, M., Sheffield, J., Cosgrove, B., Wood, E., et al. (2012). Continental-scale water and energy flux analysis and validation for the North American Land Data Assimilation System project phase 2 (NLDAS-2): 1. Intercomparison and application of model products. *Journal of Geophysical Research, 117*(D3), D03109. https://doi.org/10.1029/2011jd016048

Yang, S., Yang, D., Chen, J., Santisirisomboon, J., Lu, W., & Zhao, B. (2020). A physical process and machine learning combined hydrological model for daily streamflow simulations of large watersheds with limited observation data. *Journal of Hydrology, 590*, 125206.

Yang, T., Gao, X., Sorooshian, S., & Li, X. (2016). Simulating California reservoir operation using the classification and regression-tree algorithm combined with a shuffled cross-validation scheme. *Water Resources Research, 52*, 1626–1651.

Yilmaz, K. K., Gupta, H. V., & Wagener, T. (2008). A process-based diagnostic approach to model evaluation: Application to the NWS distributed hydrologic model. *Water Resources Research, 44*, 1–18. https://doi.org/10.1029/2007WR006716

6

USE OF NEAR-REAL-TIME SATELLITE PRECIPITATION DATA AND MACHINE LEARNING TO IMPROVE EXTREME RUNOFF MODELING

Paul Muñoz[1,2], Gerald A. Corzo Perez[3], Dimitri P. Solomatine[3,4,5], Jan Feyen[6], and Rolando Célleri[1,2]

[1] *Department of Water Resources and Environmental Sciences, University of Cuenca, Cuenca, Ecuador*
[2] *Faculty of Engineering, University of Cuenca, Cuenca, Ecuador*
[3] *IHE Delft Institute for Water Education, Delft, The Netherlands*
[4] *Water Resources Section, Delft University of Technology, Delft, The Netherlands*
[5] *Water Problems Institute of the Russian Academy of Sciences, Moscow, Russia*
[6] *Faculty of Bioscience Engineering, Catholic University of Leuven, Leuven, Belgium*

Extreme runoff modeling is hindered by the lack of sufficient and relevant ground information and the low reliability of physically based models. This chapter proposes combining precipitation Remote Sensing (RS) products, Machine Learning (ML) modeling, and hydrometeorological knowledge to improve extreme runoff modeling. The approach applied to improve the representation of precipitation is the object-based Connected Component Analysis (CCA), a method that enables classifying and associating precipitation with extreme runoff events. Random Forest (RF) is employed as an ML model. Two and a half years of near-real-time hourly RS precipitation from the

Advanced Hydroinformatics: Machine Learning and Optimization for Water Resources, Special Publications 78, First Edition. Edited by Gerald A. Corzo Perez and Dimitri P. Solomantine.

DOI: 10.1002/9781119639268.ch6

PERSIANN-CCS and IMERG-early run databases and runoff at the outlet of a basin in the tropical Andes of Ecuador were used. The developed models show the ability to simulate extreme runoff for the cases of long-duration precipitation events regardless of the spatial extent, obtaining Nash-Sutcliffe efficiencies above 0.72. On the contrary, there was an unacceptable model performance for a combination of short-duration and spatially extensive precipitation events. The strengths/weaknesses of the developed ML models are attributed to the ability/difficulty to represent complex precipitation-runoff responses.

6.1. Introduction

Physically based precipitation-runoff models used in water management describe the physical processes that occur in a system (basin) by using balance and conservation equations (Clark et al., 2017). However, those models demand extensive data and might be subject to overparameterization, limiting their operational value (Mosavi et al., 2018; Young, 2002). As a solution, during the last decades, a data-driven approach, using Machine Learning (ML) techniques, gained popularity among hydrologists (Solomatine et al., 2009; Bontempi et al., 2013; Chang et al., 2019; Galelli & Castelletti, 2013; Mosavi et al., 2018). An important reason responsible for the increasing interest is the fact that ML exploits the available and relevant information (e.g., precipitation, past runoff) to find relations to the target variable (i.e., runoff) without requiring knowledge about the underlying physical processes. However, apart from providing accurate hydrograph simulations, the use of ML for deriving inferences about the processes within a hydrological system has received considerable critical attention (Beven, 2020; Herath et al., 2021; Moreido et al., 2021; Nearing et al., 2021; Solomatine & Ostfeld, 2008). It is argued that conceptualizing and understanding forcing processes in a system would enable ML models to simulate the hydrological response beyond the range of training data, or even to transfer models to similar ungauged systems (Beven, 2020). Some strategies (also known as feature engineering) for gaining understanding can be achieved through the addition of mass and energy balance equations for the training task (Wang et al., 2019), or hypothesis testing of additional processes using additional input data such as soil types, land uses, topography, residence time distributions, among others (Huang & Lee, 2021).

Among ML techniques, the Random Forest (RF) algorithm is preferably being used by hydrologists for runoff modeling (including extreme runoff magnitudes associated with low probabilities of occurrence) (Li et al., 2020; Mosavi et al., 2018; Muñoz et al., 2021, 2018; Orellana-Alvear et al., 2020; Papacharalampous & Tyralis, 2018). This is due to RF's simplicity (few parameters to calibrate) and higher accuracy when compared with other ML techniques such as Support Vector Machines and Artificial Neural Networks, model robustness, and capacity to deal with small size samples and complex data structures (Biau & Scornet, 2016; Breiman, 2001; Contreras et al., 2021; B. Li et al., 2016; Liu et al., 2020; Muñoz et al., 2018, 2021; Papacharalampous & Tyralis, 2018; Tyralis et al., 2019; Wang et al., 2015).

In terms of data availability, in many regions, ground precipitation networks are either nonexistent or scarce and if available mostly with extremely low areal density. This is especially true for mountainous regions, such as the tropical Andes, where the remoteness of the study areas and monitoring authorities' budget constraints limits the development of accurate precipitation-runoff models. Fortunately, continuous development of Remote Sensing (RS) products, for example, space-based satellites have dramatically enhanced the quantity (spatiotemporal resolution) and quality of areal precipitation observations. However, RS precipitation obtained from a single sensor (satellite) hardly provides accurate estimations (Hong et al., 2019). This has stimulated the development of multisatellite precipitation products such as the NASA Global Precipitation Measurement (GPM) Integrated Multi-satellitE Retrievals for GPM (IMERG) (Huffman et al., 2015), and the Precipitation Estimation from Remotely Sensed Information using Artificial Neural Networks (PERSIANN) (Hsu et al., 1997). IMERG and PERSIANN products are characterized by quasi-global coverage, free access, high spatiotemporal resolutions, and continuous development (Tang et al., 2016). Given previous, both RS products are nowadays widely used in hydrometeorological applications including tracking of precipitation anomalies (Nguyen et al., 2014; Sakib et al., 2021), precipitation early warning systems (Sorooshian et al., 2014), and flood forecasting and mapping (Belabid et al., 2019; Nguyen et al., 2015). Of particular concern is the determination of the optimal RS product for a given study area, the necessity to combine two or more RS products, or even fuse RS databases to improve precipitation representation and minimize precipitation uncertainties (Soo et al., 2020; Yin et al., 2021; Yu et al., 2020).

Once the issue of data availability is solved, the arising research questions are (1) is precipitation well represented by RS data? and (2) can the assimilation of RS precipitation be improved (in the context of extreme runoff ML models)? The first research question is mandatory for the cases when the interest lies in providing accurate precipitation estimations. This can be achieved by validating RS products with ground precipitation estimations, see for instance the studies of Laverde-Barajas et al. (2019) and Li et al. (2016). For the second question, we focus on the cases where there are no ground stations for validating RS imagery. Here, the hypothesis is that precipitation uncertainties over reduced areas remain more or less constant, and, therefore, the use of an ML algorithm (such as the RF algorithm) exploiting not quantitate precipitation estimates but rather precipitation differences within the system can be used to explore strategies for improving data assimilation. Investigating strategies for improving data assimilation in ML is a continuous concern within the field of hydrological modeling (Cisty & Soldanova, 2018; Laverde-Barajas et al., 2020; Yang & Chui, 2020). The methodology adopted consists of applying a feature engineering strategy to RS precipitation data enabling a better ML precipitation assimilation during the learning process, improving ultimately extreme runoff model efficiencies. For instance, ML assimilation can be improved by building specialized runoff models able to discriminate between different precipitation event types (Laverde-Barajas et al., 2020), or by adding specific information (features) such as mean and peak precipitation intensities (Yang & Chui, 2020). This is because different precipitation events produce different runoff responses as a result of various runoff generation processes, mainly infiltration and saturation excess (Gutiérrez-Jurado et al., 2019).

Precipitation events can be distinguished by applying object-based methods to RS imagery (Davis et al., 2006; Laverde-Barajas et al., 2019; J. Li et al., 2016; Peña-Barragán et al., 2011; Vogels et al., 2020; Wang et al., 2016). A simple yet effective object-based method is the Connected Component Analysis (CCA) employed by Laverde-Barajas et al. (2019). The CCA includes a physical description of precipitation events (centroid, extension area, etc.), as well as key meteorological attributes (intensity, duration, volume, etc.). These characteristics are then used for classifying precipitation events, which can be paired with their associated runoff responses.

In this context, the objective of this study is to develop specialized (smart) ML extreme runoff models for a 3,393 km^2 basin in Ecuador. We used a feature engineering methodology to exploit readily available satellite precipitation sources to improve the areal representation of the precipitation and to maximize runoff model efficiencies by identifying and classifying precipitation events associated with extreme hydrological events.

6.2. Study Area and Data Set

6.2.1. Study Area

The Jubones basin, located in the tropical Andes of Ecuador, was selected as the study area (Fig. 6.1) and covers an area of ~3,391 km^2

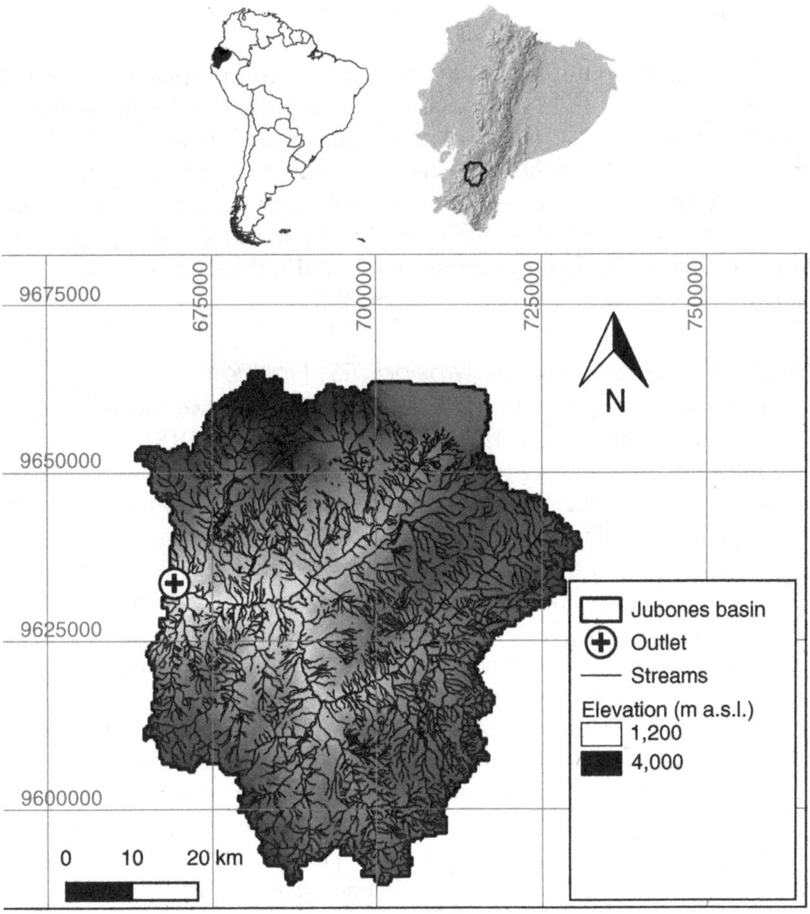

Figure 6.1 The Jubones basin in the tropical Andes of Ecuador, South America (UTM coordinates).

upstream of the Minas-San Francisco hydroelectric dam, with an elevation ranging between 1,250 and 3,920 m above sea level. The climatology of the basin is extremely variable due to the presence of the Andean mountain range, trade winds, and ocean currents from the Pacific Ocean. A distinction can be made between at least four rainfall regions, including a semi-arid region. The basin climate ranges from humid to arid, with average annual rainfall ranging spatially from 350 to 1,170 mm.

6.2.2. Data Set

The data set comprises hourly satellite-derived precipitation covering the Jubones basin, and hourly runoff data collected at the hydrological station, situated in the outlet of the basin. Runoff is measured at the entrance of the Minas-San Francisco hydroelectric dam, and although measurements are prone to dam oscillations uncertainties, they are expected to be absorbed/compensated by the use of the RF algorithm. Moreover, since the dam was completed in 2018, the study period is ~2.5 yr, from 18 November 2018 to 31 March 2021.

Precipitation From Remote Sensing (RS) Products

Precipitation information was retrieved from two near-real-time multisatellite sources, the IMERG-early run, and the PERSIANN-Cloud Classification System (CCS) precipitation subproducts. Data were derived at hourly intervals. The main difference between both precipitation sources is the spatial resolution. The PERSIANN-CCS possesses the highest spatial resolution for the region (0.04° x 0.04°), being the result of infrared imagery processing and cloud classification using artificial neuronal networks (Hong et al., 2004). Whereas the IMERG-early run approach interpolates various microwave precipitation estimates and delivers data with a spatial resolution of 0.1° x 0.1°.

Figure 6.2 shows the mean annual precipitation of the Jubones basin, measured by the PERSIANN-CCS (728.5 mm) and IMERG-early run (727.2 mm) precipitation satellite subproducts. For this plot, we used hourly information for 2019 and 2020 (see also the precipitation plot in Fig. 6.3). We found differences of 1.3 and 116 mm between the mean and the maximum annual precipitation obtained from the PERSIANN-CCS and the IMERG-early run subproducts, respectively. This is attributed to the spatial resolution difference and the measurement principle of each satellite subproduct.

Runoff at the Entrance of the MSF Hydropower Dam

Hourly time series of runoff at the outlet of the Jubones basin were derived from the server of the Corporación Eléctrica del Ecuador (CELEC EP, https://www.celec.gob.ec/), the company that manages the Minas-San Francisco hydroelectric dam. Figure 6.3 depicts the runoff information for the study period. Figure 6.3a shows the hourly time series, whereas Figure 6.3b shows the corresponding probability of exceedance from which 55 nearly independent peak flow events were selected based on peak-over-threshold values (dark dots in Fig. 6.3a). The peak flow

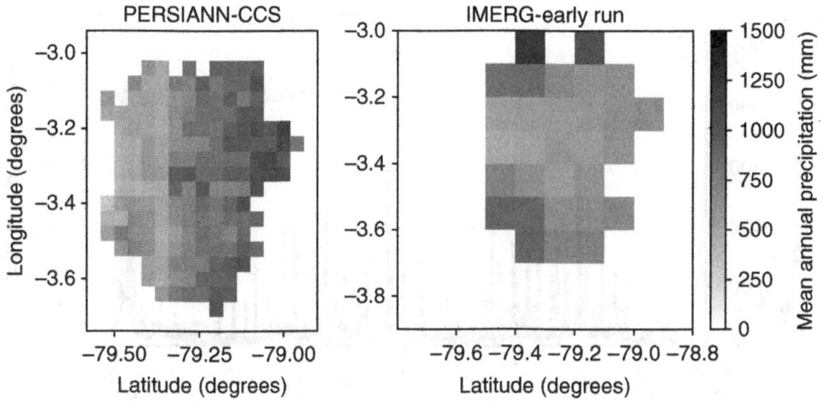

Figure 6.2 Mean annual precipitation in mm (for 2019 and 2020) measured by the PERSIANN-CCS and the IMERG-early run precipitation satellite subproducts.

events selection was done using the WETSPRO tool (Willems, 2009). Exceedance probability analysis reveals that for the study period, the runoff magnitudes of 103.5 and 159.4 $m^3.s^{-1}$ are exceeded with probabilities of 10% and 5%, respectively. These probabilities, which correspond to the 90% and 95% quartiles, served to determine extreme hydrological runoff events for the development of the extreme precipitation-runoff models.

6.3. Methodology

6.3.1. Determination of Nearly Independent Peak Hydrological Events

The extreme hydrological events from the complete runoff time series were derived by applying the following two criteria: (1) extreme hydrological events must exceed the 90% quartile values (98.8 $m^3.s^{-1}$), and (2) such events must be nearly independent. For meeting both criteria of independence, we used the WETSPRO time series tool (Willems, 2009), which splits runoff series in nearly independent peak and low flow events following a peak-over-threshold approach. The WETSPRO has two parameters to be calibrated, the interevent time and peak height. In summary, we selected extreme hydrological events with a definition of independence controlled by the recession time and peak height difference of two consecutive runoff events.

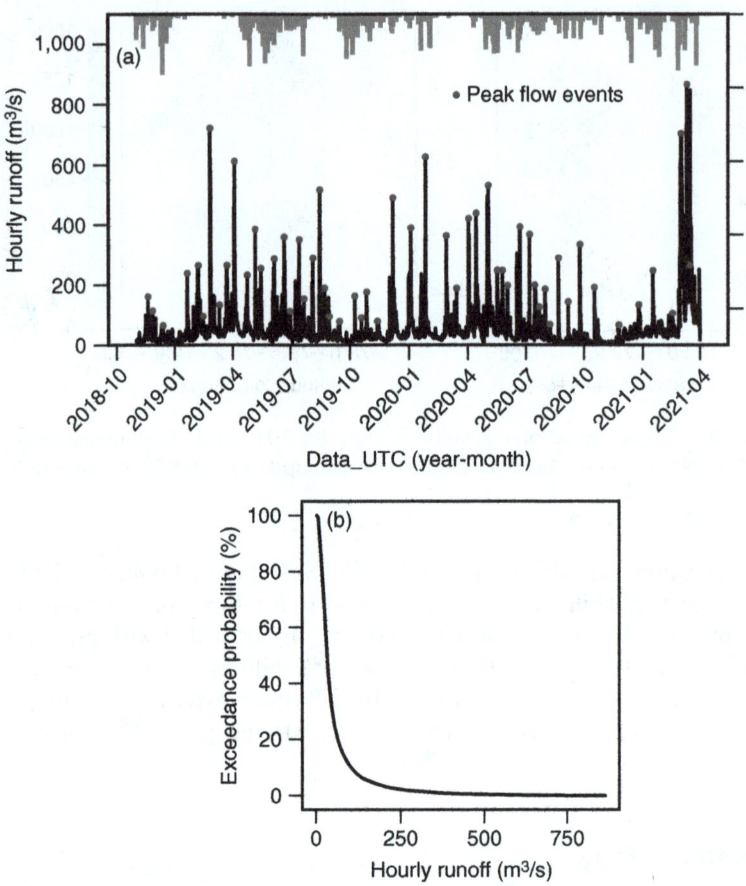

Figure 6.3 (a) Runoff and precipitation (PERSIANN-CCS) time series at the outlet of the Jubones basin. Peak flow events are displayed as grey dots. (b) Exceedance probability for the study period (18 Nov 2018 to 31 Mar 2021).

6.3.2. Object-Based Connected Component Analysis

Once extreme hydrological events were selected, the next step was to analyze their correspondent precipitation imagery from the highest-resolution satellite subproduct, the PERSIANN-CCS. The precipitation analysis was done by applying a feature engineering strategy based on an object-based Connected Component Analysis (CCA) algorithm. The CCA algorithm is fully detailed in Laverde-Barajas et al. (2019). We implemented the CCS algorithm through the scikit-image processing package in Python version 3.7 (van der Walt et al., 2014). The approach consists of the following steps (see also Fig. 6.4):

1. Clipping of the precipitation imagery to the Jubones basin (Fig. 6.4a).
2. Identification and localization of precipitation objects (latitude, longitude, see Fig. 6.4b). For this we defined a precipitation threshold

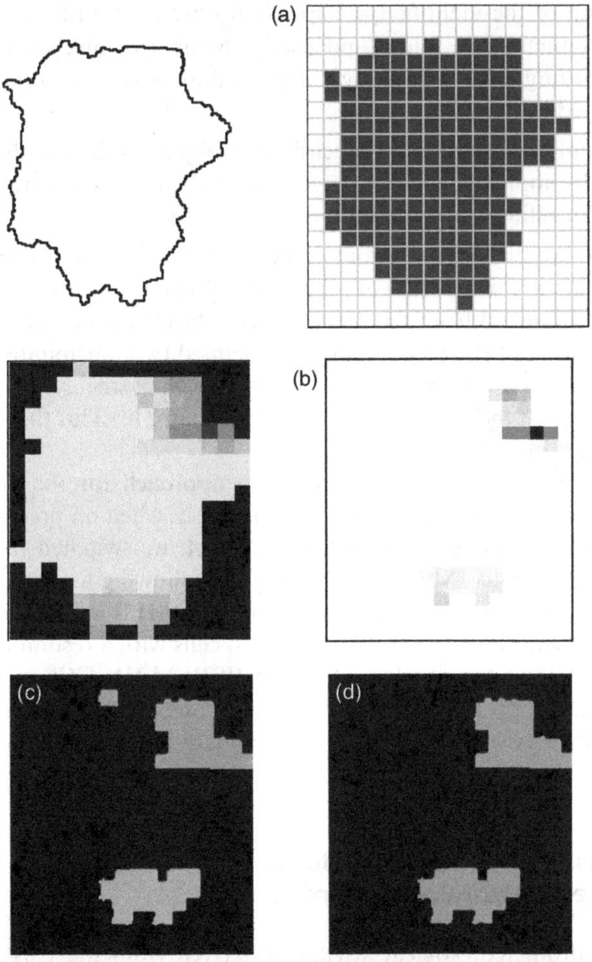

Figure 6.4 Precipitation identification with an object-based Connected Component Analysis Illustration of the PERSIAN-CCS 2021-12-25 05:00 UTC image: (a) Jubones basin clipping, (b) precipitation identification in mm from PERSIANN-CCS imagery, (c) initial identification of seven precipitation objects with CCA analysis, and (d) selection of two precipitation objects according to object size filtering and morphological closing.

volume of 0.5 mm, that is, precipitation objects with an associated precipitation volume of less than 0.5 mm are trimmed off. This was done on a trial-and-error basis validated with precipitation objects observed in randomly selected precipitation events. The target was to remove noise from precipitation imagery and keep only clear precipitation objects in the precipitation imagery (Fig. 6.4c).

3. Filtering of the identified precipitation objects according to size criteria. Similarly, we found and used a number-of-pixels threshold of 6, according to a trial-and-error procedure with the same target as employed in step 2.
4. The identified and filtered precipitation objects were morphologically closed by applying, as a final procedure, a dilatation-and-erosion algorithm for refining precipitation objects (Fig. 6.4d).
5. Retrieval of physical (centroid and extension area) and hydrometeorological attributes (volume of precipitation, maximum intensity, precipitation duration) from the precipitation objects defined in step 4. For the duration of the precipitation, we defined two precipitation objects are considered consecutive (i.e., belong to the same event) when the time between their appearance is shorter than 2 hr. This threshold was calibrated on a trial-and-error basis.

Additionally, a modular precipitation approach for the analysis of the precipitation imagery was used. For the cases when no precipitation is observed by the PERSIANN-CCS subproduct, we switched the precipitation data source to IMERG-early imagery, following a simple spatially undersampling technique. This means that an IMERG-early run cell of size 0.1 x 0.1° was directly divided into ~6.4 cells with a resolution of 0.04 x 0.04°, matching the resolution of the PERSIANN-CCS subproduct. This modular approach assures that all extreme hydrological events are trained with an existent precipitation signal, reducing noise and improving the learning process of the further developed runoff models.

6.3.3. Classification of Precipitation Events Associated With Extreme Hydrologic Events

The hydrometeorological attributes derived from the CCA analysis are used to classify precipitation events together with their associated runoff response. For this we used the following two criteria, respectively, the extension of the precipitation objects (local and spatially extensive) and the duration of the precipitation events (short and long). As a result, by defining extension and duration thresholds, we could establish four precipitation event classes: (1) Local and short extreme events (LSE), (2) local and long-duration extreme events (LLE), (3) spatially extensive extreme events (SEE), and (4) spatially extensive and long-duration extreme events (SLE).

6.3.4. Event-Based Runoff Modeling

We developed one runoff model for each precipitation event class and one model without precipitation discrimination (base model). For this, we used the ML technique known as Random Forest (RF) for regression. The RF is described in subsection 6.3.4.1. Moreover, the input feature space to each model was formed with hourly precipitation and runoff, as well as an indicator of the belonging precipitation class. In addition to current-time precipitation and runoff information, we used past lag information, which is determined according to statistical correlation analyses: partial- and auto-correlation functions for runoff, and cross-correlation functions for precipitation. The construction of the input feature space was conducted following the methodology developed by Muñoz et al. (2018), with the purpose to add only relevant information to the models and improve their efficiencies.

Random Forest for Regression

Random Forest (RF) is an ML technique of supervised learning where the main idea is to build multiple decorrelated trees (models) in which the input feature space is related to output(s) by successively applying a set of hierarchically organized conditions (Breiman, 2001). The key to the RF algorithm is the random selection of resampled data sets from the input feature space (bagging technique), which assures decorrelation between stochastically formed models.

We implemented the RF runoff models through the scikit-learn package for ML in Python version 3.7 (Pedregosa et al., 2011). A full explanation of the RF algorithm can be found in Breiman (2001) and can be summarized as follows:

1. Construction of each decision tree by randomly selecting several bootstrap samples from the input feature space. A process known as out-of-bag (OOB) is used for forming each bootstrap with roughly two thirds of the input feature space. On one hand, the OOB process serves to obtain unbiased estimates of the regression, and on the other hand, it allows to estimate the importance of each feature (predictor) of the feature space in the tree construction process.

Table 6.1 Search space (grid) of the RF runoff models.

Hyperparameter	Domain
n_trees*	40;800;10*
max_features	n_features, n_features$^{(1/2)}$, $\log_2$ (n_features)
max_depth*	40;800;10*

*Domain defined by min, max, and increment.

2. Optimally splitting of the data selected in step 1 at each node of each tree. This is done by determining a maximum number of features to perform the best split from the total number of predictors in the feature space. This also avoids overfitting by assuring variety and the nonexistence of duplicated models.
3. Growth of all the trees constructed in step 1 with the splits defined in step 2 up to a size defined either by a maximum depth parameter or a minimum number of samples expected in the final node. Control of the depth of the trees aims to reduce the structural complexity of the models, leading to model parsimony and noise reduction.
4. Determination of the output of the model as the mean response from all regression trees.

According to Contreras et al. (2021), the most-influencing RF hyperparameters for hydrological forecasting applications are the number of trees in the forest (n_trees), the maximum number of features to perform the splits of the data (max_features), and the maximum depth for pruning purposes (max_depth). For all runoff models, we determined the optimal combinations of hyperparameters following a random grid-search procedure implemented with a 10-fold cross-validation process to prevent overfitting. The measure of agreement was evaluated according to the coefficient of determination (R^2) between simulations and observations for the training subsets. Table 6.1 presents the domain of the selected hyperparameters, which forms the search space for the optimization task.

Model Evaluation

We used four goodness-of-fit metrics for evaluating the efficiencies of the four runoff models. The Nash-Sutcliffe Efficiency (NSE) coefficient was set as the reference for measuring and comparing the overall model accuracy. To complement the analysis, we relied on the Kling-Gupta Efficiency (KGE), the Percent Bias (PBIAS), and the Root Mean Square Error

(RMSE) metrics. The following equations were used:

$$\text{NSE} = 1 - \frac{\sum_{i=1}^{n}\left(Q_s(i) - Q_o(i)\right)^2}{\sum_{i=1}^{n}\left(Q_o(i) - \overline{Q_o}\right)^2}, \tag{6.1}$$

$$\text{KGE} = 1 - \sqrt{(r-1)^2 + (\alpha - 1)^2 + (\beta - 1)^2}, \tag{6.2}$$

$$\text{PBIAS} = \frac{\sum_{i=1}^{n}\left(Q_o - Q_s\right)}{\sum_{i=1}^{n} Q_o}, \tag{6.3}$$

and

$$\text{RMSE} = \sqrt{\frac{1}{n}\sum_{i=1}^{n}\left(Q_s - Q_o\right)^2}, \tag{6.4}$$

where n is the number of instances, Q_s is the simulated runoff, Q_o is observed runoff, $\overline{Q_o}$ is the mean observed runoff, $\overline{Q_s}$ is the mean simulated runoff, r is the correlation coefficient between Q_s and Q_o, $\alpha = \frac{\sigma_s}{\sigma_o}$ is the variability ratio, $\beta = \frac{\overline{Q_s}}{\overline{Q_o}}$ is the bias ratio, and σ is the standard deviation.

The NSE is dimensionless and ranges between $-\infty$ and 1.0, NSE = 1 being the optimal value. A limitation of NSE is the underestimation of peak flows and overestimation of low flows; in such cases, the KGE is suggested (Gupta et al., 2009), with KGE = 1 as the optimal value. Additionally, the optimal value of PBIAS is 0, positive values indicate model underestimation bias, and negative values overestimation bias. Finally, RMSE measures how model residuals are spread out from the best fit between simulations and observations, being RMSE = 0 the optimal value.

6.4. Results

6.4.1. Determination of Nearly Independent Peak Hydrological Events

The WETSPRO tool for the Jubones basin was calibrated using the following parameters: inter-event time of 120 hr (i.e., consecutive extreme hydrological events must be separated by a time frame of at least 5 days), and a maximum ratio of runoff drop down of 0.6 (i.e., runoff, q, drops down in between two consecutive events to a ratio $\frac{q_{min}}{q_{max}} < 0.6$). Moreover, we considered only events exceeding the 90% quartile values of the runoff time series ($98.8\ m^3.s^{-1}$). With these criteria, we obtained 55 nearly independent peak hydrological events (see Fig. 6.3a).

6.4.2. Object-Based Connected Component Analysis

For the 55 peak hydrological events, we first retrieved hourly precipitation maps from the PERSIANN-CCs and the IMERG-early run subproducts. Then, we applied the CCA algorithm with the precipitation threshold volume of 0.5 mm to derive the meteorological attributes and classify the precipitation event. The step-by-step application of the CCA algorithm for the map corresponding to the PERSIAN-CCS 2021-12-25 05:00 UTC is presented in Figure 6.4 (see section 6.3, Methodology).

CCA results showed that, for 15 extreme hydrological events, there was nearly or even a nonexistent precipitation signal from the PERSIANN-CCS subproduct. For these 15 cases, we performed the CCA algorithm on the IMERG-early run data set, and this resulted in a reduction of 40% of the events without any precipitation signal. In other words, although we used two precipitation satellite sources, we encountered nine hydrological events where either no precipitation at all was observed or any precipitation object was identified according to the CCA algorithm. Therefore, these events were trimmed off, leaving 46 events available for further analysis.

The validity of the precipitation modular approach is demonstrated in two extreme hydrological events (see Fig. 6.5). For instance, for the

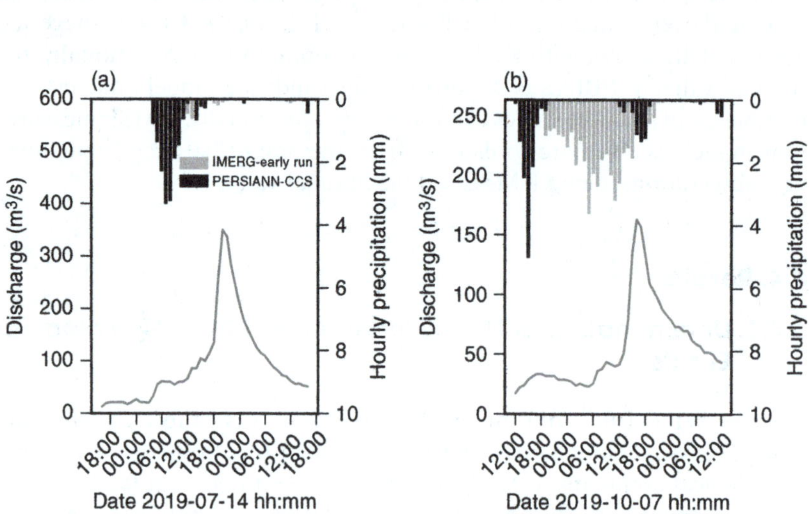

Figure 6.5 Illustration of the precipitation-retrieval modular approach using PERSIANN-CCS and IMERG-early run data sources, respectively, for the events from (a) 2019-07-13 18:00 to 2019-07-14 18:00 UTC, and (b) from 2019-10-07 12:00 to 2019-10-08 12:00 UTC.

event from 2019-07-13 20:00 to 2019-07-14 20:00 UTC, it seems evident that the highest resolution of the PERSIANN-CCS subproduct leads to a clearer precipitation-runoff relation when compared with precipitation obtained from the IMERG-early run subproduct. The opposite happened for the event from 2019-10-07 at 16:00 to 2019-10-08 at 16:00 UTC, where the PERSIANN-CCS signal was practically nonexistent, and the IMERG-early run signal was used to relate precipitation with runoff.

Moreover, the precipitation objects identified with the CCA algorithm for each one of the 46 extreme hydrological events were tracked down. From this analysis, the following information was retrieved: quantity, localization (centroids) and extension of precipitation objects, precipitation duration, total precipitation volume, and precipitation maximum intensity. This information is summarized in Figure 6.6 and served to infer duration and extension thresholds of 7 hr and 50 km^2, respectively. These thresholds were used in the following subsection to classify the precipitation events.

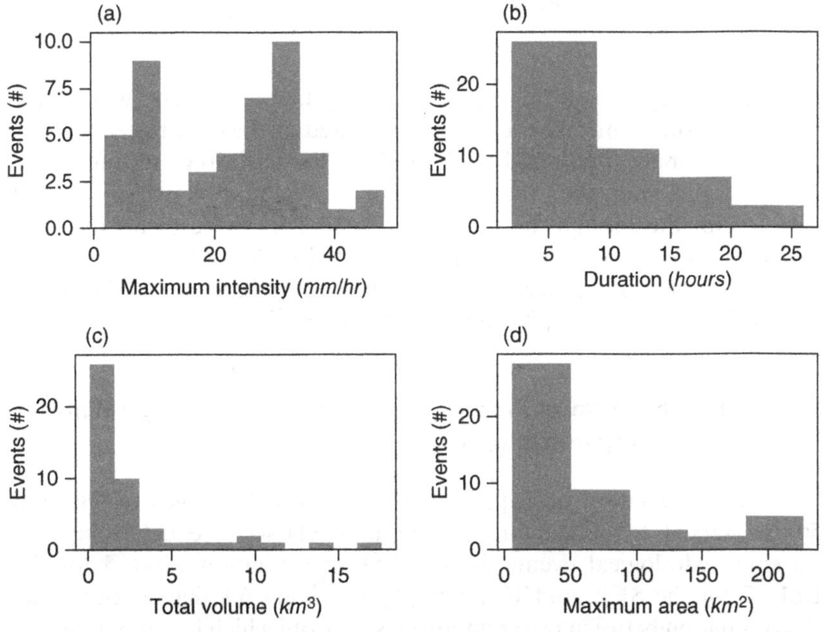

Figure 6.6 Meteorological precipitation information retrieved from 47 extreme hydrological events: (a) maximum intensity, (b) duration, (c) total volume, and (d) maximum area.

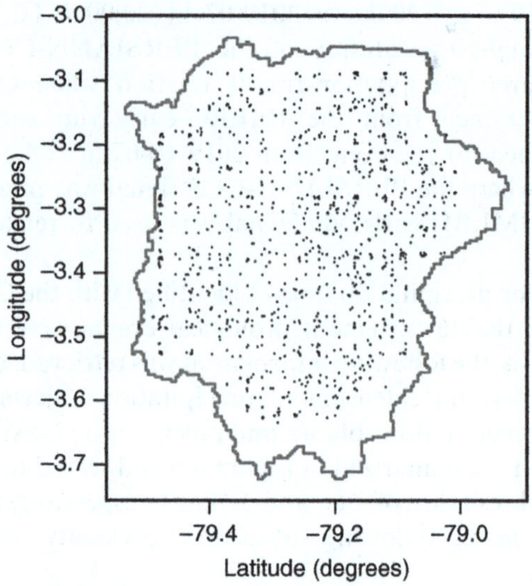

Figure 6.7 Localization of precipitation object centroids (dots) associated with extreme hydrological events in the Jubones basin.

Concerning the localization of precipitation objects within the Jubones basin, centroid occurrence appeared to be unaffected by any physical attribute that could be derived for the basin (i.e., altitude, land use, etc.). It is interesting that no hotspot of precipitation occurrence was detected for the Jubones basin (see Fig. 6.7). This suggests, for instance, that there is no evident orographic precipitation enhancement, and that the runoff generation process is rather driven by infiltration and saturation mechanisms before precipitation becomes streamflow.

6.4.3. Classification of Precipitation Events Associated With Extreme Hydrologic Events

The combination of duration and extension thresholds of 7 hr and 50 km^2 served to define four precipitation classes. We determined 24 extreme hydrological events for the LSE precipitation class, 5 for the LLE, 7 for the SEE, and 10 for the SLE. Figure 6.8 depicts the visual discrimination between precipitation classes, from which it is apparent that the majority of extreme hydrological events occurred as a result of short duration and spatial local (LSE) precipitation events, and long duration and spatially extensive events (SLE).

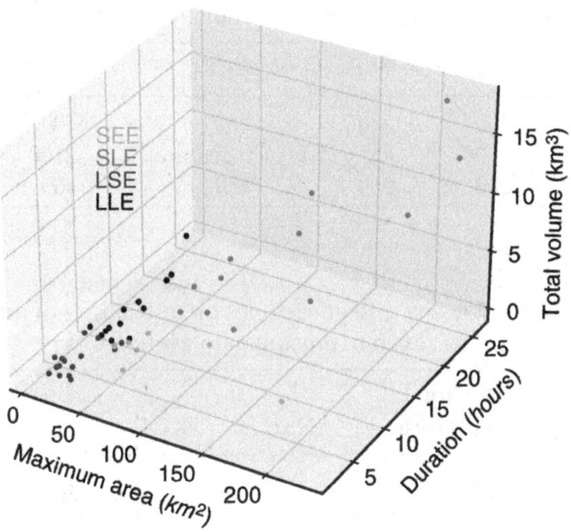

Figure 6.8 Precipitation classes associated with extreme hydrological events: Local and short extreme events (LSE), local and long-duration extreme events (LLE), spatially extensive extreme events (SEE), and spatially extensive and long-duration extreme events (SLE).

6.4.4. Event-Based Runoff Modeling

First, we defined the dimension of the input feature space of all extreme runoff models as a combination of current time precipitation together with past precipitation and past runoff data influencing current rime runoff. In this regard, results from partial- and auto-correlation functions for runoff suggest using past lags (hours) from 1 up to 12 lags, with a 95% confidence level for both correlation functions. Similarly, the cross-correlation function for precipitation determined 13 past lags (hours) of precipitation with correlations higher than 0.2. These results are congruent with the concentration time of the Jubones basin, which was estimated at 11 hr by averaging the concentration times found with the equations of Giandotti, Johnstone, and the U.S. Army Corps of Engineers (equations recommended for the basin area, see de Almeida et al., 2014).

Once the input feature space was defined, we constructed RF models for each precipitation class and the base model. For the model training and testing of each model, we assigned 70% of the events for training and the remaining 30% for testing. For instance, there were 46 events available for the LSE precipitation class; therefore, we assigned 32 events for training and 14 for testing. Moreover, since the objective was to simulate

Table 6.2 RF hyperparameterization of extreme runoff models.

Hyperparameter	None	LSE	LLE	SLE	SEE
n_trees*	300	280	250	300	300
max_features	$2100^{(1/2)}$	$\log_2(2100)$	$\text{n\_features}^{(1/2)}$	$2100^{(1/2)}$	$\log_2(2100)$
max_depth*	200	200	150	180	200

Table 6.3 Number of events and efficiencies on test subsets of runoff models specifically developed for different precipitation events.

Precipitation class	No. total events (test)	NSE	KGE	PBIAS	RMSE
None	46 (14)	0.83	0.85	4.49	55.38
LSE	24 (7)	0.67	0.71	−1.45	35.00
LLE	5 (2)	0.72	0.74	−23.94	41.76
SEE	7 (3)	−1.93	−0.48	−61.44	60.44
SLE	10 (3)	0.90	0.94	−2.72	69.09

the hydrographs corresponding to each event, we used a time frame of 24 hr before and after peak events. Concerning RF hyperparameterization, Table 6.2 presents the optimized combination of hyperparameters for each runoff model. The coefficient of determination between simulations and observations for the training subsets of each model was always higher than 0.91.

Table 6.3 summarizes the number of events used for developing extreme runoff models, and a comparison of the NSE coefficients obtained for each precipitation class and the base model. It is apparent from this table that LSE and especially SEE precipitation events are causing decay in the overall NSE-value of 0.83 (see also Fig. 6.9b and d). Surprisingly, LSE presents the majority of extreme hydrological events, and it seems contradictory that for LSE events, the higher number of events for training did not result in a higher NSE. This suggests that there are physical processes not well represented in the input feature space that disturbs the learning process of the RF models, as further discussed.

From the data in Figure 6.9, we can infer the spectrum of the runoff magnitudes modeled for each precipitation class. What is striking from the subfigures in Figure 6.9 is that regardless of the spatial extension, short-duration precipitation events (LSE and SEE classes) caused the lowest extreme runoff magnitudes at the outlet of the Jubones basin.

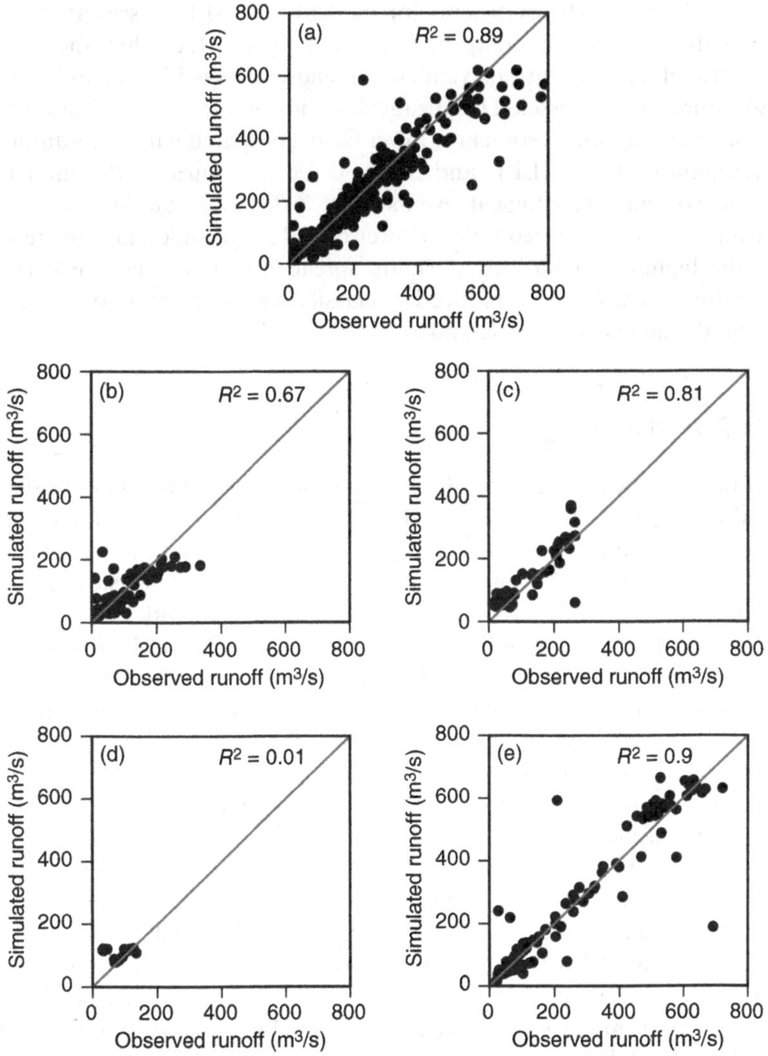

Figure 6.9 Scatter plot between extreme runoff observations and simulations for (a) no-precipitation event classification, (b) LSE events, (c) LLE events, (d) SEE events, and (e) SLE events.

Now, since we developed models for extreme runoff, we maximized the efficiencies for the highest runoff magnitudes. Therefore, it is evident that the lowest NSE coefficients for the LSE and SEE classes are found. Physically, this finding may be explained by the fact that the runoff response of short-duration events is somehow softened by the infiltration and saturation processes. This means that the volume of precipitation that becomes streamflow is somehow lower when compared with long-duration precipitation classes (LLE and SLE). If we now turn to the modeling of all extreme hydrological events (Fig. 6.9a), we can infer that the learning process is biased toward lower runoff magnitudes, and the results for the highest magnitudes are more spread out. However, the bias for long-duration events was reduced by classifying precipitation types before the modeling task (Fig. 6.9c and e).

6.5. Discussion

In this study, specialized (smart) extreme runoff models were developed for a 3,391 km^2 representative basin of the Ecuadorian tropical Andes. The efficiencies of the developed ML models are comparable and outperformed the ones obtained with traditional physically based models such as HEC-RAS (see the study of Belabid et al., 2019), wflow-sbm (see Laverde-Barajas et al., 2020), and the hydrologic-hydraulic HiResFlood-UCI model (see Nguyen et al., 2015). Particular to this finding is that, unlike physically based models, data-driven runoff models exploit precipitation satellite data without prior ground validation. Therefore, this study represents a solution for the cases when ground precipitation networks are scarce or even nonexistent.

The specificities of our extreme runoff models were delineated for four precipitation-event types based on a combination of their duration and spatial extension (LSE, LLE, SEE, and SLE). Developing specialized models served to identify the hidden strong-and-weak points of the base runoff model without precipitation classification. For instance, this approach could be used in the study of Belabid et al. (2019), where they obtained, in some cases, unacceptable runoff efficiencies (negative NSE).

For the Jubones basin, the vast majority of extreme hydrological events are the result of local and short-duration (LSE) precipitation events. In addition, we found that the centroids of LSE-associated objects were well distributed across the Jubones basin. These results indicate that small precipitation volumes are concentrated on many small different land-use areas, characterized by a variety of specific runoff generation processes. Therefore, even for a discriminated LSE precipitation event, multiple precipitation-runoff responses can mislead the learning process of RF models. This explains the lower model efficiencies of LSE events (NSE = 0.67) in comparison with SLE (0.90) and LLE (0.72) events. The opposite occurred in the case of long-duration and spatially extensive events (SLE), which were associated with the most extreme runoff magnitudes. For such events, even though we had less than half of the events available for LLE, model efficiencies reached the maximum (NSE = 0.90). The LLE runoff model was optimized for extreme runoff magnitudes (KGE = 0.94). Physically, this is explained by the fact that the RF learning process becomes straightforward after a greater portion of the basin is saturated, and any additional precipitation volume is directly converted into streamflow. The major difficulty comes from the modeling of extreme runoff triggered by spatially extensive and short-duration precipitation events (SEE).

The efficiencies of the developed and tested models highlighted the advantage of developing specialized extreme runoff models but also revealed the need to include additional information on antecedent soil saturation and its dynamic along with extreme hydrological events. This is particularly required for short-duration precipitation events (SEE and LSE), where the runoff generation process is governed by the antecedent saturation state of the basin. Foregoing is the reason why short-duration and nonextreme precipitation intensities can trigger extreme hydrological events. Given this, we encourage the approach employed by Massari et al. (2018) where they used satellite soil moisture observations to improve extreme runoff forecasting. Moreover, unveiling the limitations of runoff modeling for the Jubones basin opens the path for future research focused on exploring additional ML algorithms. We recommend, for instance, exploration of additional ML algorithms for the modeling of LSE and SEE events, and come up with a superior model consisting of an ensemble of specialized runoff models.

6.6. Conclusions

This study exploits the possibility of using two near-real-time satellite precipitation sources (without ground validation) for the development of smart extreme runoff models for a 3,391 km^2 basin. Smart models are characterized by the use of an ML algorithm with prior data assimilation enhancement under hydrometeorological criteria. For dealing with the complex precipitation-runoff response and the optimization of the runoff model efficiency, a straightforward feature engineering methodology was used. The major finding emerging from this study is that improvement of the representation of precipitation maximizes the efficiency of extreme runoff models. In addition, precipitation discrimination also served to unveil the precipitation-runoff scenarios misleading the learning process of RF extreme models.

In general, we found that the spatial extension of precipitation events made no significant difference in the learning process of RF models when they occurred for long-duration periods. These particular events produced the highest runoff magnitudes at the outlet of the basin. Physically, the success in modeling such precipitation events is attributed to a clear precipitation-runoff signal resulting from a gradual soil saturation process before precipitation is turned into a runoff. This signal served to improve the learning process of RF models by reducing noise and maximizing model efficiencies. In terms of input data, the present study intentionally used and tested two near-real-time precipitation satellite sources, the PERSIAN-CCS and IMERG-early run subproducts. We used a modular framework of precipitation data acquisition that reduced 40% of precipitation events with nearly or even nonexistent precipitation signals.

A natural progression of this work is to test the proposed methodology in other study areas with similar hydrological conditions. The contribution of this study is to inform ML models about the functioning of a hydrological system. Therefore, caution must be taken in hydrological systems where there is the contribution of groundwater; here, additional analyses might be needed to conceptualize and implement appropriate feature engineering strategies.

All in all, the knowledge gained from the functioning of the basin, the proposed feature engineering methodology, and the evaluation of nearly real-time satellite precipitation sources provides hydrologists with the tools for the future development of real-time runoff forecasting models. In addition, this study can be used to assist decision makers in the fields of flood forecasting, water resources management, optimization of hydropower generation, and many more.

References

Belabid, N., Zhao, F., Brocca, L., Huang, Y., & Tan, Y. (2019). Near-real-time flood forecasting based on satellite precipitation products. *Remote Sensing*, *11*(3), 252. https://doi.org/10.3390/rs11030252.

Beven, K. (2020). Deep learning, hydrological processes and the uniqueness of place. *Hydrological Processes*, *34*, 3608–3613. https://doi.org/10.1002/HYP.13805

Biau, G., & Scornet, E. (2016). A random forest guided tour. *Test*, *25*, 197–227. https://doi.org/10.1007/s11749-016-0481-7

Bontempi, G., Ben Taieb, S., & Le Borgne, Y. A. (2013). Machine learning strategies for time series forecasting. In M. A. Aufaure & E. Zimányi (Eds.), *Business intelligence.* eBISS 2012. Lecture Notes in Business Information Processing, 138. Berlin, Heidelberg: Springer. https://doi.org/10.1007/978-3-642-36318-4_3

Breiman, L. (2001). Random forests. *Machine Learning*, *45*, 5–32. https://doi.org/10.1023/A:1010933404324

Chang, L.-C., Chang, F.-J., Yang, S.-N., Kao, I.-F., Ku, Y.-Y., Kuo, C.-L., et al. (2019). Building an intelligent hydroinformatics integration platform for regional flood inundation warning systems. *Water*, *11*, 9. https://doi.org/10.3390/w11010009

Cisty, M., & Soldanova, V. (2018). Flow prediction versus flow simulation using machine learning algorithms. In P. Perner (Eds.), *Machine learning and data mining in pattern recognition.* MLDM 2018. Lecture Notes in Computer Science, 10935. Springer, Cham. https://doi.org/10.1007/978-3-319-96133-0_28

Clark, M. P., Bierkens, M. F. P., Samaniego, L., Woods, R. A., Uijlenhoet, R., Bennett, K. E., et al. (2017). The evolution of process-based hydrologic models: Historical challenges and the collective quest for physical realism. *Hydrology and Earth System Sciences*, *21*, 3427–3440. https://doi.org/10.5194/hess-21-3427-2017

Contreras, P., Orellana-Alvear, J., Muñoz, P., Bendix, J., & Célleri, R. (2021). Influence of random forest hyperparameterization on short-term runoff forecasting in an Andean mountain catchment. *Atmosphere*, *12*, 238. https://doi.org/10.3390/atmos12020238

Davis, C., Brown, B., & Bullock, R. (2006). Object-based verification of precipitation forecasts. Part I: Methodology and application to mesoscale rain areas. *Monthly Weather Review*, *134*, 1772–1784. https://doi.org/10.1175/MWR3145.1

de Almeida, I. K., Almeida, A. K., Anache, J. A. A., Steffen, J. L., & Alves Sobrinho, T. (2014). Estimation on time of concentration of overland flow in watersheds: A review. *Geociencias*, *33*, 661–671.

Galelli, S., & Castelletti, A. (2013). Assessing the predictive capability of randomized tree-based ensembles in streamflow modeling. *Hydrology and Earth System Sciences*, *17*, 2669–2684. https://doi.org/10.5194/hess-17-2669-2013

Gupta, H. V., Kling, H., Yilmaz, K. K., & Martinez, G. F. (2009). Decomposition of the mean squared error and NSE performance criteria: Implications for improving hydrological modeling. *Journal of Hydrology*, *377*, 80–91. https://doi.org/10.1016/j.jhydrol.2009.08.003

Gutiérrez-Jurado, K.Y., Partington, D., Batelaan, O., Cook, P., & Shanafield, M. (2019). What triggers streamflow for intermittent rivers and ephemeral streams in low-gradient catchments in Mediterranean climates. *Water Resources Research*, *55*, 9926–9946. https://doi.org/10.1029/2019WR025041

Herath, H. M. V. V., Chadalawada, J., & Babovic, V. (2021). Hydrologically informed machine learning for rainfall-runoff modelling: Towards distributed modelling. *Hydrology and Earth System Sciences*, *25*, 4373–4401. https://doi.org/10.5194/HESS-25-4373-2021

Hong, Y., Hsu, K.-L., Sorooshian, S., & Gao, X. (2004). Precipitation estimation from remotely sensed imagery using an artificial neural network cloud classification system. *Journal of Applied Meteorology and Climatology*, *43*, 1834–1853. https://doi.org/10.1175/JAM2173.1

Hong, Y., Tang, G., Ma, Y., Huang, Q., Han, Z., Zeng, Z., et al. (2019). Remote sensing precipitation: Sensors, retrievals, validations, and applications. *Observation and Measurement of Ecohydrological Processes*, *107–128*. https://doi.org/10.1007/978-3-662-47871-4_4-1

Hsu, K., Gao, X., Sorooshian, S., & Gupta, H. V, (1997). Precipitation estimation from remotely sensed information using artificial neural networks. *Journal of Applied Meteorology and Climatology*, *36*, 1176–1190. https://doi.org/10.1175/1520-0450(1997)0362.0.CO;2

Huang, P. C., & Lee, K. T. (2021). Influence of topographic features and stream network structure on the spatial distribution of hydrological response. *Journal of Hydrology*, *603*, 126856. https://doi.org/10.1016/j.jhydrol.2021.126856

Huffman, G. J., Bolvin, D. T., Braithwaite, D., Hsu, K., Joyce, R., Xie, P., et al. (2015). NASA global precipitation measurement (GPM) integrated multi-satellite retrievals for GPM (IMERG). *Algorithm Theoretical Basis Document (ATBD) version 4*.4, 26.

Laverde-Barajas, M., Corzo, G., Bhattacharya, B., Uijlenhoet, R., & Solomatine, D. P. (2019). Spatiotemporal analysis of extreme rainfall events using an object-based approach. In *Spatiotemporal analysis of extreme hydrological events* (pp. 95–112). Elsevier. https://doi.org/10.1016/B978-0-12-811689-0.00005-7

Laverde-Barajas, M., Perez, G. A. C., Chishtie, F., Poortinga, A., Uijlenhoet, R., & Solomatine, D. P. (2020). Decomposing satellite-based rainfall errors in flood estimation: Hydrological responses using a spatiotemporal object-based verification method. *Journal of Hydrology*, *591*, 125554. https://doi.org/10.1016/j.jhydrol.2020.125554

Li, B., Yang, G., Wan, R., Dai, X., & Zhang, Y. (2016). Comparison of random forests and other statistical methods for the prediction of lake water level: A case study of the Poyang Lake in China. *Hydrology Research*, *47*, 69–83. https://doi.org/10.2166/nh.2016.264

Li, J., Hsu, K.-L., AghaKouchak, A., & Sorooshian, S. (2016). Object-based assessment of satellite precipitation products. *Remote Sensing*, *8*, 547. https://doi.org/10.3390/rs8070547

Li, M., Zhang, Y., Wallace, J., & Campbell, E. (2020). Estimating annual runoff in response to forest change: A statistical method based on random forest. *Journal of Hydrology*. https://doi.org/10.1016/j.jhydrol.2020.125168

Liu, D., Fan, Z., Fu, Q., Li, M., Faiz, M. A., Ali, S., et al. (2020). Random forest regression evaluation model of regional flood disaster resilience based on the whale optimization algorithm. *Journal of Cleaner Production*, *250*, 119468.

Massari, C., Camici, S., Ciabatta, L., & Brocca, L. (2018). Exploiting satellite-based surface soil moisture for flood forecasting in the Mediterranean area: State update versus rainfall correction. *Remote Sensing*, *10*(2), 292. https://doi.org/10.3390/rs10020292

Moreido, V., Gartsman, B., Solomatine, D. P., & Suchilina, Z. (2021). How well can machine learning models perform without hydrologists? Application of rational feature selection to improve hydrological forecasting. *Water*, *13*, 1696. https://doi.org/10.3390/w13121696

Mosavi, A., Ozturk, P., & Chau, K.W. (2018). Flood prediction using machine learning models: Literature review. *Water (Switzerland)*, *10*, 1–40. https://doi.org/10.3390/w10111536

Muñoz, P., Orellana-Alvear, J., & Célleri, R. (2021). Application of a machine learning technique for developing short-term flood and drought forecasting models in tropical mountainous catchments. In *Integrated research on disaster risks* (pp. 11–35). Springer. https://doi.org/10.1007/978-3-030-55563-4_2

Muñoz, P., Orellana-Alvear, J., Willems, P., & Célleri, R. (2018). Flash-flood forecasting in an Andean mountain catchment-development of a step-wise methodology based on the random forest algorithm. *Water (Switzerland)*, *10*. https://doi.org/10.3390/w10111519

Nearing, G. S., Kratzert, F., Sampson, A. K., Pelissier, C. S., Klotz, D., Frame, J. M., et al. (2021). What role does hydrological science play in the age of machine learning? *Water Resources Research*, *57*. https://doi.org/10.1029/2020WR028091

Nguyen, P., Sellars, S., Thorstensen, A., Tao, Y., Ashouri, H., Braithwaite, D., et al. (2014). Satellites track precipitation of super typhoon Haiyan. *Eos, Transactions American Geophysical Union*, *95*, 133–135. https://doi.org/10.1002/2014EO160002

Nguyen, P., Thorstensen, A., Sorooshian, S., Hsu, K., & AghaKouchak, A. (2015). Flood forecasting and inundation mapping using HiResFlood-UCI and near-real-time satellite precipitation data: The 2008 Iowa flood. *Journal of Hydrometeorology*, *16*, 1171–1183. https://doi.org/10.1175/JHM-D-14-0212.1

Orellana-Alvear, J., Célleri, R., Rollenbeck, R., Muñoz, P., Contreras, P., & Bendix, J. (2020). Assessment of native radar reflectivity and radar rainfall estimates for discharge forecasting in mountain catchments with a random forest model. *Remote Sensing*, *12*, 1986. https://doi.org/10.3390/rs12121986

Papacharalampous, G. A., & Tyralis, H. (2018). Evaluation of random forests and prophet for daily streamflow forecasting. *Advances in Geosciences*, *45*, 201–208. https://doi.org/10.5194/adgeo-45-201-2018

Pedregosa, F., Varoquaux, G., Gramfort, A., Michel, V., Thirion, B., Grisel, O., et al. (2011). Scikit-learn: Machine learning in Python. *Journal of Machine Learning Research*, *12*, 2825–2830.

Peña-Barragán, J. M., Ngugi, M. K., Plant, R. E., & Six, J. (2011). Object-based crop identification using multiple vegetation indices, textural features and crop phenology. *Remote Sensing of Environment*, *115*, 1301–1316. https://doi.org/10.1016/j.rse.2011.01.009

Sakib, S., Ghebreyesus, D., & Sharif, H. O. (2021). Performance evaluation of IMERG GPM products during tropical storm Imelda. *Atmosphere*, *12*(6), 687. https://doi.org/10.3390/atmos12060687

Solomatine, D., & Ostfeld, A. (2008). Data-driven modeling: Some past experiences and new approaches. *Journal of Hydroinformatics*, *10*(1), 3–22. doi: https://doi.org/10.2166/hydro.2008.015

Solomatine, D., See, L. M., & Abrahart, R. J. (2009). Data-driven modelling: Concepts, approaches and experiences. In R. J. Abrahart et al. (Eds.), *Practical hydroinformatics*. Berlin, Heidelberg: Springer. https://doi.org/10.1007/978-3-540-79881-1_2

Soo, E. Z. X., Jaafar, W. Z. W., Lai, S. H., Othman, F., Elshafie, A., Islam, T., et al. (2020). Precision of raw and bias-adjusted satellite precipitation estimations (TRMM, IMERG, CMORPH, and PERSIANN) over extreme flood events: Case study in Langat River basin, Malaysia. *Journal of Water and Climate Change*, *11*, 322–342. https://doi.org/10.2166/WCC.2020.180

Sorooshian, S., Nguyen, P., Sellars, S., Braithwaite, D., AghaKouchak, A., & Hsu, K. (2014). Satellite-based remote sensing estimation of precipitation for early warning systems. In *Extreme natural hazards, disaster risks and societal implications* (pp. 99–112). Cambridge University Press.

Tang, G., Long, D., & Hong, Y. (2016). Systematic anomalies over inland water bodies of High Mountain Asia in TRMM precipitation estimates: No longer a problem for the GPM era? *IEEE Geoscience and Remote Sensing Letters*, *13*(12), 1762–1766. https://doi.org/10.1109/LGRS.2016.2606769

Tyralis, H., Papacharalampous, G., & Langousis, A. (2019). A brief review of random forests for water scientists and practitioners and their recent history in water resources. *Water (Switzerland)*. https://doi.org/10.3390/w11050910

van der Walt, S., Schuonberger, J. L., Nunez-Iglesias, J., Boulogne, F., Warner, J. D., Yager, N., et al. (2014). Scikit-image: Image processing in Python. *PeerJ Computer Science*, *2*, e453. https://doi.org/10.7717/peerj.453

Vogels, M. F. A., de Jong, S. M., Sterk, G., Wanders, N., Bierkens, M. F. P., & Addink, E. A. (2020). An object-based image analysis approach to assess irrigation-water consumption from MODIS products in Ethiopia. *International Journal of Applied Earth Observation and Geoinformation*, *88*, 102067. https://doi.org/10.1016/j.jag.2020.102067

Wang, L., Zhou, X., Zhu, X., Dong, Z., & Guo, W. (2016). Estimation of biomass in wheat using random forest regression algorithm and remote sensing data. *Crop Journal*. https://doi.org/10.1016/j.cj.2016.01.008

Wang, N., Zhang, D., Chang, H., & Li, H. (2019). Deep learning of subsurface flow via theory-guided neural network. *Journal of Hydrology*, *584*. https://doi.org/10.1016/j.jhydrol.2020.124700

Wang, Z., Lai, C., Chen, X., Yang, B., Zhao, S., & Bai, X. (2015). Flood hazard risk assessment model based on random forest. *Journal of Hydrology*, *527*, 1130–1141. https://doi.org/10.1016/j.jhydrol.2015.06.008

Willems, P. (2009). A time series tool to support the multi-criteria performance evaluation of rainfall-runoff models. *Environmental Modelling & Software*, *24*, 311–321. https://doi.org/10.1016/j.envsoft.2008.09.005

Yang, Y., & Chui, T. (2020). *Uncertainties of machine learning in predicting the hydrological responses of LID practices*. Proceedings of the 22nd IAHR-APD Congress. Sapporo, Japan.

Yin, J., Guo, S., Gu, L., Zeng, Z., Liu, D., Chen, J., et al. (2021). Blending multi-satellite, atmospheric reanalysis and gauge precipitation products to facilitate hydrological modelling. *Journal of Hydrology*, *593*, 125878. https://doi.org/10.1016/J.JHYDROL.2020.125878

Young, P. C. (2002). Advances in real-time flood forecasting. *Philosoophical Transactions Royal Society of London. Series A, Mathematical and Physical Sciences*, *360*, 1433–1450.

Yu, L., Ma, L., Li, H., Zhang, Y., Kong, F., & Yang, Y. (2020). Assessment of high-resolution satellite rainfall products over a gradually elevating mountainous terrain based on a high-density rain gauge network. *International Journal of Remote Sensing*, *41*(14), 5620–5644. https://doi.org/10.1080/01431161.2020.173425541

Part II

Forecasting Water Resources

7

FORECASTING WATER LEVELS USING MACHINE (DEEP) LEARNING TO COMPLEMENT NUMERICAL MODELING IN THE SOUTHERN EVERGLADES, USA

Courtney S. Forde[1], Biswa Bhattacharya[2], Dimitri P. Solomatine[2,3,4], Eric D. Swain[5], and Nicholas G. Aumen[6]

[1] *Caribbean Institute for Meteorology and Hydrology, Saint Michael, Barbados*
[2] *IHE Delft Institute for Water Education, Delft, The Netherlands*
[3] *Water Resources Section, Delft University of Technology, Delft, The Netherlands*
[4] *Water Problems Institute of the Russian Academy of Sciences, Moscow, Russia*
[5] *Caribbean-Florida Water Science Center, United States Geological Survey, Lutz, Florida, USA*
[6] *Southeast Region, United States Geological Survey, Boynton Beach, Florida, USA*

Water level is an important guide for water resource management and wetland ecosystems, defining one of the most basic processes in hydrology. This research seeks to investigate the possibility of complementing numerical modeling with a Machine Learning (ML) model to forecast daily water levels in the southern Everglades in Florida, USA. An exact analytical solution to water level may not be possible, but using the computational methods afforded by ML, the traditional numerical techniques may be enhanced to generate more robust, scalable predictions. Five locations were chosen for application of the Time-Delayed Neural Network (TDNN) and Long-Short Term Memory Recurrent

Advanced Hydroinformatics: Machine Learning and Optimization for Water Resources, Special Publications 78, First Edition. Edited by Gerald A. Corzo Perez and Dimitri P. Solomantine.

DOI: 10.1002/9781119639268.ch7

Neural Network (LSTM-RNN) ML models, which were built to estimate water level with 1, 2, 3, 7 and 10 day forecasts using a simulation step of 1 day. The results showed that rainfall forecasts from weather models could improve water-level forecasts if the accuracy and performance of the weather models can be improved. The ML models presented here improve water-level predictions from a historical hydrologic model for a 24 hour forecast horizon.

7.1. Introduction

Water level is widely used in flood risk management, monitoring of ground and surface water, and ecosystem and habitat monitoring and restoration in wetland environments (Forde, 2020). The Florida Everglades is considered one of the most significant wetland ecosystems in the world, particularly within Everglades National Park (ENP), the southern-most part of the Greater Everglades (Haman & Svendsen, 2006) established in 1947, which was designated as an International Biosphere Reserve in 1976, a World Heritage Site in 1979, and a Ramsar Convention Wetland of International Importance in 1987 (Maltby & Dugan, 1997). Prior to large-scale alterations of south Florida's watershed, water flowed unhindered from the Kissimmee River into Lake Okeechobee, which periodically overflowed its southern bank during the wet season and drove sheet flow of surface water through the Everglades into the Gulf of Mexico, Florida Bay, and Biscayne Bay (Ingebritsen et al., 1999) (Fig. 7.1).

After severe flooding from hurricanes, a levee was constructed around almost all of Lake Okeechobee to prevent the overflow of water and avoid further flooding. A network of canals, levees, water control structures, and pumps was built throughout south Florida to provide water supply and flood protection to an ever-increasing population. Improvements and expansion of channel networks, development of the historical Everglades wetlands into surface reservoirs called Water Conservation Areas, and the construction of water-control structures further altered the watershed. As a result, the natural environment significantly deteriorated due to altered hydrodynamics as it relates to flow, water level, seasonal timing, and spatial distribution (Light & Dineen, 1994). The altered flows resulted in shifts in vegetation composition, as well as dramatic reductions in wading bird numbers due to habitat loss and reduced aquatic production, both of which were a result of ecosystem degradation and shortened hydroperiods (Light & Dineen, 1994).

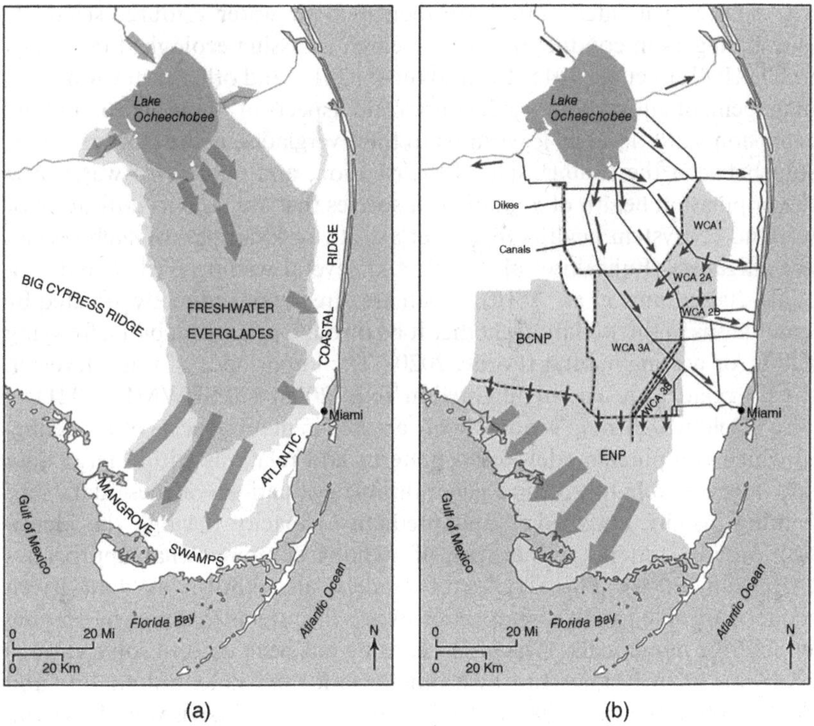

Figure 7.1 (a) Historical flow through the Everglades before drainage and (b) the current flow pattern with implementation of the water-control system of canals, levees, and structures. BCNP = Big Cypress National Preserve; WCA = Water Conservation Area; ENP = Everglades National Park (image from Telis et al., 2015).

With more than 1,800 km of canals and levees, and more than 2,200 water-control structures managed by the South Florida Water Management District (SFWMD), the water system is substantially altered to provide flood control and water conservation for water supply. The Comprehensive Everglades Restoration Plan (CERP) was created to rectify the issue of the severely degraded wetland ecosystem due to reduced quality and quantity of water, without sacrificing water supply and flood protection (Perry, 2008).

A significant tool for assistance in implementing the CERP is the Everglades Depth Estimation Network (EDEN, https://sofia.usgs.gov/eden). This network consists of more than 240 gauging stations, a water surface model, and a ground elevation model, which are used to monitor and simulate continuous water depths throughout the Everglades.

The EDEN-generated values are then used by water resource scientists and managers in conducting their research assessing ecological responses to CERP (Telis et al., 2015; Conrads et al., 2014) and other restoration and management efforts. One of the important aspects of water depth, and, by extension water level as it pertains to the Everglades, is the close relationship between the timing, spatial distribution, and volume of water, and the population health of a number of species that are used as indicators of wetland ecosystem health (Brandt et al., 2016). Examples of such species are alligators (Fujisaki et al., 2009) and several wading birds (Frederick et al., 2009; Lantz et al., 2010), which are directly or indirectly affected by water levels in the wetland, whether it be through prey availability, foraging habits, or nesting habitat (Forde, 2020). The importance of water levels in the Everglades are also highlighted in Telis (2006) and SFWMD (2021).

Machine Learning is now widely applied in many aspects of modeling, in which computer models learn patterns and relations within large data sets, and are able to produce generalizable evaluations of these data sets, gaining insight, knowledge, and predictive capacity through the identification, utilization, and testing of various computational approaches (Alpaydin, 2009). These types of models, also known as data-driven models, are able to solve complex problems in systems that are challenging to describe numerically. Machine Learning has been used to solve various problems in hydrology and hydraulics, see for example, Solomatine and Dulal (2003), Bhattacharya and Solomatine (2005), Kasiviswanathan et al. (2016), Xiang et al. (2020), Kao et al. (2020), and Moreido et al. (2021). A historical overview of Artificial Neural Networks (ANN) for hydrological forecasting is provided by Abrahart et al. (2012). Additionally, there are multiple examples of using ML for water-level forecasting (e.g., Zhu et al., 2020). Cruz et al. (2018) successfully used a multilayer ANN (Aggarwal, 2018) in predicting water level using real-time rainfall, water level, and soil moisture data. Amaranto et al. (2018) demonstrated the use of ANN (multilayer perceptron) for groundwater-level forecasting. Another study by Ma et al. (2019) showed promising results using a model based on LSTM networks to investigate whether the accuracy of water-level forecasting could be improved in the Rhine River Basin in Germany. A LSTM network, one form of Recurrent Neural Networks, is a deep-learning architecture that includes memory cells that store information over long periods of time and is known to be the optimal contemporary network architecture for predicting sequences such as in time-series, speech recognition, text prediction, and others (Kratzert et al., 2018). Unlike traditional approaches for time-series forecasting like Autoregressive Integrated Moving Average (ARIMA), LSTM has the ability to adapt to features of the data such as seasonality (Siami-Namini, 2018). Additionally,

LSTM can utilize more hyperparameters and, because the extent of the short-term memory can be defined, LSTM tends to outperform SARIMA (Seasonal-ARIMA).

We could find only two published research papers presenting explicit water-level modeling in the Everglades. The first was a study by Conrads and Roehl (2007) where an ANN was used to hindcast water levels. The second and more applicable to this study focused on prediction boundaries and forecasting of nonlinear hydrological stage data in the A.R.M. Loxahatchee National Wildlife Refuge, the northern-most extent of the remnant Everglades (Park et al., 2005). An ANN model was developed to forecast stage over a 16 yr period from just 2 yr of training data, with forecast horizons of 1, 2, 8, and 14 days. This model was compared with linear models that forecast stages, and the results showed that the ANN model outperformed the linear models, most notably at the longer lead times. Additional methods for evaluating machine-learning models, such as visualization tools, were not applied in this study.

Apart from these examples, the models that currently exist for forecasting water level are numerical models, which simulate overall regional hydrological conditions in the entire, or specific parts, of the Everglades and are used by water managers. One of these models is the Regional Simulation Model (RSM), a surface water-groundwater model that solves for water stages and flows under natural (predrainage system) conditions as well as under controlled conditions such as the operation of water-control structures. The RSM quantifies changes in groundwater levels, groundwater and overland flow, and wetland hydroperiods, among other things (SFWMD, 2006). The RSM is solved for an irregular triangular mesh grid, which is considered an improvement on the South Florida Water Management Model (SFWMM), which has a square grid size of 3.2 x 3.2 km (SFWMD, 2008). The SFWMM simulates the hydrology and management of the water resources system from Lake Okeechobee to Florida Bay and is a tool for assessing operational changes to the complex water management system in South Florida (SFWMD, 2005).

Another regional simulation model, and the one used for comparison in this study, is the Biscayne and Southern Everglades Coastal Transport (BISECT) model, which simulates hydrologic conditions in various scenarios (Swain et al., 2019). This model uses dynamic equations so that surface water is fully represented, making it suitable for comparison in this study. One of the disadvantages of the BISECT and possibly other regional models is the averaging of inputs and outputs over each grid cell, which reduces the extent of peaks and troughs in the predictions. As a result, the BISECT model is more suitable for long-term regional analysis of hydrological conditions, than it is for local, shorter-term predictions (Swain et al., 2019).

Many ML models use real-time data in forecasting, but there is a lack of research as it relates to the use of Numerical Weather Prediction (NWP) (Pu & Kalnay, 2019) in predicting over longer forecast horizons. Additionally, from the characteristics of the numerical models mentioned as well as the issues faced by both water managers and ecologists, improved forecasts could assist operational water management and ecological assessments, particularly under the CERP.

The main objective of this study is to apply and test ML models to potentially complement conventional numerical models in an effort to improve water-level forecasting in the southern Everglades. Reaching this objective involved choosing suitable predictors for forecasting water levels. Two ML model architectures used to predict water level were tested for various forecast horizons and compared with a numerical model. A novel aspect of this study is analyzing the effect of NWP on the accuracy of the water-level predictions.

7.2. Methods

The southern Everglades is an important resource because ENP is one of the last remaining natural parts of the Everglades. The area just south of Tamiami Trail (Fig. 7.2) was identified as an area where water-level forecasts would be beneficial. Because restoring hydrologic conditions in the natural ecosystem is a primary objective of the CERP, water-level forecasts are essential to inform and aid restoration efforts as well as water resources management.

Five stations in the area were selected (Fig. 7.2). Station P33 was identified as an index station, considered to be representative of water levels and hydroperiods in Shark Slough, a low-lying area that is the main drainage through ENP. Station NP201 was selected from a map of the EDEN network (https://sofia.usgs.gov/eden) on the basis of its relative proximity to Tamiami Trail, as was station BCA11 (also selected to provide spatial representation along the trail). Finally, two stations influenced by control structures S12A_T and S12D_T (Fig. 7.2) were selected to allow for the consideration of discharges from water-control structures.

For each station, rainfall, evapotranspiration, and water-level data were available, while discharge data were available for the control structures S12A and S12D (vertical lift gates), all at a daily resolution. The rainfall, evapotranspiration, and water-level data had an 18 yr period of record (1 Jan 2002 to the present), 23 yr (1 Jun 1995–31 Dec 2018), and 30 years (1 Jan 1990 to the present), respectively, for all the stations. An exception was BCA11, which had a water-level record from 28 Dec 1990 to present

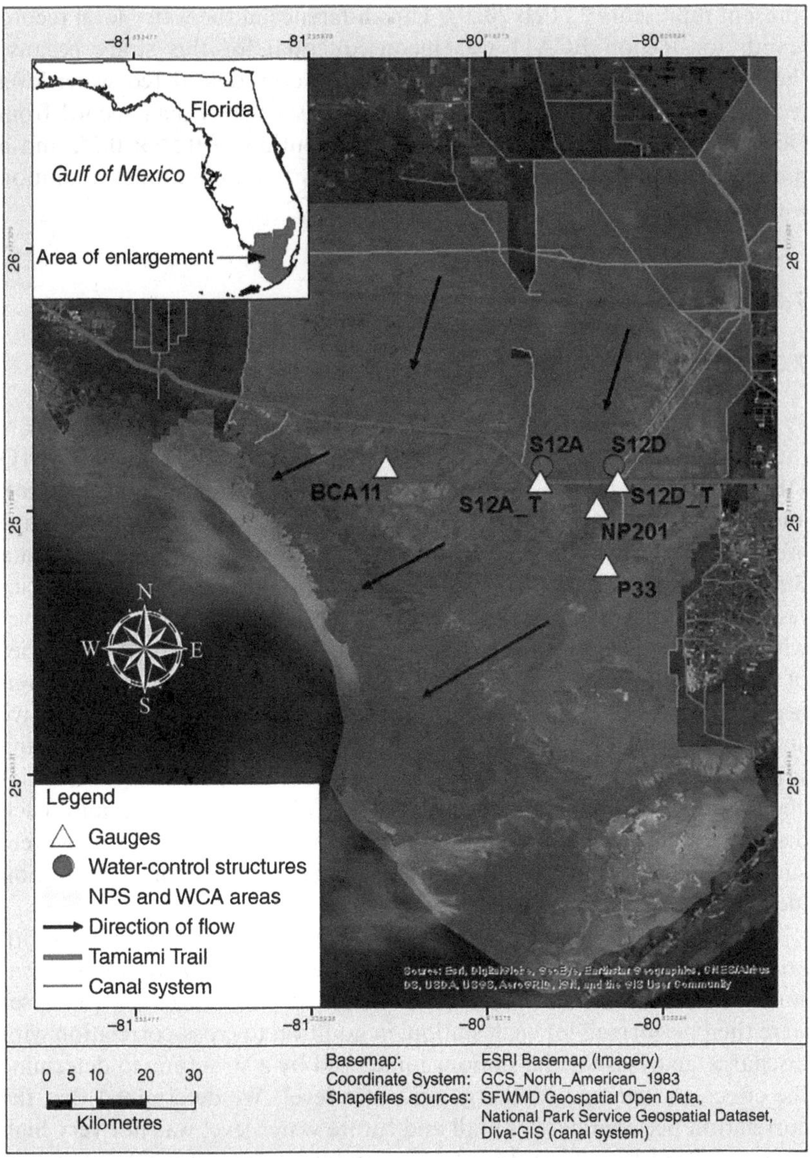

Figure 7.2 Map showing the study area in the southern Everglades, Florida, and the gauges used in this study. The general direction of water flow is from north to south. Water-control structures S12A and S12D are just upstream of the gauges S12A_T and S12D_T, respectively. The T designation for the station names indicates a tailwater (downstream from the structure). Gauges BCA11, NP201, and P33 are not directly influenced by a structure. NPS = National Park Service; WCA = Water Conservation Areas (map adapted from Forde (2020); copyright Courtney Forde 2020; used with permission).

(present represents 22 Feb 2020). This difference in the water-level record length for station BCA11 was inconsequential for this study because the rainfall record started in 2002, and we only considered water-level records starting in 2002. The rainfall forecasts had a data record from October 2006 to present with a spatial resolution of 0.5° × 0.5°, and a minimum temporal resolution of 6 hr. A daily temporal model resolution was selected.

7.3. Workflow

7.3.1. Data Acquisition and Processing

The data used were rainfall and water level from USGS EDEN (https://sofia.usgs.gov/eden/eve/index.php), discharge from SFWMD DBHYDRO database (https://www.sfwmd.gov/science-data/dbhydro), and rainfall forecasts from the European Centre for Medium-Range Weather Forecasts (ECMWF) (https://apps.ecmwf.int/datasets). Data were subjected to extensive quality assurance measures by the agencies responsible for collection. The discharge data set was the only data set with missing values, which were gap-filled using two additional time series of discharge collected by the SFWMD and the U.S. Army Corps of Engineers. The USGS time series for discharge was the most comprehensive and was selected for use by ML models, but it also contained some missing values. To gap-fill USGS discharge, values from the other two time series were used to fill missing values of the USGS time series. Because each discharge time series represents data from the same location, values were similar and in some cases identical, confirming that this substitution method was adequate.

Once the rainfall, water-level, and discharge data were continuous with no missing values, each time series was graphed to observe trends and seasonal patterns. Cross-correlation with rainfall and autocorrelation analyses were then performed for each station, in addition to cross-correlation with discharge analyses for the stations influenced by a structure, to determine the effect of structure discharge on water level. We determined that the correlation between past rainfall and future water level was not very high (Fig. 7.3a and Fig. 7.4a). Due to the random nature of rainfall together with the influence of control structure operation, the relationship between water level and rainfall is nonlinear and may not always be captured by a linear statistic such as correlation coefficient.

The high autocorrelation between past and future water levels was expected, because past water level largely influences future water level

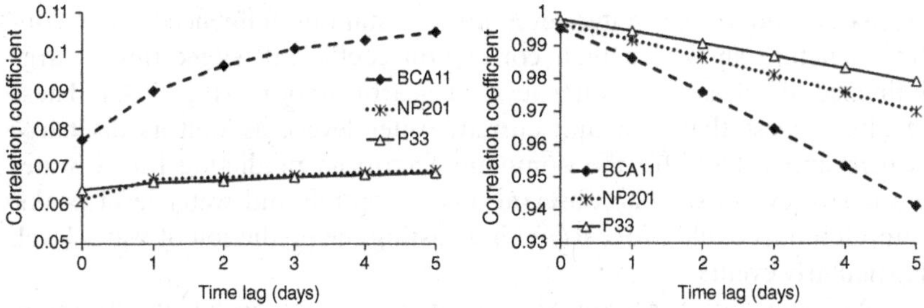

Figure 7.3 Correlation analyses for the three stations *not* influenced by a control structure: (a) Cross-correlation analyses (past rainfall–future water level); (b) autocorrelation analyses (past water level–future water level).

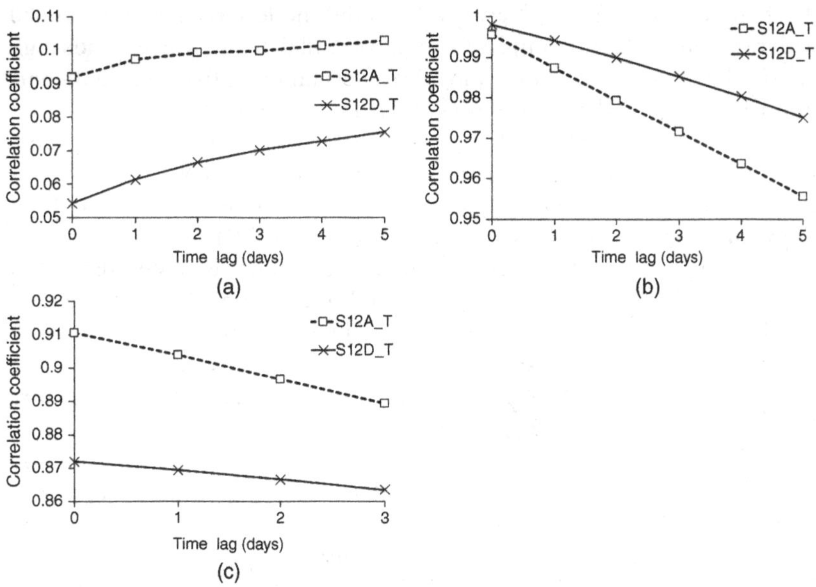

Figure 7.4 Correlation analyses for the stations influenced by a control structure: (a) Cross-correlation analysis (past rainfall–future water level); (b) autocorrelation analysis (past water level–future water level); (c) cross-correlation analysis (past structure discharge–future water level).

(Fig. 7.3b and Fig. 7.4b). The additional cross-correlation between past discharge and future water level for the stations influenced by a control structure showed a high correlation coefficient, suggesting a large influence of releases on water level at nearby stations (Fig. 7.4c). These results suggest that past and current water levels, as well as discharge (where applicable), are the dominant factors in predicting future water level. However, from the observations of rainfall and water-level peaks, rainfall inputs could prove useful in assisting the prediction of water level, particularly events.

An important step in any ML exercise is partitioning data into training, cross-validation, and testing subsets. There are various approaches for data partitioning, from simple methods like random sampling, to more complex methods aimed at optimization of the resulting models (e.g., Genetic Algorithm, Self Organizing Maps; Bowden et al., 2002). The most widely used and simplest method is random sampling, which was used in this study for the Time-Delayed Neural Network (TDNN) models. Each subset is filled with randomly selected samples from the overall data set, which avoids high bias in the performance of the model (Reitermanova, 2010). The data partitioning for the LSTM model is explained in a later section.

The NWP rainfall data were from the Interactive Grand Global Ensemble data set, and downloaded from the ECMWF (https://apps.ecmwf.int/datasets/). This download was performed using a Python script and Application Programming Interface (API), which allows for connecting to the ECMWF server and downloading requested data. The script specified the class of data preferred, the type of data set, the date required, grid size, the parameter(s) required, the time step(s) desired, and specific grid coordinates for selecting a desired region. The data were stored in gridded format and these files were processed using the software Panoply, a data viewing tool developed by National Aeronautics and Space Administration, which allows for the visualization of Network Common Data Form (NetCDF), General Regularly Distributed Information in Binary form (GRIB), and other data sets.

For building ML models, Google's TensorFlow (Abadi et al., 2016a; Abadi et al., 2016b) was used. It is used widely for its deep learning capabilities, and the high-level API Keras (Chollet, 2015) makes building and evaluating ML models easy. The ANN was the Machine Learning technique used, and two model architectures were developed: a Multi-Layer Perceptron (MLP) TDNN (Luk et al., 2001) as well as a LSTM (Sherstinsky, 2020) network. Both architectures were designed with a single hidden layer (see experimental setup for number of nodes), and to generate a single output, and one of each architecture was developed for each location under study. The performance indicators used to assess the models were

Root Mean Squared Error (RMSE) and Coefficient of Determination (R^2). Python was employed as the main programming language.

7.4. Experimental Setup

7.4.1. Selecting Model Inputs and Data Preparation

Based on the correlation analyses and the usefulness of rainfall as one of the predictors, a number of trials were conducted to determine the most suitable combination of predictors and their lags. Models were set up to initially run using the inputs of rainfall (referred to as R) ([R_{t-1}, R_t]), water level (referred to as WL) (combinations of [WL_{t-1}, WL_t], [WL_{t-2}, WL_{t-1}, WL_t], and [WL_{t-3}, WL_{t-2}, WL_{t-1}, WL_t]), and discharge (referred to as FLOW), where applicable (combinations of [$FLOW_t$], and [$FLOW_{t-1}$, $FLOW_t$]. The dependent variable was [WL_{t+1}]. Similar to the correlation analyses, the tests using the WL inputs provided the best results and were used as a basis for further experiments. For the stations not influenced by a water-control structure, the rainfall inputs were added to the combinations of WL inputs initially tested, resulting in the combinations [R_{t-1}, R_t, WL_{t-1}, WL_t], [R_{t-1}, R_t, WL_{t-2}, WL_{t-1}, WL_t], and [R_{t-1}, R_t, WL_{t-3}, WL_{t-2}, WL_{t-1}, WL_t]. The test for the stations not influenced by a water-control structure were conducted first, and then the combination that provided the best model performance was used for tests for the stations influenced by a water-control structure, adding the additional input(s) of discharge. The combinations tested here were [R_{t-1}, R_t, WL_{t-2}, WL_{t-1}, WL_t, $FLOW_t$] and [R_{t-1}, R_t, WL_{t-2}, WL_{t-1}, WL_t, $FLOW_{t-1}$, $FLOW_t$]. Table 7.1 shows the improvement of model performance at the majority of lead times with the addition of the rainfall inputs. This improvement is observed in the RMSE values in Table 7.1. All tests were conducted using the TDNN model architecture.

Additional experiments were conducted with inputs, such as the 7 day moving average of rainfall and change in water level, but rendered no improvement in model performance and were not considered further. The final combination of inputs was decided based on the results of the tests. The models for the stations (one for each gauging station location) not directly impacted by a control structure were trained with the predictors [R_{t-1}, R_t, WL_{t-2}, WL_{t-1}, WL_t]. For those that were influenced by the discharge from a control structure [R_{t-1}, R_t, WL_{t-2}, WL_{t-1}, WL_t, $FLOW_t$] were the predictors used. Table 7.2 shows the inputs for each forecast lead time presented in this study. The measured predictors used were R_{t-1}, R_t, WL_{t-2}, WL_{t-1}, WL_t, and all FLOW predictors. All remaining predictors

Table 7.1 Root mean squared error of example model.

Lead time	RMSE (m) (TDNN model) Input variables: WL	Input variables: Rainfall & WL
1 day	0.02	0.02
2 days	0.04	0.03
3 days	0.07	0.04
7 days	0.11	0.07
10 days	0.13	0.09

Note: Root mean squared error of example model tested with only the water-level predictors, and the selected combination of water-level and rainfall predictors, for gauge NP201. The average water level is 1.8 m. TDNN = Time-delayed neural network; WL = water level.

Table 7.2 Details for short-term and medium-range forecast horizons.

Lead Time	Inputs: Gauges not influenced by structures	Inputs: Gauges influenced by structures	Output
1 day	R_{t-1}, R_t, WL_{t-2}, WL_{t-1}, WL_t	R_{t-1}, R_t, WL_{t-2}, WL_{t-1}, WL_t, FLOW_t	$\mathbf{WL_{t+1}}$
2 days	R_t, R_{t+1}, WL_{t-1}, WL_t, $\mathbf{WL_{t+1}}$	R_t, R_{t+1}, WL_{t-1}, WL_t, $\mathbf{WL_{t+1}}$, FLOW_{t+1}	$\mathbf{WL_{t+2}}$
3 days	R_{t+1}, R_{t+2}, WL_t, $\mathbf{WL_{t+1}}$, $\mathbf{WL_{t+2}}$	R_{t+1}, R_{t+2}, $\mathbf{WL_t}$, $\mathbf{WL_{t+1}}$, $\mathbf{WL_{t+2}}$, FLOW_{t+2}	$\mathbf{WL_{t+3}}$
7 days	R_{t+5}, R_{t+6}, $\mathbf{WL_{t+4}}$, $\mathbf{WL_{t+5}}$, $\mathbf{WL_{t+6}}$	R_{t+5}, R_{t+6}, $\mathbf{WL_{t+4}}$, $\mathbf{WL_{t+5}}$, $\mathbf{WL_{t+6}}$, FLOW_{t+6}	$\mathbf{WL_{t+7}}$
10 days	R_{t+8}, R_{t+9}, $\mathbf{WL_{t+7}}$, $\mathbf{WL_{t+8}}$, $\mathbf{WL_{t+9}}$	R_{t+8}, R_{t+9}, $\mathbf{WL_{t+7}}$, $\mathbf{WL_{t+8}}$, $\mathbf{WL_{t+9}}$, FLOW_{t+9}	$\mathbf{WL_{t+10}}$

Note: Details for short-term forecast horizons (1–3 days) and medium-range forecast horizons (7 and 10 days). The numerical weather prediction inputs in the table are underlined, the forecasted water level variables are in bold, and the remaining variables are measured. R = rainfall; WL = water level.

for further time steps are forecasted values. As the table of predictors indicates, each prediction was used as input for the prediction at the subsequent forecast time step. In other words, the water-level prediction for the 1 day forecast horizon was used as an input in predicting the 2 day forecast horizon, and so on, up to 10 days. The five forecast horizons assessed were grouped into short-term forecasts (1–3 days) and medium-range forecasts (7 and 10 days). The rainfall forecasts from the NWP were used at every lead time following the 1 day prediction. Measured discharge was used at each time step because future pump schedules were not available at the time of the study.

Eighteen years (2002–2019) of rainfall, water level, and discharge data were used and these data were split initially into subsets of 82% and 18%. The 18% represents approximately 3 yr (2008–2010) of data, which were selected in order to have a sequential time series for visualization, analysis, and comparison against numerical model output, and was considered the test set. The 82% was then randomly split into subsets of 66% (80% of the 82%) for training, and 16% (20% of the 82%) as a randomly selected test set. This additional randomly selected test set was used to ensure there was a test set with comparable statistics (mean and standard deviation) to the randomly selected training set.

To ensure these two subsets had comparable statistics, a Python script was written to randomly split the 82% proportion of the data set continuously into the subset proportions mentioned above, until the statistics of the R_t input on both sets were respectively similar (difference of 0.02). The R_t predictor was used because the rainfall inputs had the largest range in data and it was important to have distribution representative of the entire data set in both the training and test sets. The reason was to ensure the optimal training process and fair testing of the model. The 3 yr data period 2008–2010 was not included in the training, cross-validation, nor the randomly selected test subsets, but was reserved as an additional test set to predict a sequential time series after training and initial testing occurred. This process of data partitioning resulted in the final proportions being 53% for training, 13% for cross-validation, and 34% for testing (combination of the 3 yr sequential time series, 18%, and the randomly selected test set, 16%). For the LSTM network, the random sampling could not be applied due to the nature of the architecture, which requires sequences of data as opposed to a single input value of each variable. As a result, the same 3 yr sequential period (2008–2010) used for final validation for the TDNN was used for testing of the LSTM, but the remaining data were split into training and cross cross-validation subsets. The final subsets for the LSTM network were 66.6% for training (12 yr), 16.7% for cross-validation (3 yr), and 16.7% for testing (3 yr; same period as for the TDNN).

To normalize the data for better learning, the Z-score normalization was performed (Eesa & Arabo, 2017). The equation for the Z-score is as follows:

$$z_i = \frac{x_i - \mu}{\sigma}, \tag{7.1}$$

where x_i is a single sample, μ is the sample mean, and σ is the sample standard deviation.

7.5. Comparison of Measured Rainfall and NWP Rainfall Forecasts

The ECMWF rainfall forecasts were used as predictors for the models built in this study, and their accuracy (i.e., to what extent they can be assumed to represent the rainfall) is an important issue. Rainfall forecasts are prone to error due to the random nature of rainfall, particularly in the tropics where high temperatures as well as reduced influence of the Earth's Coriolis forcing on atmospheric circulation at these latitudes, result in challenges in weather forecasting (Sobel, 2012). Despite the presence of error in daily forecasted rainfall amounts, seasonal averages provide an understanding of the comparability of the forecasted rainfall to the measured rainfall. The maps in Figure 7.5 show that the differences in average rainfall between the gauged measurements and the forecasts are not very significant. The dry season (approximately Nov–Apr) rainfall total (sum of all the totals) for the gauges used in this study was 3,883 mm over the dry periods of the 3 yr prediction period 2008–2010. The sum of the forecasted precipitation for the 4 ECMWF grid cells covering the general locations of the gauges (Fig. 7.5b) was 3,725 mm. For the wet season (approximately May–Oct), the sum of total rainfall from the gauges was 15,972 mm, while the sum for the four forecast grid cells was 14,316 mm (Fig. 7.5d).

In addition to the inherent error in rainfall forecasts, the forecasts are averaged over a ~50 × 50 km grid whereas a gauge records point rainfall. The effect of error in the daily rainfall amounts being used as model predictors will be analyzed in the water-level predictions, but it is expected that error will be greater as the forecast lead time increases.

7.5.1. Model Development

The architecture of the TDNN model is a multilayer perceptron feed forward network. The nodes in the input layer have the input features of rainfall and water level, but additionally has those same inputs delayed by one or more time steps. More detail on the TDNN can be found in Lingras et al. (2008). The LSTM model architecture is similar to a standard Recurrent Neural Network (RNN), but the difference is the presence of memory blocks in the LSTM allowing it to store long-term information and later access it. Within these memory blocks are multiplicative units called gates, which facilitate the storage and access of long-term information (Xiang et al. 2020). A further description of LSTM networks and memory cells can be found in Hochreiter and Schmidhuber (1997) and Graves (2012).

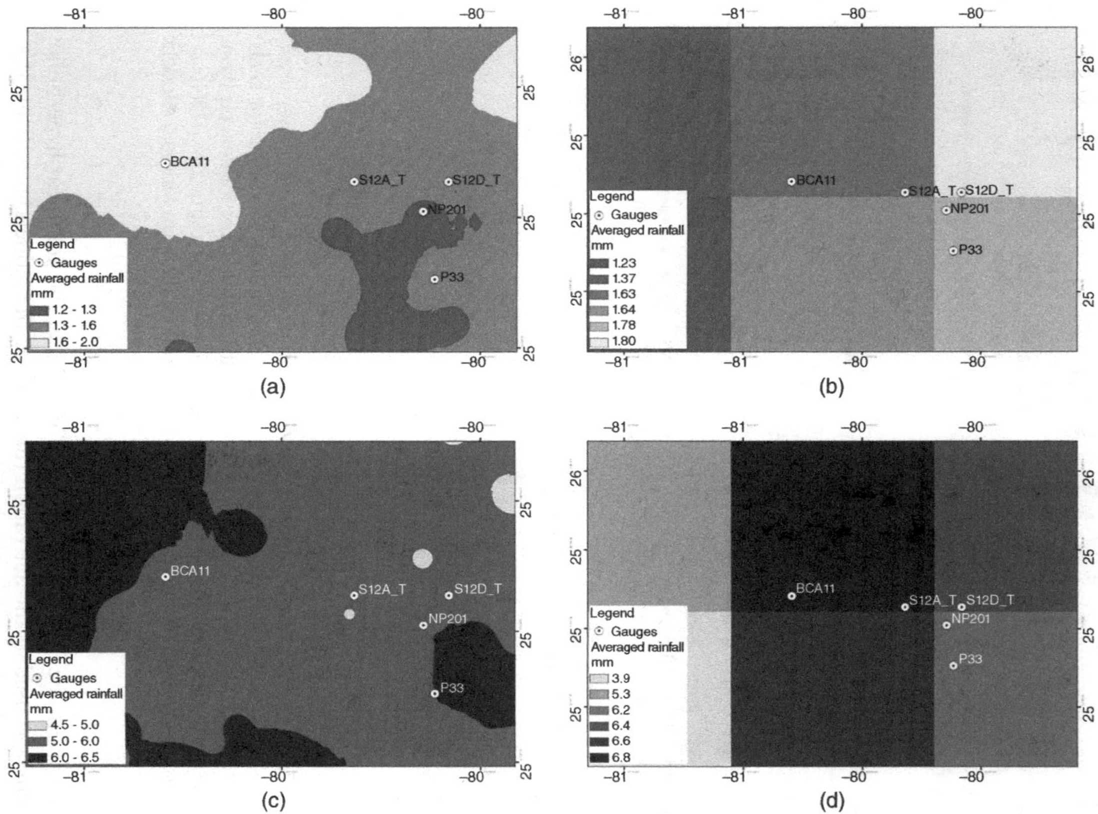

Figure 7.5 Averaged rainfall (mm) for the dry season and wet season in the study area for the period 2008–2010. The figure presents (a) interpolated average rainfall of 93 gauges in the study area for the dry season; (b) the dry season average for each cell of the European Centre for Medium-Range Weather Forecasts (ECMWF) gridded rainfall forecast covering the study area; (c) the interpolated average rainfall of 93 gauges for the wet season; and (d) wet season average for each cell of the ECMWF gridded rainfall forecast.

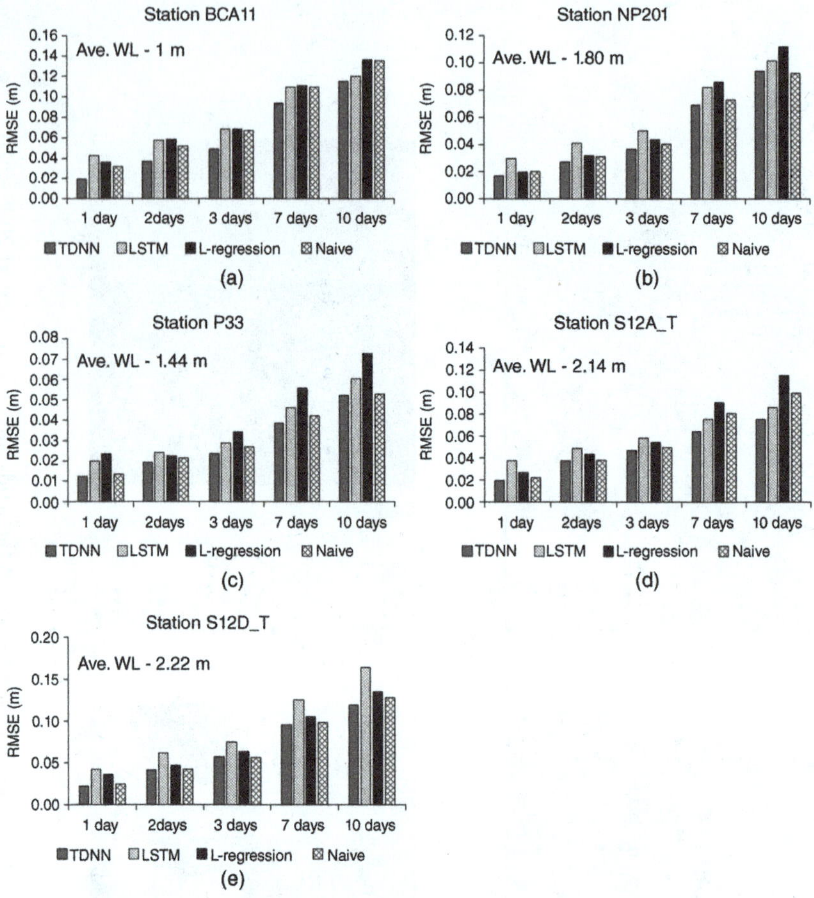

Figure 7.6 Charts comparing the Root Mean Squared Error of the water-level (WL) predictions for each data-driven model at each forecast lead time for each station studied. Charts (a), (b), and (c) relate to the stations *not* influenced by a water-control structure; (d) and (e) are for those influenced by a structure. TDNN = Time-Delayed Neural Network Model; LSTM = Long-Short Term Memory Model; L-regression = Linear Regression Model; Naive = Naive Model.

For both the TDNN and the LSTM models, the nonlinear hyperbolic tangent was used as the activation function for the hidden layer, while the linear transfer function was used as the activation function for the output layer. The sigmoid function was used for the recurrent activation in the LSTM models. The number of hidden nodes for the TDNN was determined by trial and error based on the minimization of the cross-validation error, resulting in six hidden nodes. Trials were made using 6, 8, 16, and 32 nodes for the hidden layer of the LSTM models. The LSTM models for BCA11 and NP201 have 32 nodes in the hidden layer while those for the remaining stations have 8 nodes, based on the trials conducted. The Root Mean Squared Propagation (RMSProp) optimizer was used in this study, and the loss (error) function was the mean squared error. The optimizer is a gradient descent optimization algorithm, which divides the gradient by the moving average of the root mean squared gradient, first proposed and explained in an online course entitled "Neural Networks for Machine Learning" (Tieleman & Hinton, 2012), and is thought to be first used in Graves (2013). More recently, it was used in a study on stochastic optimization and was shown to be beneficial in MLP ANNs, although it is typically used in recurrent neural networks (Huk, 2020).

The LSTM was built similarly to the TDNN using the same number of nodes in the hidden layer, the same optimizer, and the same activation functions. The length of the historical sequence that the model was trained on was 10 days. Trials were made for sequence window lengths of 5, 7, 10, and 20 days.

Cross-validation was carried out using the Keras *EarlyStopping* "callback" (https://www.tensorflow.org/api_docs/python/tf/keras/callbacks/EarlyStopping) feature, which monitors the error (loss) on the cross-validation subset, and terminates training when the error no longer improves after a consecutive number of training loops or epochs. To ensure training did not stop before error was minimized, a "patience" of 100 epochs was specified. This 'patience' means that 100 consecutive epochs need to pass without the error of the cross-validation set improving, before training is terminated. The callback was set up to ensure that the weights that resulted in the lowest cross-validation error were restored after training was completed.

Once configured, each model could be trained and the total number of epochs specified for training was 500. After training was completed, the TDNN models were initially tested using the randomly selected test set, and then were further tested to predict the 3 yr sequential time series. The LSTM models were tested only on the 3 yr sequential time series. The simulation time step for the models is 24 hr. In order to forecast lead times of 7 and 10 days, forecasts had to be made for lead times 4–6 days and 8–9 days, respectively, in order to get the necessary WL inputs as indicated in Table 7.2.

For model comparison, two benchmark models were employed: linear regression and naive models. Linear regression models were developed for each case using the same inputs and loss function. Random walk forecasts were also made based on the naive method. This type of forecast assumes that at every point, a random step is taken away from the previous step, and the distribution of the steps is independent (Pwasong & Sathasivam, 2015). The type of random walk forecast used in this study was the random walk without drift, which means that the mean value of the steps is zero. Simply put, the forecast assumes that the next value is the same as the current value. The performances of these benchmark models were compared with those of the ML models.

7.6. Results and Discussion

7.6.1. Performance of All Data-Driven Models

In comparing all data-driven models (TDNN, LSTM, linear regression and naive), the ML models generally outperformed the simpler models (linear regression and naive) (Fig. 7.6). The TDNN model performed the best in predicting water level at each lead time investigated for station BCA11. This observation is the same for station NP201 with the exception of the 10 day water-level prediction, where the naive model showed better performance than the TDNN model. Similar to station BCA11, the TDNN model for station P33 outperforms the other models at each forecast lead time. The charts of the remaining two stations show similar characteristics. The performance of the TDNN model on the short-term forecasts was similar to or better than the naive model, while it also performed better on the medium-range forecasts. For each lead time prediction for each station, the linear regression model was outperformed by both the TDNN model and the naive model, while the TDNN model generally performed better overall. The LSTM model outperformed the linear regression model in predicting the medium-range forecast horizons for each station with the exception of S12D_T. For stations BCA11 and P33, the LSTM model also performs better than the naive model. The results of the medium-range predictions of the LSTM model, and those of the TDNN model in particular, however, indicate that ML is useful in this study.

7.6.2. Short-Term Forecasts

Station NP201 was used as an example to present the results of the water-level forecasts. The TDNN model accurately predicts the water level 1 day ahead, over the 3 yr sequential test data, with an RMSE of 0.017 m (average water level = 1.8 m) and an R^2 of 0.997. The peaks and continuous changes in water level are well predicted in the first 200 days of the

3 yr prediction, particularly the peaks labeled 1 and 2 (Fig. 7.7c). Up to 3 days lead time, these peaks are well-predicted temporally as well as in magnitude, and can be linked to accurate forecasted rainfall at these points (Fig. 7.7a). Peak 3 is not as accurately predicted as peaks 1 and 2, resulting from small differences in the forecasted rainfall, particularly for the 2 day prediction. This error in the forecasted rainfall leads to a small lag, which is then propagated to the 3 day prediction. These observations show the influence of the rainfall predictors, but more important, how improved rainfall forecasts could result in more accurate predictions.

RMSE for the overall prediction for the 2 day and 3 day lead times were 0.027 m and 0.037 m, respectively, with an R^2 greater than 0.98 in both cases. There is some lag in the prediction to the end of the graph (Fig. 7.7c) resulting in differences between the measured water level and predictions to the magnitude of 0.25 m and 0.20 m in the case of the 3 day prediction at days 143 and 149. This error can be accounted for when considering the error in the forecasted rainfall used. Additionally, the small prediction error in the 1 day forecast would increase the error of the 2 day prediction, as well as the influence of the highly autoregressive water-level inputs, which are the main predictors in these forecasts.

At day 505, a rainfall peak of 37.8 mm resulted in a peak in the water-level prediction, which was not observed in the measured data (Fig. 7.7d). The influence of peaks in rainfall on the water-level prediction appears to be greater when the water level is indicative of a dry season (lowest values in Fig. 7.7c and d) than when the water level is higher during the wet season (indicated by the three highest periods in Fig. 7.7b). This observation also suggests that the observed rainfall was absorbed by the natural system, resulting in no major increase in the measured water level. There are a number of mechanisms by which the system can receive rainfall and not experience water-level increases. The water-control structures are designed to mitigate water levels, and drainage to the coast can vary seasonally. The major trough in the 3 yr prediction, magnified in Figure 7.7d, indicates a particularly dry period as confirmed by the 2009 South Florida Weather Year in Review report (National Weather Forecast Office, 2009) issued by the weather forecast office of the National Weather Service along with the National Oceanographic and Atmospheric Administration (NOAA). The report indicated that the winter (January–April) of 2009 was the driest dry season on record up to that date.

In the 3 day water-level prediction, the water level is overpredicted on days 503 and 504 by magnitudes of 0.45 m and 0.38 m. This overprediction could be a result of rainfall forecasts for day 502 and 503 of 30.2 mm and 27.8 mm, respectively, causing two successive peaks. Note that in the dry season of 2008 (approximately days 0–130) when there is not much

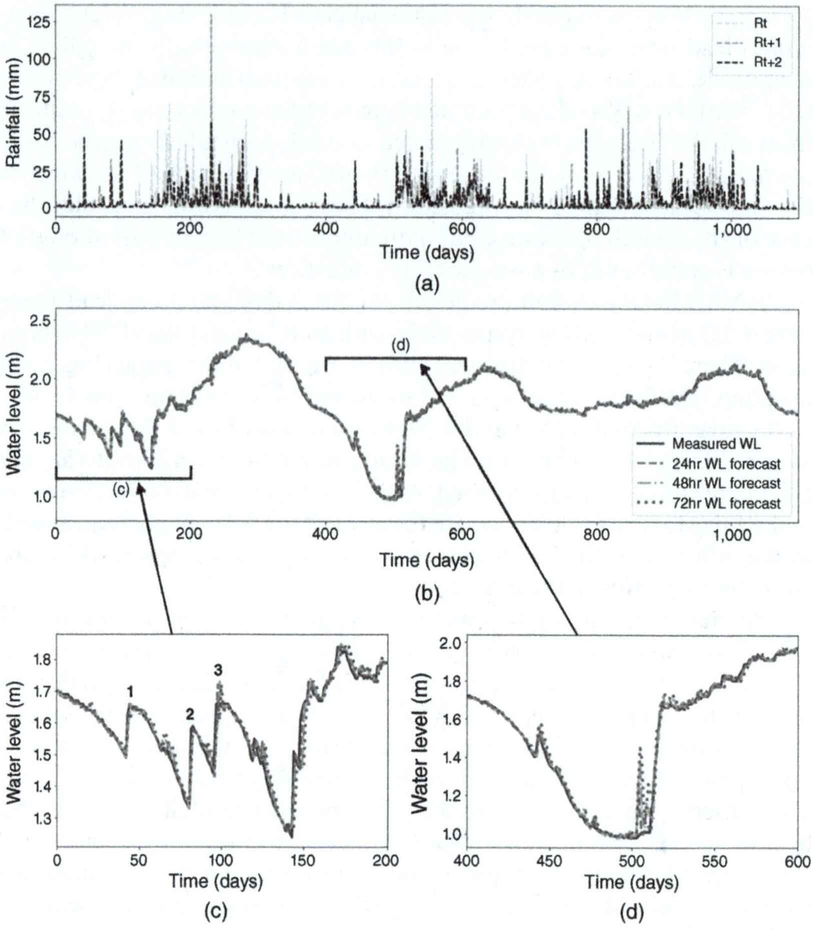

Figure 7.7 Short-term forecasts for station NP201. (a) Rainfall used in the water-level predictions: R_t is measured rainfall; R_{t+1} and R_{t+2} are forecast rainfall; (b) three-year water-level prediction from TDNN model; (c) first 200 days magnified; (d) days 400–600 magnified.

rainfall, the system reacts more to rainfall occurrences and the water level changes somewhat proportionally. During the wet season, this is not the case (Fig. 7.7b). This greater influence of rainfall inputs on the water level during the dry season is likely to have been the case in the training data as well, which would explain why the model may tend to predict a water-level peak where there was no measured peak (Fig. 7.7d). Given that longer-term antecedent conditions affect the response to rainfall, water levels representative of the dry season of 2009 were the lowest values presented to the model, and the characteristics of the reaction of the system during the dry season described prior could be the reason for the overprediction. Apart from this outlier in the predictions, each forecast of the 2009 dry season was good, particularly in capturing the increase in water level entering into the wet season from days 514 to 519 (Fig. 7.7d).

The results of the LSTM model show a lag in the water-level prediction of the 1 day forecast horizon. With each lead time following, the lag increases, suggesting the dominance of the autoregressive water-level predictors on the LSTM model's prediction. This observation specifically occurs around day 150 in Figure 7.8b. The three peaks analyzed in the results of the TDNN model are predicted with less accuracy by the LSTM model, but the timing of peaks 1 and 2 are the same for each short-term forecast lead time, although lagged, compared with the measured data. Because only four trials were made in finding a suitable number of hidden nodes for the LSTM model, a more extensive optimization of the model parameters could potentially improve the accuracy of the water-level forecasts. Characteristics similar to those observed in the predictions of the TDNN model are observed in the predictions of the LSTM model.

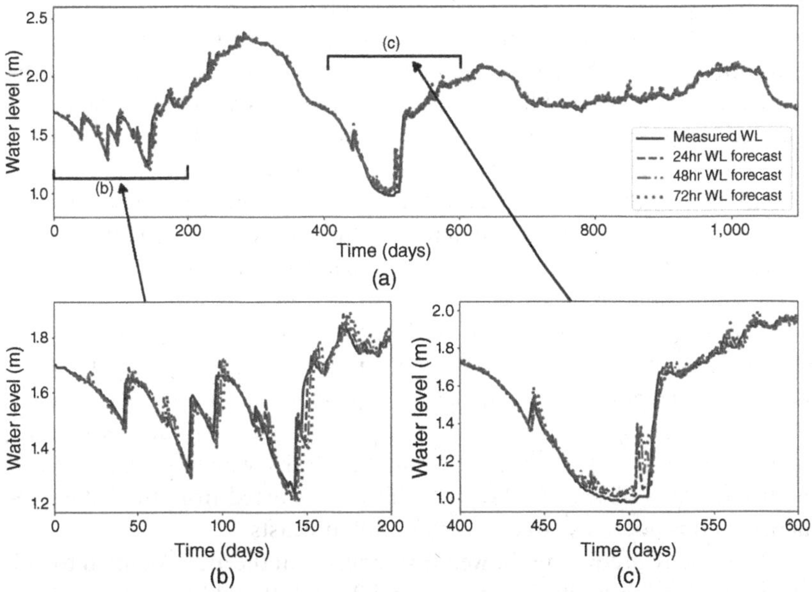

Figure 7.8 Short-term forecasts for station NP201. (a) Three-year water-level prediction from LSTM model; (b) first 200 days magnified; (c) days 400–600 magnified.

7.6.3. Medium-Range Forecasts

As expected with longer forecast horizons, the accuracy is notably reduced in the 7 day and 10 day water-level predictions. Similar to the 2 day and 3 day forecasts, peaks in medium-range predicted water level can be linked to peaks in the rainfall forecast. A combination of the error in the rainfall forecast used to predict both forecast horizons, as well as the propagation of error from the prediction at each lead time, could contribute to the larger error at these lead times. The two medium-range forecasts had RMSE values of 0.069 m and 0.094 m, and R^2 of 0.94 and 0.89, respectively. The general pattern of the measured data is still preserved in the forecasts despite some overprediction of the two latter peaks (Fig. 7.9b).

Even at such a long lead time, some peaks are still well predicted due to reasonably accurate rainfall forecast points. For the 7 day water-level prediction, peak 1 is delayed in comparison to the measured data. Table 7.3 indicates there is no forecasted rainfall to drive the prediction of the peak on time and as a result, it arrives late. Peak 2 occurred a day late, which coincides with a peak in forecasted rainfall being forecasted a day late. Peak 3 was well predicted due to accurate timing and comparable magnitude of the forecasted rainfall to the measured rainfall (Table 7.3). From day 139 to 200, there is significant lag, which causes point errors as large as 0.35 m at around day 143 (Fig. 7.9c). There is very little forecasted rainfall at the point where the measured water level increased and resulted in the model not predicting the increase at that point. This error would have grown from the 2 day forecast horizon, which did not have a comparable forecasted rainfall peak in the measured data. With each lead, the consequent error in the water-level prediction would have grown as a result of prediction errors due to inaccurate rainfall forecasts, and the inherent prediction errors with larger lead time forecasts.

The error of the outlier peak predicted in the dry season of 2009 propagated through each time step, resulting in point prediction errors of as high as 0.70 m at the point where the largest peak in predicted water level occurs, just before the measured water level begins to increase (7 day water-level forecast in Fig. 7.9d). There are two rainfall peaks, which result in the two predicted peaks observed. However, this error started from the 1 day prediction and is not solely due to the rainfall forecasts.

The 10 day prediction showed further error in the areas pointed to in 7 day prediction for similar reasons (Fig. 7.9c and d). Additionally, the forecasts were notably more variable over the first 200 days than in the same

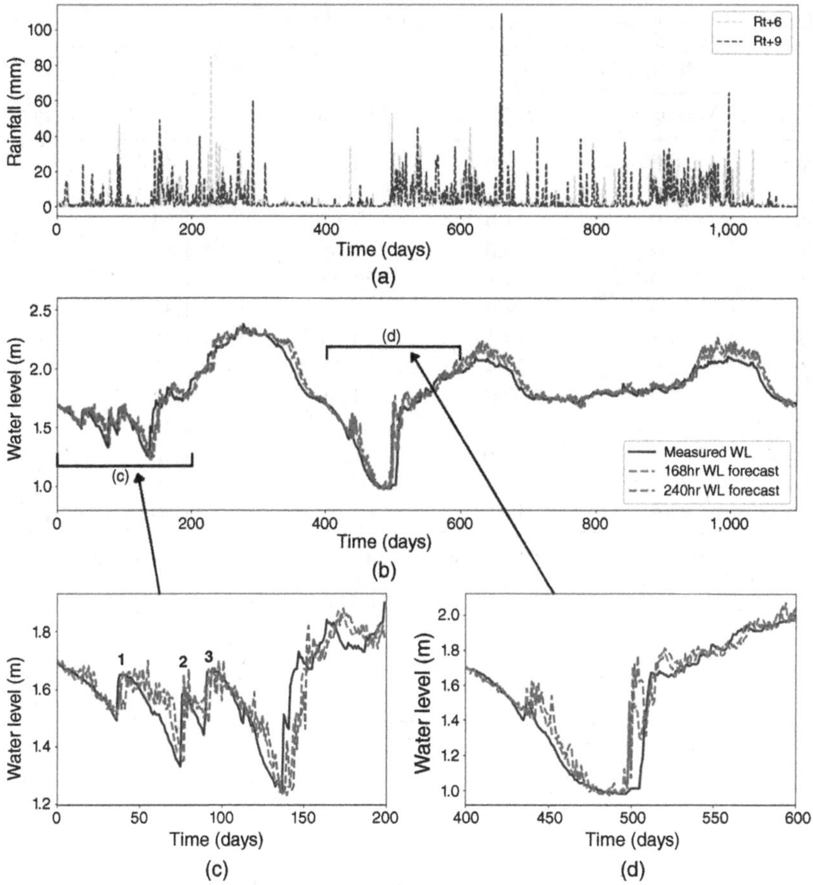

Figure 7.9 Medium-range forecasts for station NP201. (a) Rainfall used in the water-level predictions; (b) three-year water-level prediction for TDNN model; (c) first 200 days magnified; (d) days 400–600 magnified.

time period of the short-term forecasts as a result of more rainfall forecasted (Fig. 7.9c).

The medium-range water-level forecasts of the LSTM model shows more variability in the predictions (also observed in the short-term forecasts for the LSTM), possibly due to the complexity of that model or to the hyperparameters of the model compared with the TDNN. However, the general trends observed in the predictions of the TDNN model are also seen in the predictions of the LSTM model (Fig. 7.10).

Table 7.3 Measured and forecasted rainfall and water level.

	Rainfall at t + 6 (mm)		Water level at t + 7 (m)		
Day	Measured	Forecasted	Measured	Predicted	Peak
37	0.0	0.0	1.49	1.49	1
38	*19.6*	5.3	*1.63*	1.51	
39	18.0	1.9	1.65	1.54	
40	0.0	0.6	1.65	1.6	
76	0.0	1.1	1.33	1.36	2
77	*25.9*	4.6	*1.58*	1.38	
78	4.3	19.1	1.59	1.65	
79	0.0	21.4	1.58	1.53	
80	0.0	0.0	1.57	1.54	
91	0.0	0.0	1.52	1.49	3
92	*26.7*	30.0	*1.66*	1.60	
93	*39.4*	46.3	*1.66*	1.66	
94	0.0	13.1	1.66	1.60	

Note: Measured and forecasted rainfall and water level corresponding to the peaks in water level for the 7-day prediction. The underlined values show the data points of high rainfall, and the corresponding water-level peaks. The shaded values show forecasted rainfall peaks and the corresponding water-level peaks except for peak 1.

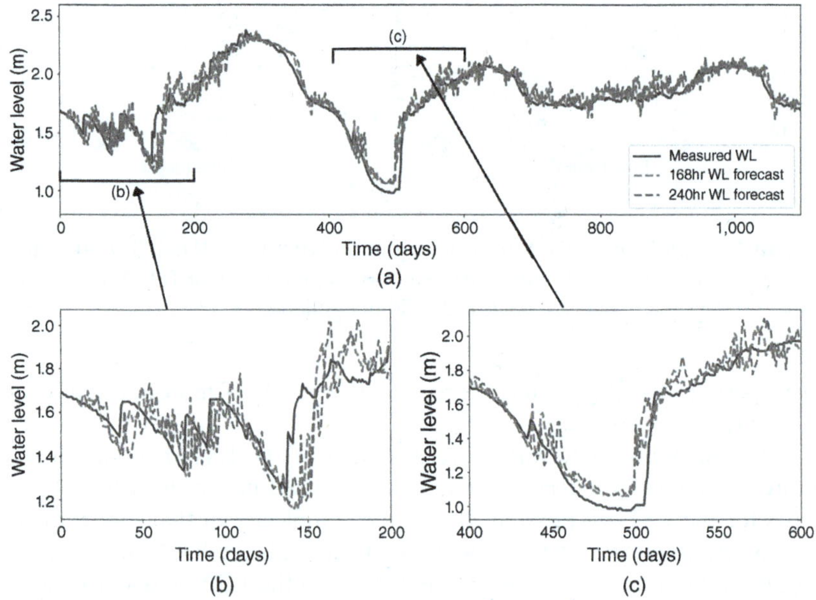

Figure 7.10 Medium-range forecasts for station NP201. (a) Three-year water-level prediction for LSTM model; (b) first 200 days magnified; (c) days 400–600 magnified.

Since the two ML models have different training sets based on the data partitioning, the references made between the two do not represent a true comparison. As the main aim of the research is to compare them with a numerical model and not each other, the references made were for the purpose of discussing the performance of both models and not to determine which ML model is better.

The characteristics observed in the results presented for both the short-term and medium-range water-level forecasts at station NP201 were also observed in the results for the other stations. The performances of these models are presented in Appendix Tables 7.A1 and 7.A2.

7.6.4. Comparison of the ML Model With the Numerical Model

The BISECT model (Swain & Lohmann, 2019) simulates surface water on a 10 min time step, so daily average water-level predictions are used for comparison in this study. When comparing the daily water-level prediction from each model, the results are promising as they relate to water-level forecasting in the southern Everglades. The ML models are able to predict peaks and troughs much better than the numerical model (Fig. 7.11). The sections of the BISECT model water-level prediction that correspond to troughs in the graphs of the measured data appear to be flattened. This flattening of the troughs is because only the surface water component of the BISECT model was considered for comparison, because the ML models also are solely based on surface water. Regardless, the peaks are much more accurately predicted by the ML models and would suggest that, were the groundwater component of the BISECT model added, its overall performance would improve; but the ML models may still predict more accurately. The BISECT model can have a bias in water levels, which varies from station to station, such as the low overall mean values seen at station NP201 in Figure 7.11. The causes of some of these biases are unknown, but others are from uncertainties in the numerical grid formulation, where a good model fit at one location can correspond to a poor fit at another location. The ML models represent locations independent of each other, so adjusting for a better fit at one location does not affect other locations. The BISECT model is able to capture the general seasonal variations and fluctuations in water level, however it is less precise than the ML models, due to the properties of the numerical model described earlier, and how its inputs and outputs are processed.

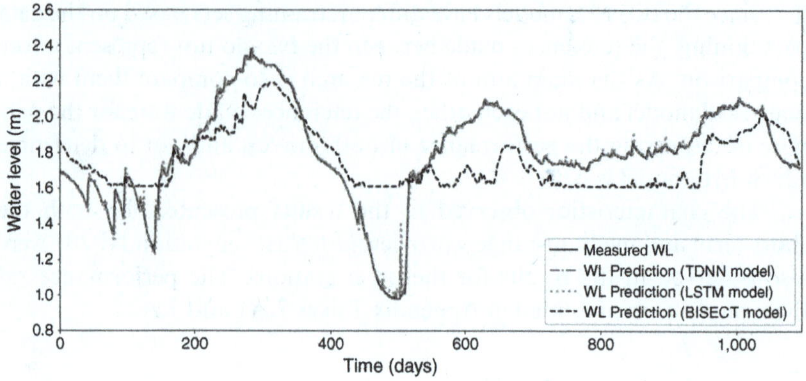

Figure 7.11 A 24 hr water-level forecast comparison of ANN and BISECT model outputs to measured water level at station NP201.

Similar characteristics are observed when comparing the output of the BISECT model to the ML water-level prediction for the remaining four stations investigated. In each case, the ML models outperformed the BISECT model in predicting water level 1 day ahead.

An improvement in water-level forecasting could be of significant importance in the southern Everglades where such a vast water-control system exists. The ML models proved to be better than the BISECT model in predicting water level 1 day ahead for each location investigated and could complement the BISECT model in monitoring and analyzing the region's hydrology. Applications where water-level forecasts are required could benefit from these more accurate local forecasts, increasing the quality and accuracy of ecological studies and assessments, and aid in the CERP.

7.7. Conclusions and Future Directions

The following itemizes conclusions and future directions.

1. Machine Learning is a good complement to the conventional numerical models used in the southern Everglades for predicting water levels and thus hydroperiods for water manager and ecologic studies. The accuracy of water-level predictions from both ML models improved when compared with an existing numerical model.

2. A rational choice of predictors that affect water level positively contributed to the performance of ML models. Also, the water level inputs are the most influential variables on the predictions.
3. Incorporation of weather predictions from numerical models as inputs to ML models improved water-level forecasting, particularly at greater lead times (7 and 10 days). Skill and accuracy of the numerical weather models is a limiting factor in achieving accuracy in water-level forecasts.
4. Incorporation of volumetric discharge releases from water-control structures into the ML models (where applicable) had a positive effect on the accuracy of water-level forecasts.

A future focus on improving the accuracy of weather forecast models or selecting weather models that perform well at forecasting rainfall could be beneficial. A local model for south Florida based on the Weather Research and Forecasting (WRF) code (https://www.mmm.ucar.edu/weather-research-and-forecasting-model) could provide some improvement in the accuracy of the rainfall forecasts because this is a mesoscale model and as a result has a higher resolution than the global ECMWF model. Rainfall forecasts from the WRF model could be corrected using gauged data through different methods. Additionally, using future scheduled releases instead of measured ones at each forecast horizon could lead to improvements in system management through better tailoring of these releases based on the predictions. Finally, medium-range forecasts could explore the use of ensemble (committee) models, as were used by Fenicia et al. (2007), Kayastha et al. (2013), and Zhang et al. (2022). Ensemble models can be created by classifying water levels into several classes (e.g., low, medium, and high), training several ML models, each responsible for its own class of data, and then combining them using a dynamic weighting scheme (Zhang et al., 2018; Zhang et al, 2022, equation (6)). Such an approach typically leads to higher performance from a combined ensemble of many ML models with different strengths and weaknesses.

Disclaimer

Any use of trade, firm, or product names is for descriptive purposes only and does not imply endorsement by the U. S. government.

Appendix

Table 7.A1 Performance indicators for each station specific TDNN model investigated.

Lead time (hr)	BCA11 R2	BCA11 RMSE (m)	NP201 R2	NP201 RMSE (m)	P33 R2	P33 RMSE (m)	S12A_T R2	S12A_T RMSE (m)	S12D_T R2	S12D_T RMSE (m)
24	0.994	0.021	0.997	0.017	0.994	0.012	0.993	0.021	0.997	0.022
48	0.981	0.039	0.991	0.027	0.986	0.019	0.977	0.038	0.990	0.039
72	0.968	0.051	0.983	0.037	0.979	0.024	0.964	0.047	0.981	0.052
168	0.892	0.093	0.940	0.069	0.944	0.039	0.933	0.065	0.939	0.094
240	0.835	0.115	0.888	0.094	0.897	0.052	0.907	0.076	0.903	0.119

Table 7.A2 Performance indicators for each station specific LSTM model investigated.

Lead time (hr)	BCA11 R2	BCA11 RMSE (m)	NP201 R2	NP201 RMSE (m)	P33 R2	P33 RMSE (m)	S12A_T R2	S12A_T RMSE (m)	S12D_T R2	S12D_T RMSE (m)
24	0.977	0.043	0.989	0.030	0.985	0.020	0.978	0.038	0.988	0.042
48	0.958	0.058	0.979	0.041	0.978	0.024	0.962	0.049	0.974	0.061
72	0.941	0.069	0.968	0.050	0.970	0.028	0.946	0.058	0.961	0.075
168	0.851	0.109	0.915	0.082	0.919	0.046	0.909	0.076	0.890	0.126
240	0.819	0.121	0.870	0.102	0.873	0.058	0.895	0.082	0.812	0.164

Note: The average water level for each station over the prediction period at every lead time was BCA11 = 1.00 m; NP201 = 1.80 m; P33 = 1.44 m; S12A_T = 2.14 m; S12D_T = 2.22 m.

References

Abadi, M., Agarwal, A., Barham, P., Brevdo, E., Chen, Z., Citro, C., et al. (2016a). TensorFlow: Large-scale machine learning on heterogeneous distributed systems. *ArXiv*, abs/1603.04467. https://doi.org/10.48550/arXiv.1603.04467

Abadi, M., Barham, P., Chen, J., Chen, Z., Davis, A., Dean, J., et al. (2016b). TensorFlow: A system for large-scale machine learning. *12th USENIX Symposium on Operating Systems Design and Implementation, OSDI '16* (pp. 265–283), Savannah, Georgia, USA, 2–4 November 2016. https://dl.acm.org/doi/10.5555/3026877.3026899

Abrahart, R. J., Anctil, F., Coulibaly, P., Dawson, C. W., Mount, N. J., See, L. M., et al. (2012). Two decades of anarchy? Emerging themes and outstanding challenges for neural network river forecasting. *Progress in Physical Geography: Earth and Environment*, *36*(4), 480–513. https://doi.org/10.1177/0309133312444943

Aggarwal, C. C. (2018) An introduction to neural networks. In *Neural networks and deep learning* (pp. 1–51). Switzerland: Springer, Cham.

Alpaydin, E. (2009). *Introduction to machine learning* (2d ed.). MIT Press.

Amaranto, A., Munoz-Arriola, F., Corzo, G., Solomatine, D. P., & Meyer, G. (2018) Semi-seasonal groundwater forecast using multiple data-driven models in an irrigated cropland. *Journal of Hydroinformatics*, *20*(6), 1227–1246. https://doi.org/10.2166/hydro.2018.002

Bhattacharya, B., & Solomatine, D. P. (2005). Neural networks and M5 model trees in modeling water level discharge relationship. *Neurocomputing*, *63*, 381–396. https://doi.org/10.1016/j.neucom.2004.04.016

Bowden, G. J., Maier, H. R., & Dandy, G. C. (2002). Optimal division of data for neural network models in water resources applications. *Water Resources Research*, *38*(2), 2-1–2-11. https://doi.org/10.1029/2001WR000266

Brandt, L. A., Beauchamp, J. S., Jeffery, B. M., Cherkiss, M. S., & Mazzotti, F. J. (2016). Fluctuating water depths affect American alligator (*Alligator mississippiensis*) body condition in the Everglades, Florida, USA. *Ecological Indicators*, *67*, 441–450. https://doi.org/10.1016/j.ecolind.2016.03.003

Chollet, F. (2015). Keras,*GitHub respository.* https://github.com/keras-team/keras

Conrads, P.A., & Roehl, E. A., Jr. (2007). Hydrologic record extension of water-level data in the Everglades Depth Estimation Network (EDEN) using artificial neural network models, 2000–2006. U.S. Geological Survey Open-File Report 2007-1350. https://doi.org/10.3133/ofr20071350

Conrads, P. A., Petkewich, M. D., O'Reilly, A. M., & Telis, P. A. (2014). *Hydrologic record extension of water-level data in the Everglades Depth Estimation Network (EDEN), 1991–99.* U. S. Geological Survey Scientific Investigations Report 2014–5226. https://doi.org/10.3133/sir20145226

Cruz, F. R. G., Binag, M. G., Ga, M. R. G., & Uy, F. A. A. (*2018*). Flood prediction using multi-layer artificial neural network in monitoring system with rain gauge, water level, soil moisture sensors. *TENCON* 2018–2018 *IEEE Region 10 Conference*, Jeju, *Korea (South)*, 2018. https://doi.org/10.1109/TENCON.2018.8650387

Eesa, A., & Arabo, W. (2017). A normalization method for backpropagation: A comparative study. *Science Journal of University of Zakho*, *5*(4), 319–323. https://doi.org/10.25271/2017.5.4.381

Fenicia, F., Solomatine, D. P., Savenije, H. H. G., & Matgen, P. (2007). Soft combination of local models in a multi-objective framework. *Hydrology and Earth System Sciences*, *11*(6), 1797–1809. https://doi.org/10.5194/hess-11-1797-2007

Forde, C. S. (2020). Forecasting water level using artificial neural networks to support water resource management and ecosystem restoration: a case study in the

Florida Everglades. Master of Science Thesis, IHE Delft Institute for Water Education, Delft, the Netherlands, April 2020.

Frederick, P., Gawlik, D. E., Ogden, J. C., Cook, M. I., & Lusk, M. (2009). The White Ibis and Wood Stork as indicators for restoration of the Everglades ecosystem. *Ecological Indicators*, *9*(6), S83–S95. https://doi.org/10.1016/j.ecolind.2008.10.012

Fujisaki, I., Rice, K. G., Pearlstine, L. G., & Mazzotti, F. J. (2009). Relationship between body condition of American alligators and water depth in the Everglades, *Florida. Hydrobiologia*, *635*(1), 329–338. https://doi.org/10.1007/s10750-009-9925-3

Graves, A. (2012). Long short-term memory. In *Supervised sequence labeling with recurrent neural networks* (pp. 5–13). Berlin, Heidelberg: Springer. https://doi.org/10.1007/978-3-642-24797-2_2

Graves, A. (2013). Generating sequences with recurrent neural networks. *ArXiv:1308*.0850. https://doi.org/10.48550/arXiv.1308.0850

Haman, D., & Svendsen, M. (2006). Managing the Florida Everglades: Changing values, changing policies. *Irrigation and Drainage Systems*, *20*(2), 283–302. https://doi.org/10.1007/s10795-006-9008-9

Hochreiter, S., & Schmidhuber, J. (1997). Long short-term memory. *Neural Computation*, *9*(8), 1735–1780. https://doi.org/10.1162/neco.1997.9.8.1735

Huk, M. (2020). Stochastic optimization of contextual neural networks with RMSprop. In N. T. Nguyen, et al. (Eds.), *Intelligent information and database systems. ACIIDS 2020*. Lecture Notes in Computer Science, vol. *12034*. Springer, Cham. https://doi.org/10.1007/978-3-030-42058-1_29

Ingebritsen, S. E., McVoy, C., Glaz, B., & Park, W. (1999). Florida Everglades: Subsidence threatens agriculture and complicates ecosystem restoration. In D. Galloway et al. (Eds.), *Land subsidence in the United States* (pp. 95–106). U.S. Geological Survey Circular 1182, U.S. Department of the Interior, U.S. Geological Survey, Reston, Virginia, USA. https://pubs.usgs.gov/circ/circ1182/

Kao, I., Zhou, Y., Chang, L. & Chang, F. (2020). Exploring a long short-term memory based encoder-decoder framework for multi-step-ahead flood forecasting. *Journal of Hydrology*, *583*, 124631. https://doi.org/10.1016/j.jhydrol.2020.124631

Kasiviswanathan, K. S., He, J., Sudheer, K. P., & Tay, J. (2016). Potential application of wavelet neural network ensemble to forecast streamflow for flood management. *Journal of Hydrology, 536*, 161–173. https://doi.org/10.1016/j.jhydrol.2016.02.044

Kayastha, N., Ye, J., Fenicia, F., Kuzmin, V., & Solomatine, D. P. (2013). Fuzzy committees of specialized rainfall-runoff models: further enhancements and tests. *Hydrology and Earth System Sciences*, *17*(11), 4441–4451. https://doi.org/10.5194/hess-17-4441-2013

Kratzert, F., Klotz, D., Brenner, C., Schulz, K. & Herrnegger, M. (2018). Rainfall-runoff modeling using long short-term memory (LSTM) networks. *Hydrology and Earth System Sciences*, *22*(11), 6005–6022. https://doi.org/10.5194/hess-22-6005-2018

Lantz, S. M., Gawlik, D. E., & Cook, M. I. (2010). The effects of water depth and submerged aquatic vegetation on the selection of foraging habitat and foraging success of wading birds. *The Condor*, *112*(3), 460–469. https://doi.org/10.1525/cond.2010.090167

Light, S. S., & Dineen, J. W. (1994). Water control in the Everglades. In S. M. Davis & J. C. Ogden (Eds.), *Everglades: The ecosystem and its restoration* (pp. 47–84). Boca Raton, FL: CRC Press.

Lingras, P., Zhong, M., & Sharma, S. (2008). Evolutionary regression and neural imputations of missing values. In *Soft computing applications in industry* (pp. 151–163). Berlin, Heidelberg: Springer. https://doi.org/10.1007/978-3-540-77465-5_8

Luk, K. C., Ball, J. E., & Sharma, A. (2001). An application of artificial neural networks for rainfall forecasting. *Mathematical and Computer Modeling*, *33*(6–7), 683–693. https://doi.org/10.1016/S0895-7177(00)00272-7

Ma, Y., Matta, E., Meißner, D., Schellenberg, H., & Hinkelmann, R. (2019). Can machine learning improve the accuracy of water-level forecasts for inland navigation? Case study: Rhine River Basin, Germany. Proceedings of the 38th IAHR World Congress Panama City 2019, Water-Connecting the World, Panama City, Panama, 1–6 September 2019. https://doi.org/10.3850/38WC092019-0274

Maltby, E., & Dugan, P. J. (1997). Wetland ecosystem protection, management, and restoration: an international perspective. In S. M. Davis & J. C. Ogden (Eds.), *Everglades: The ecosystem and its restoration* (pp. 29–46). Boca Raton, FL: CRC Press.

Moreido, V., Gartsman, B., Solomatine, D. P., & Suchilina, Z. (2021). How well can machine learning models perform without hydrologists? Application of rational feature selection to improve hydrological forecasting. *Water*, *13*(12), 1696. https://doi.org/10.3390/w13121696

National Weather Forecast Office (2009). 2009 south Florida weather year in review. https://www.weather.gov/media/mfl/news/2009WeatherSummary.pdf

Park, J., Obeysekera, J., & VanZee, R. (2005). Prediction boundaries and forecasting of non-linear hydrologic stage data. *Journal of Hydrology*, *312*(1–4), 79–94. https://doi.org/10.1016/j.jhydrol.2005.02.022

Perry, W. B. (2008). Everglades restoration and water quality challenges in south Florida. *Ecotoxicology*, *17*(7), 569. https://doi.org/10.1007/s10646-008-0240-y

Pu, Z., & Kalnay, E. (2019). Numerical weather prediction basics: Models, numerical methods, and data assimilation. In Q. Duan et al. (Eds.), *Handbook of hydrometeorological ensemble forecasting* (pp. 67–97). Berlin Heidelberg: Springer. https://doi.org/10.1007/978-3-642-39925-1_11

Pwasong, A. D., & Sathasivam, S. (2015). Forecasting performance of random walk with drift and feed forward neural network models. *International Journal of Intelligent Systems and Applications (IJISA)*, *7*(9), 49–56. https://doi.org/10.5815/ijisa.2015.09.07

Reitermanova, Z. (2010). Data splitting. WDS'10 Proceedings of Contributed Papers, Part 1, Prague, Czech Republic, 1–4 June 2010. extension://elhekieabhbkpmcefcoobjddigjcaadp/viewer.html?pdfurl=https%3A%2F%2Fwww.mff.cuni.cz%2Fveda%2Fkonference%2Fwds%2Fproc%2Fpdf10%2FWDS10_105_i1_Reitermanova.pdf&clen=155228&chunk=true

Sherstinsky, A. (2020). Fundamentals of Recurrent Neural Network (RNN) and Long Short-Term Memory (LSTM) network. *Physica D: Nonlinear Phenomena, 404*, 132306. https://doi.org/10.1016/j.physd.2019.132306

Siami-Namini, S., Tavakoli, N., & Namin, A. S. (2018). *A comparison of ARIMA and LSTM in forecasting time series*. 2018 17th IEEE International Conference on Machine Learning and Applications (ICMLA), 17–20 December 2018. https://doi.org/10.1109/ICMLA.2018.00227

Sobel, A. H. (2012). Tropical weather. *Nature Education Knowledge*, *3*(12) 2. https://www.nature.com/scitable/knowledge/library/tropical-weather-84224797/

Solomatine, D. P., & Dulal, K. N. (2003). Model trees as an alternative to neural networks in rainfall-runoff modeling. *Hydrological Sciences Journal*, *48*(3), 399–411. https://doi.org/10.1623/hysj.48.3.399.45291

South Florida Water Management District (SFWMD) (2005). Documentation of the South Florida Water Management Model Version 5.5. https://www.sfwmd.gov/sites/default/files/documents/sfwmm_final_121605.pdf

South Florida Water Management District (SFWMD) (2006). Regional simulation model (RSM) fact sheet. https://www.sfwmd.gov/sites/default/files/documents/rsmsplash_2006june1.pdf

South Florida Water Management District (SFWMD). (2008). Regional simulation model (RSM), model overview. https://www.sfwmd.gov/sites/default/files/documents/rsm_module_1_0.pdf

South Florida Water Management District (SFWMD) (2021). 2020 statewide annual report on total maximum daily loads, basin management action plans, minimum flows or minimum water levels, and recovery or prevention strategies. https://fdep.maps.arcgis.com/apps/MapSeries/index.html?appid=07030d9388134673a538ad27455b3029

Swain, E. D., & Lohmann, M. A. (2019). *FTLOADDS (combined SWIFT2D surface-water model and SEAWAT groundwater model) simulator used to assess proposed sea-level rise response and water-resource management plans for the hydrologic system of the South Florida Peninsula for the Biscayne and Southern Everglades Coastal Transport (BISECT) Model*. U.S. *Geological Survey Data Release*. https://doi.org/10.5066/P9MDUQPK

Swain, E. D., Lohmann, M. A., & Goodwin, C. R. (2019). The hydrologic system of the South Florida Peninsula: Development and application of the Biscayne and Southern Everglades Coastal Transport (BISECT) Model. *U.S. Geological Survey Scientific Investigations Report 2019–5045*. https://doi.org/10.3133/sir20195045

Telis, P. A. (2006). The Everglades Depth Estimation Network (EDEN) for support of ecological and biological assessments. *U.S. Geological Survey Fact Sheet 2006–3087*. https://doi.org/10.3133/fs20063087

Telis, P. A., Xie, Z., Liu, Z., Li, Y., & Conrads, P. (2015). The Everglades Depth Estimation Network (EDEN) surface-water model, version 2. *U.S. Geological Survey Scientific Investigations Report 2014–5209*. https://doi.org/10.3133/sir20145209

Tieleman, T., & Hinton, G. (2012). Lecture 6.5-rmsprop: Divide the gradient by a running average of its recent magnitude. *COURSERA: Neural Networks for Machine Learning*, *4*(2), 26–31.

Xiang, Z., Yan, J., & Demir, I. (2020). A rainfall-runoff model with LSTM-based sequence-to-sequence learning. *Water Resources Research*, *56*, e2019WR025326. https://doi.org/10.1029/2019WR025326

Zhang, C., Brodylo, D., Rahman, M., Rahman, M. A., Douglas, T. A., & Comas, X. (2022). Using an object-based machine learning ensemble approach to upscale evapotranspiration measured from Eddy Covariance Towers in a subtropical wetland. *Science of the Total Environment*, *831*, 154969. https://doi.org/10.1016/j.scitotenv.2022.154969

Zhang, C., Denka, S., Cooper, H., & Mishra, D. R. (2018). Quantification of sawgrass marsh aboveground biomass in the coastal Everglades using object-based ensemble analysis and Landsat data. *Remote Sensing and Environment*, *204*, 366–379. https://doi.org/10.1016/j.rse.2017.10.018

Zhu, S., Hrnjica, B., Ptak, M., Choianski, A., & Sivakumar, B. (2020). Forecasting of water level in multiple temperate lakes using machine learning models. *Journal of Hydrology*, *585*, 124819. https://doi.org/10.1016/j.jhydrol.2020.124819

8

APPLICATION OF A MULTILAYER PERCEPTRON ARTIFICIAL NEURAL NETWORK (MLP-ANN) IN HYDROLOGICAL FORECASTING IN EL SALVADOR

Jose Valles

IHE Delft Institute for Water Education, Delft, The Netherlands

The Grande de San Miguel catchment in El Salvador generates floods almost every year and produces significant damage to transport infrastructure, houses, and agricultural land, as well as causing people to evacuate. To anticipate flood conditions, an early warning system (EWS) has been in place since 2001 in collaboration with local inhabitants, civil protection, and the National Hydrometeorological Service, but a rainfall-runoff forecast model has not been implemented in the catchment to predict these flood events. In this study, a Multilayer Perceptron Artificial Neural Network (MLP-ANN) model is proposed as an operational flood forecast model for this catchment to address the short-term flood forecast. The model results show excellent performance for lead time no larger than 4 hours and good results for lead time between 5 and 9 hours, but an unsatisfactory result for a forecast lead time of 12 hours since there are some high-flow events that the model underestimates.

Advanced Hydroinformatics: Machine Learning and Optimization for Water Resources, Special Publications 78, First Edition. Edited by Gerald A. Corzo Perez and Dimitri P. Solomantine.

DOI: 10.1002/9781119639268.ch8

8.1. Introduction

El Salvador has been implementing the Flood Early Warning System since the impact of Hurricane Mitch in 1998. After this hydrometeorological event, many governmental and nongovernmental organizations recognized the flood vulnerability of the country and the lack of a suitable hydrometeorological monitoring system. As a result of that, many automatic hydrometeorological gauges were installed, flood emergency plans were implemented, and a flood forecast center was created in order to both monitor and predict the floods impacts.

One the major components of the early warning system (WMO, 2018) is the monitoring and the flood forecasting, which allows to anticipate their impacts and reduce the properties damage and the loss of lives. A hydrological model constitutes the basis of a flood forecasting system. For this reason, many hydrological models have been built to predict the hydrological response in important catchment in El Salvador. For instance, Häggström and Lindström (1990) calibrate the conceptual HBV model to six rivers in Central America, which includes Grande de San Miguel catchment in El Salvador. They found that the calibration of the model was not satisfactory for Grande de San Miguel catchment, although the model has succeeded in simulating most of the flood peaks. In 2004, the Hydrological National Service of El Salvador (2004) reported a rainfall-runoff model for forecasting purposes in the Lempa catchment, which includes the four hydropower reservoir located in the main stream. Additionally, Riverside Technology (2012) implemented a semidistributed Sacramento rainfall-runoff model in Lake Ilopango and Jiboa Catchment, which was implemented in a Delft-FEWS standalone version, and integrated it with real-time and forecast data. In this research, the implementation of the Multilayer Perceptron Artificial Neural Network (MLP-ANN) technique is proposed as operational forecast model in order to address the short-range hydrological flood forecast in Grande de San Miguel river basin. This catchment generates floods almost every year and generates significant damage to transport infrastructure, houses, crop areas, and people evacuated. There is not an operational rainfall-runoff forecast model in this catchment that allows local forecast operators, stakeholders, and decision makers to foresee the flood conditions and increase the lead time for communities located in flood-prone areas to respond efficiently.

The implementation of the MLP-ANN forecast model in Grande de San Miguel catchment is outlined as follows. Section 8.2 provides a general introduction of data-driven modeling, in particular the Multilayer Perceptron Artificial Neural Network (MLP-ANN) technique. Section 8.3 describe the Grande de San Miguel river basin in El Salvador. Section 8.4

provides the methodology used in this research. Section 8.5 defines the experimental setups of the MLP-ANN models for different lead times. The results and discussion are presented in section 8.6. Moreover, the conceptual design of the operational MLP-ANNs model is presented in section 8.7. Finally, the relevant conclusion of this study is presented in section 8.8.

8.2. Data-Driven Modeling Technique

Solomatine and Wagener (2011) defined a Data-Driven Model (DDM) as a model connecting the system states variables (input, internal, and output variable) and built on the basis of collected data, with only a limited number of assumptions about the physical behavior of the system. These models can be used to solve prediction problems, reconstruct highly nonlinear functions, perform classification, cluster data, and build rule-based systems. However, a general disadvantage of DDM is that the model performance depends on the amount of input data. (Chacon-Hurtado, 2012). Furthermore, many techniques in DDM are taken from various areas, such as Machine Learning (ML), Soft Computing (SC), Data Mining, Statistical Methods, Computational Intelligence, and knowledge discovery in database.

The modeling process in a Data-Driven Model (DDM) follows the same principles as the conceptual and physically based models: study the problem, collect the data, select model structure, build the model, test the model. However, Solomatine and Wagener (2011) mentioned that the model parameters and structure are often subject to optimization, which is a difference between DDM and physically based models.

Dawson and Wilby (2001) detailed a number of decisions that have to be made before applying Artificial Neural Network. First, one must select the appropriate ANN type. Secondly, one must select the training algorithm. Thirdly, one must decide how the input and output data should be processed.

Many applications of Data-Driven Models are known in the flow forecasting and rainfall-runoff modeling area, for example Hsu et al. (1995); Minns and Hall (1996), Abrahart and See (2000), Coulibaly et al. (2000), Dawson and Wilby (2001), Rajurkar et al. (2002), Solomatine and Dulal (2003), Wang et al. (2006), Abrahart and See (2007), Hu et al (2007), Solomatine et al. (2008), Wu and Chau (2010), Govindaraju and Rao (2013), Badrzadeh et al. (2015), Solomatine and Siek (2006), Ahani et al. (2018). Most of these authors used Artificial Neural Network (ANN), as Data-Driven method for rainfall-runoff modeling allows reproducing better the nonlinear relationship between rainfall and discharge, as well as handling incomplete, noisy, and ambiguous data (Maier & Dandy, 1996).

In the following sections, a brief description is provided about the Artificial Neural Network architecture, the training algorithm, and the choice of model variable (i.e., input and output).

8.2.1. Artificial Neural Network (ANN) Architecture

One of the primary mainstream technologies for data-driven modeling is Artificial Neural Networks (ANN). This method can be used in the fields of clustering, classification, and prediction. ANNs loosely imitate the functionality of neurons in a human's brain and combines these neurons to reproduce any multivariable or multivalue function (Solomatine, 2016). Figure 8.1 presents the Artificial Neural Network architecture (a) and the structure of an individual neuron (b), which consists of a number of inputs (x), an activation function (i.e., nonlinear function) computed based on the weighted sum of all the inputs and outputs. This activation function may be a logistic sigmoid, Gaussian or hyperbolic tangent function. The selection of the function depends on the network type and training algorithm employed (Dawson & Wilby, 2001). In the majority of Artificial Neural Network studies in hydrology, the logistic sigmoid (equation (8.1)) function and the hyperbolic tangent functions (equation (8.2)) are used, in which u represents the weighted sum of inputs to the neuron and f(x) is the neuron output.

$$f(\mathrm{u}) = \frac{1}{1 + \mathrm{e}^{-\mathrm{u}}}, \tag{8.1}$$

$$f(\mathrm{u}) = \tanh^{-1}(\mathrm{u}) = \frac{\mathrm{e}^{\mathrm{u}} - \mathrm{e}^{-\mathrm{u}}}{\mathrm{e}^{\mathrm{u}} + \mathrm{e}^{-\mathrm{u}}}. \tag{8.2}$$

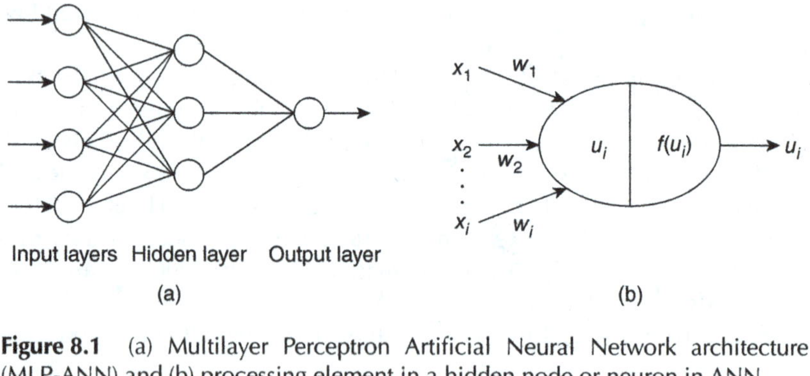

Figure 8.1 (a) Multilayer Perceptron Artificial Neural Network architecture (MLP-ANN) and (b) processing element in a hidden node or neuron in ANN.

The connection between neurons that flows in one direction is called feed-forward networks. This flow direction goes from the input layer, through hidden layers that contain the neurons, to the output layers (see Fig. 8.1). There are two feed-forward network types have been used by many authors in rainfall-runoff modeling: The Multilayer Perceptron (MLP) and the Radial-Basis function (RBF). The MLP is defined by Solomatine and Wagener (2011) as a device that consists of several layers of mutually interconnected neurons, which transform the inputs using a multiparameter nonlinear transformation so the resulting model is capable of approximate complex input-output relationships. The objective of the training process is to optimize the number of neurons and hidden layers to which the network is applied using a trial and error procedure (Shamseldin, 1997). If there are too few neurons in the hidden layer, the Artificial Neural Network may not be capable of reproducing the output function. On the contrary, if there are too many neurons, the ANN may overfit the output function, losing the ability to generalize (Dawson & Wilby, 2001). The Radial-Basis function is a feed-forward ANN, similar in topology to the MLP (Fernando & Jayawardena, 1998), which tries to approximate a function $y = f(x)$ by a superposition of other functions $F(x,\sigma)$, where σ is a parameter that characterizes the width of the function in the input space (Solomatine & Wagener, 2011). Typically, Gaussian functions are used as RBF functions. More details on RBF networks can be seen in Dayhoff (1990), Fernando and Jayawardena (1998), Dawson and Wilby (2001), Sudheer and Jain (2003), and Solomatine and Wagener (2011).

8.2.2. Data Partitioning

Data analysis and preparation is an important procedure in developing hydrological models. In particular, a DDM Rainfall-Runoff model requires that the complete data set should be separated into three different data sets: training (i.e., calibration), cross-validation, and verification. This particular data partitioning in DDM models is quite different from the standard procedure for calibration in conceptual and physically based models since the data are split into two data sets: calibration and verification. The cross-validation data set serves as a small test set, which avoids overfitting of the function and stops the training process. The verification data set serves as testing set, which allows seeing how the model would perform if new data were fed into it. Nevertheless, Lachtermacher and Fuller (1994) identified the following problems in three data sets partitioning.

1. Difficulties in creating three independent data sets in cases where limited data are encountered. In hydrology, it is difficult to split the data in

independent and statistically similar data sets, so the modeler should try to ensure similarity in the distribution.

2. The model results can be affected by the method of partitioning.
3. When using a cross-validation data set, it is not always clear when the model is learning the noise of the data.

8.2.3. Choice of Model Input Variable

The data cleansing and the selection of proper input variables are another important step in DDM, which take more time and effort that the modeling part. It is common that real-life data have a lot of errors, noise, or missing data. Additionally, outliers may also pose a problem since they could be an indicator of extreme event or erroneous measurement. For this reason, the modeler should be able to identify these atypical values (i.e., outliers) before building any kind of hydrological model.

Another important step is the choice of the model variable that has to consider the physics of the underlying process. In DDM, the physics is introduced by the choice of the relevant input variable that contributes to predicting the output value. Bowden et al. (2005) stated that many DDM modelers often present a large number of inputs and rely on the network to identify the inputs. This deficiency in the modeling process can increase the complexity, reduce the learning process, and model performance.

In literature, five approaches for input determination in DDM modeling have been addressed by Bowden et al. (2005): (1) Methods that rely upon the use of a priori knowledge, (2) methods based on linear cross-correlation, (3) methods that utilize a heuristic approach, (4) methods that extract the knowledge contained within trained ANN, and (5) a combination of the above four approaches.

In this study, the most appropriate inputs for predicting future discharge were estimated using the linear correlation and the average mutual information (AMI) as it was done by Solomatine and Dulal (2003) and Corzo et al. (2009).

The Average Mutual Information (Shannon, 1948) is used to investigate linear and nonlinear relationships and lag effects between the time series variables. It is the measure of information available from one set of data having knowledge of another data set. In other words, it represents the measure of information that can be learned from one data set (X) having the knowledge another data set (Y).

$$AMI = \sum_{i,j} P_{XY}\left(x_i y_j\right) \log_2 \left[\frac{P_{XY}\left(x_i, y_j\right)}{P_X\left(x_i\right) P_Y\left(y_j\right)}\right]. \tag{8.3}$$

Equation (8.3) presents the Average Mutual Information formula where $P_X(x)$ and $P_Y(y)$ are the marginal probability density functions of X and Y, respectively; and $P_{XY}(x,y)$ is the joint probability density function of X and Y. A high value of AMI would indicate a strong dependency between two variables. Conversely, if X is completely independent of Y, then AMI is equal to zero.

Another way to identify the relevant input variables is the analysis of linear correlation between the predicted discharge and the input variable. The correlation analysis is a measure of only linear dependency where a value +1 indicates a total positive linear correlation, 0 indicates that a linear relationship does not exist between two variables, and −1 indicates total negative linear relationship. Overall, the input variables with high correlation and Average Mutual Information (AMI) to the output should be selected in the experimental setup of Data-Driven Models. These metrics are good indicators for detecting the input variable in a DDM. (Solomatine, 2016).

8.3. Characterization of The Grande De San Miguel Catchment

The Grande de San Miguel river is one of the most important rivers in El Salvador. It is located in the eastern part of El Salvador (see Fig. 8.2) between the Pacific Ocean and the mountain range Tecapa and Chinameca. In the catchment headwater, many fast-moving streams are encountered. These streams join and form the Grande de San Miguel river, which pours its water into Bahia de Jiquilisco, a mangrove estuary in the Pacific Ocean (Häggström & Lindström, 1990). The catchment has a total drainage area of 2,397 km^2, which includes two lakes: the Olomega lake (24.2 km^2), which has a damping influence on the flow in the river, and the Jocotal lake with an area of 2.8 km^2. The length of the river from its source to the Pacific Ocean is about 120 km and the major tributaries are the San Francisco, Yamabal, and Yaguabar rivers.

In recent years, flood impacts produced by the Grande de San Miguel river have increased due to various factors: increased urbanization and land use change, low infiltration capacity, erosion, poor or nonexistent water management, and changes in temporal and spatial distribution of rainfall (Servicio Hidrologico Nacional, 2004).

Rainfall data are available in the catchment (see Fig. 8.2) and are provided by the Ministry of Environment and Natural Resources of El Salvador (MARN for its acronym in Spanish). In addition to this, water level and discharge data at La Canoa station are available. The

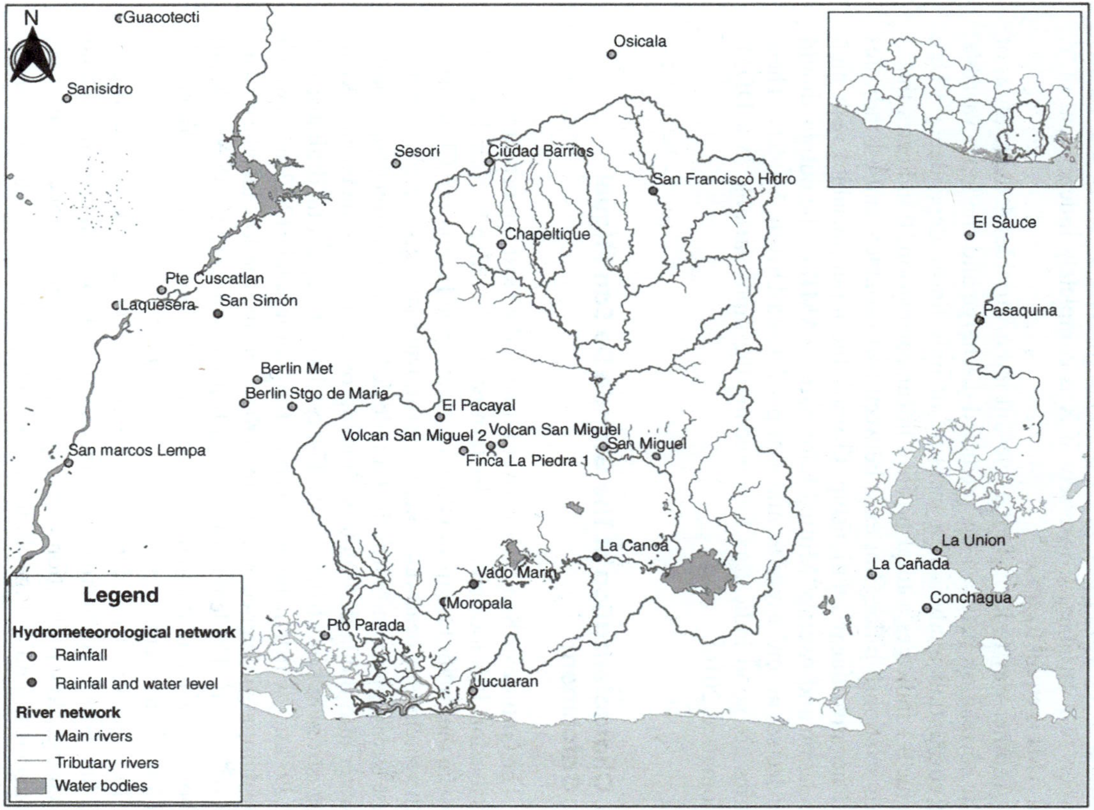

Figure 8.2 Rainfall and water level stations in Grande de San Miguel catchment.

Table 8.1 Potential evapotranspiration-elevation relationships.

Month	Regression equation y = P_e (mm), x = elevation (m)
January	$y = -7E\text{-}0.6x^2 - 0.0170x + 152.95$
February	$y = -8E\text{-}0.6x^2 - 0.0164x + 156.04$
March	$y = -1E\text{-}0.5x^2 - 0.0171x + 190.42$
April	$y = -9E\text{-}0.6x^2 - 0.0202x + 194.52$
May	$y = -7E\text{-}0.6x^2 - 0.0237x + 186.89$
June	$y = -7E\text{-}0.6x^2 - 0.0181x + 165.79$
July	$y = -9E\text{-}0.6x^2 - 0.0127x + 175.28$
August	$y = -8E\text{-}0.6x^2 - 0.0122x + 170.12$
September	$y = -8E\text{-}0.6x^2 - 0.0126x + 150.42$
October	$y = -1E\text{-}0.5x^2 - 0.0094x + 149.56$
November	$y = -9E\text{-}0.6x^2 - 0.0119x + 142.18$
December	$y = -7E\text{-}0.6x^2 - 0.0168x + 144.91$

Source: Adapted from SNET (2005).

evapotranspiration data are estimated by using an evapotranspiration-elevation relationship shown in Table 8.1. These relationships were determined using the Hargreaves method and linear regression equations. For this study, we train a MLP-ANN Rainfall-Runoff forecast model for the part of the catchment that drains to La Canoa station, which has an area of 900 km^2. This station is used for stakeholders and decision makers to monitor the flooding situation.

8.4. Methodology

8.4.1. Data Exploration

In this study, a Multilayer Perceptron Artificial Neural Network (MLP-ANN) architecture with one hidden layer with a sigmoid function and a linear model as an output layer is used to predict discharge at different lead times in Grande de San Miguel catchment. The first part of the study was related to the data cleansing and exploration. In Data-Driven models this is an important step since any wrong data can generate noise in the models; so the data were reviewed using a MATLAB code to detect outliers and remove erroneous data. The outliers were evaluated to determine whether or not they should be removed for the model training. For instance, if a rainfall intensity of 150 mm/hr was recorded by the gauge, then it was compared with the values of the closest gauges in order to determine if the rainfall intensity should be removed for the analysis or kept in the model building. Thereafter, the rainfall point data were averaged in the catchment using the Inverse Distance Weighting (IDW) method. Furthermore, the water level data set at La Canoa station was also reviewed to detect spikes, noise, or suspicious values. After this process, all the water level data points were transformed into discharge values using a rating curve equation provided by MARN. Finally, the discharge and rainfall values were aggregated into hourly time steps since the time of concentration of the catchment is in hours.

8.4.2. Input Variable Selection

In the second part of the study, a maximum forecast horizon was defined according to the maximum time in which it is possible to predict using observed rainfall and discharge. For this study, an analysis of the physical characteristic of the basin was carried out in order to identify the average response time in the catchment. After this process, an analysis of interdependencies between the rainfall, evapotranspiration, and discharge was done using the correlation coefficient and the average mutual information (AMI). The aim of this process was to estimate which past rainfall (RF), evapotranspiration (ET), and discharge (Q) values have a major influence on the model output. Hence, the main forecast formulation can be represented as follows:

$$Q_{(t+i)} = f\left(RF_t, RF_{t-1}, RF_{t-2} \ldots RF_{t-L}, Q_t, Q_{t-1}, Q_{t-2} \ldots Q_{t-M}, ET_t, ET_{t-1}, ET_{t-2} \ldots ET_{t-N}\right), \tag{8.4}$$

where the optimal lags L for rainfall, M for potential evapotranspiration, and N for discharge are obtained through model optimization for different forecast lead time (i).

8.4.3. Data Partitioning and Model Developing

The available data were partitioned into training, cross-validation, and verification. The aim of the training data set is to reduce the difference between the observed and simulated discharge. This procedure is performed in such a way as to ensure that the maximum and minimum values for each input variable are included in the training data set to reduce possible extrapolation. All the MLP-ANN models for different lead times were trained with the Levenberg-Marquardt optimization algorithm (Levenberg, 1944; Marquardt, 1963) with a learning rate of 0.1, one hidden layer with sigmoid function, and one linear function in the output layer. The ANN-MLP models for different lead times have been optimized using a cross-validation data set, which allows to stop the training process, determine the number of neurons (i.e., nodes) using a trial-and-error optimization, and avoid overfitting (i.e. the model tries to follow all the training data points). Last, the verification data set allows to see how the model would perform with a data that it never seen.

The results of the each trained MLP-ANN forecast model for different lead times were evaluated using goodness-of-fit functions and visual inspection analysis between the observed and simulated discharge. The following error functions are used as goodness-of-fit function between the observed discharge (Q_{obs}) and the simulated discharge (Q_{sim}).

The Mean Error (ME)

The Mean Error function is calculated using the mean difference between the observed (Q_{obs}) and simulated discharge (Q_{sim}). The problem related to this error is that it could display a low value as well as low accuracy and precision.

$$ME = \frac{\Sigma\left(Q_{sim} - Q_{obs}\right)}{n}. \tag{8.5}$$

The Nash-Sutcliffe Efficiency (NSE)

The Nash-Sutcliffe efficiency (1970) is defined as one minus the sum of the absolute square difference between the simulated and observed values normalized by the variance of the observed values. The NSE range varies from $-\infty$ to 1 where a value of 1 indicates a perfect reproduction between the observed and simulated discharge. Conversely, if the NSE value is equal or less than 0, it means that the model is not a better predictor than the mean observed discharge ($\overline{Q_{obs}}$). In literature, a value larger than 0.6 indicate a satisfactory result of the model performance. (Gupta et al., 1998). The major disadvantage of this goodness-of-fit function was stated by Legates and McCabe (1999) and Krause et al. (2005) in which the NSE value overestimates larger values and neglects lower ones.

$$NSE = 1 - \frac{\Sigma\left(Q_{sim} - Q_{obs}\right)^2}{\Sigma\left(Q_{obs} - \overline{Q_{obs}}\right)^2}. \tag{8.6}$$

The Root Mean Square Error (RMSE)

The Root Mean Square Error (RMSE) is a widely used goodness-of-fit function for verifying hydrological model performance ensures Maximum Likelihood Estimation (MLE). This error function punishes large deviation, which is usually found in high flows but the bias and shape might be not contemplated (Corzo Perez, 2009).

$$\mathrm{RMSE} = \sqrt{\frac{\sum\left(Q_{sim} - Q_{obs}\right)^2}{n}}. \tag{8.7}$$

The Normalized Root Mean Square Error (NRMSE)

The normalized Root Mean Square Error (NRMSE) compares the classical RMSE with the standard deviation of the measurement (σ_{obs}). This error function can be used to compare different time series (Corzo Perez, 2009).

$$\mathrm{NRMSE} = \frac{\mathrm{RMSE}}{\sigma_{obs}}. \tag{8.8}$$

The Kling-Gupta Efficiency (KGE)

The Kling-Gupta efficiency (KGE) was proposed by Gupta et al. (2009). It is based on the decomposition of the Nash-Sutcliffe efficiency into three distinctive components representing the linear correlation (r), the variability bias (α), and mean bias (β).

$$\mathrm{KGE} = 1 - \sqrt{(r-1)^2 + (\alpha-1)^2 + (\beta-1)^2}. \tag{8.9}$$

Equation (8.9) can be represented using the following formula:

$$\mathrm{KGE} = 1 - \sqrt{(r-1)^2 + \left(\frac{\sigma_{sim}}{\sigma_{obs}} - 1\right)^2 + \left(\frac{\mu_{sim}}{\mu_{obs}} - 1\right)^2}, \tag{8.10}$$

where σ_{obs} is the standard deviation in observations, σ_{sim} the standard deviation in simulations, μ_{sim} the simulation mean, and μ_{obs} the observation mean. Similar to the NSE, a value of KGE equal to 1 indicates a perfect agreement between the observed and simulated discharge.

The MLP-ANN forecast model for different lead times was compared with a naïve reference model, which allows one to benchmark model performance and to evaluate how much a sophisticated forecast model (MLP-ANN) is improving the results based on a simple autoregressive

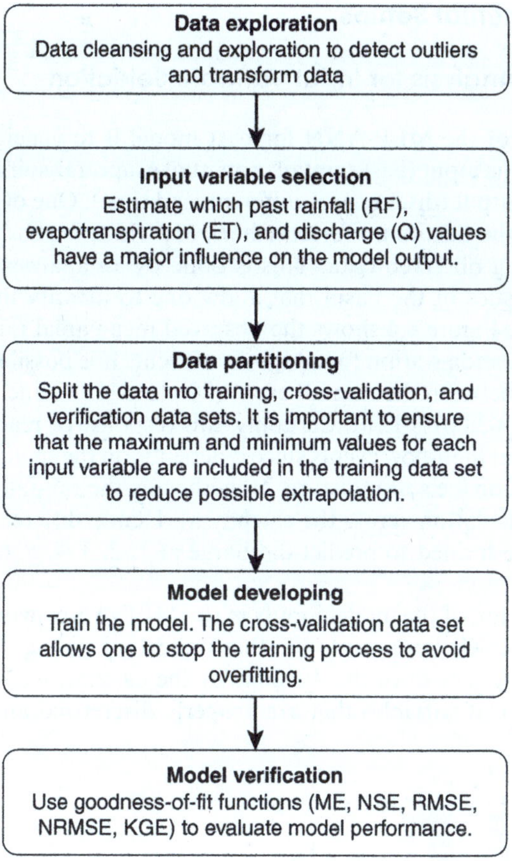

Figure 8.3 Overall methodology to build an MLP-ANN model.

forecast model. The naïve reference model is a simple model that assumes the forecast discharge at lead time i is equal to the latest available observed discharge (see equation (8.11)).

$$Q_{(t+i)} = Q_t. \tag{8.11}$$

As a summary, Figure 8.3 shows the overall methodology with the previous steps used to build a Multilayer Perceptron Artificial Neural Network (MLP-ANN) in Grande de San Miguel catchment.

8.5. Experimental Setups

8.5.1. Data Analysis for Input Variable Selection

The goal of the MLP-ANN forecast model is to generalize the relation between the input (past rainfall, potential evapotranspiration, and discharge) and output (discharge at different lead time). One of the first tasks is to identify the maximum forecast horizon, which is possible to predict discharge using observed data. This is done by an analysis of the physical characteristics of the basin that allow one to identify the reaction or response time. Figure 8.4 shows the observed mean areal rainfall and discharge in La Canoa station for a high-flow event. It is possible to conclude that the time between the rainfall event (20 Nov, 15:00) and the discharge peak (21 Nov, 03:00) is 12 hr. Certainly, this response or reaction time can vary in different high-flow events since it depends on the spatial distribution of the rainfall or the soil moisture. Nonetheless, the rainfall is considered an average calculation across the catchment. Hence, different MLP-ANN models will be trained to predict discharge at 1, 2, 3, 4, 5, 6, 9, and 12 hr ahead.

The selection of the model inputs in the MLP-ANN hydrological forecasting is extremely important. Another important task is the selection of the appropriate lags since the dynamic of the catchment can be approximated by a set of variables that are properly discretized and lagged with

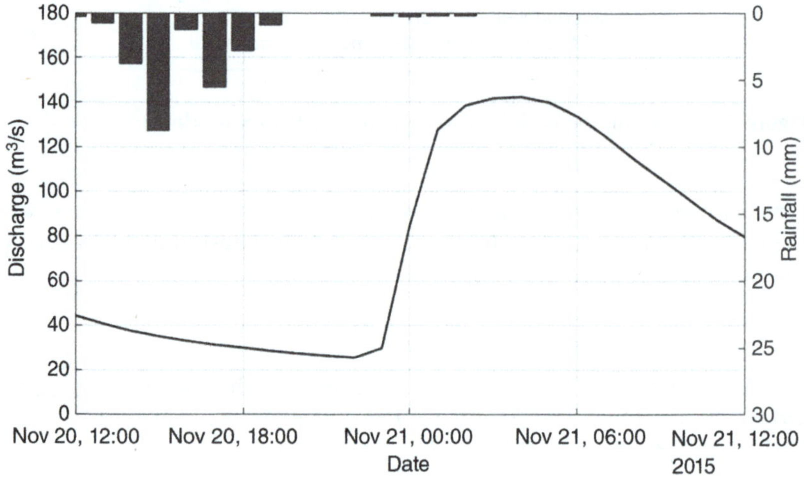

Figure 8.4 Response time based on the physical characteristic of the catchment.

respect to the output discharge (Solomatine & Dulal, 2003). For this study, the hydrometeorological variables that influence the predicted discharge at La Canoa station are the rainfall, evapotranspiration, and discharge itself. However, it is important to state that not all of these variables have a substantially influence on the output discharge. The analysis of interdependency of variable allows one to identify the hydrometeorological variables and lags that have a major influence on the output prediction. In this way, Figure 8.5 presents the Correlation Coefficient (CoC) and the Average Mutual Information (AMI) analysis between the discharge at 1 hr ahead and the rainfall, evapotranspiration, and discharge itself. It can be seen that the past discharge provides the maximum AMI and CoC at a lag 1 hr and then start decreasing, which means that past discharge has a major effect on predicting the discharge itself. Additionally, the AMI and CoC for the rainfall variable start rising at a lag around 10 hr, and then start decreasing at a lag around 14 hr; this behavior means that the rainfall data from 10 to 14 time steps back can be used as a model input. Furthermore, the

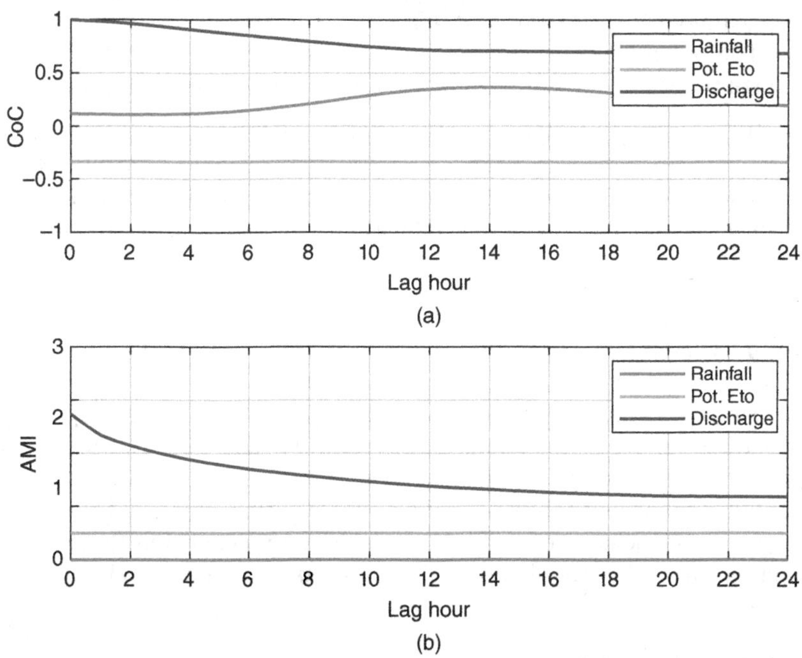

Figure 8.5 Correlation coefficient (a) and AMI (b) analysis between predicted discharge (Qt +1) and rainfall (RF), potential evapotranspiration (ET), and discharge (Q) itself.

Table 8.2 MLP-ANNs experimental summary for flow forecasting in Grande de San Miguel basin.

Input variable	Output variable
$RF_{t-14}, RF_{t-13}, RF_{t-12}, RF_{t-11}, RF_{t-10}, Q_{t-2}, Q_{t-1}, Q_t$	Q_{t+1}
$RF_{t-13}, RF_{t-12}, RF_{t-11}, RF_{t-10}, RF_{t-9}, Q_{t-2}, Q_{t-1}, Q_t$	Q_{t+2}
$RF_{t-12}, RF_{t-11}, RF_{t-10}, RF_{t-9}, RF_{t-8}, Q_{t-2}, Q_{t-1}, Q_t$	Q_{t+3}
$RF_{t-11}, RF_{t-10}, RF_{t-9}, RF_{t-8}, RF_{t-7}, Q_{t-2}, Q_{t-1}, Q_t$	Q_{t+4}
$RF_{t-10}, RF_{t-9}, RF_{t-8}, RF_{t-7}, RF_{t-6}, Q_{t-2}, Q_{t-1}, Q_t$	Q_{t+5}
$RF_{t-9}, RF_{t-8}, RF_{t-7}, RF_{t-6}, RF_{t-5}, Q_{t-2}, Q_{t-1}, Q_t$	Q_{t+6}
$RF_{t-6}, RF_{t-5}, RF_{t-4}, RF_{t-3}, RF_{t-2}, Q_{t-2}, Q_{t-1}, Q_t$	Q_{t+9}
$RF_{t-3}, RF_{t-2}, RF_{t-1}, Q_{t-2}, Q_{t-1}, Q_t$	Q_{t+12}

evapotranspiration does not provide significant information on the prediction of the discharge, which means that the evapotranspiration has little effect on the discharge prediction in the catchment. Therefore, the rainfall and discharge are used as input variables for the MLP-ANN hydrological forecast model for different lead times.

The presented consideration makes it possible to present the general form of the MLP-ANN forecast model for 1 hr ahead in the following form:

$$Q_{(t+1)} = f\left(RF_{t-14}, RF_{t-13}, RF_{t-12}, RF_{t-11}, RF_{t-10}, Q_{t-2}, Q_{t-1}, Q_t\right). \tag{8.12}$$

The previous analysis was also done for the remaining forecast lead times of 2, 3, 4, 5, 6, 9, and 12 hr. The summary of these results appears in Table 8.2.

8.5.2. Data Partitioning

Hourly rainfall and discharge data from 2013 to 2016 were separated into training, cross-validation, and verification data set. The statistical properties of the discharge data set in La Canoa station and mean areal rainfall are shown in Tables 8.3 and 8.4, where the maximum and minimum rainfall and discharge of the data available were included in the training data set to avoid extrapolation of the results in operational mode.

8.6. Results and Discussion

In this study, a Multilayer Perceptron Artificial Neural Network (MLP-ANN) architecture with a one hidden layer with a sigmoid function

Table 8.3 Statistical properties for training, cross-validation, and verification discharge data set.

Statistical properties	Training	Cross-validation	Verification
Data points (hours)	15,058	6,453	5,000
Maximum (m^3/s)	167.17	163.87	157.75
Minimum (m^3/s)	0.84	1.78	1.36
Average (m^3/s)	14.45	14.12	12.64
Standard deviation (m^3/s)	21.05	23.11	19.15

Table 8.4 Statistical properties for training, cross-validation, and verification rainfall data set.

Statistical properties	Training	Cross-validation	Verification
Data points (hours)	15,058	6,453	5,000
Maximum (mm)	28.23	20.67	19.07
Minimum (mm)	0.00	0.00	0.00
Average (mm)	0.14	0.18	0.16
Standard deviation (mm)	0.88	1.03	0.93

and a linear model as an output layer is used to predict discharge with a maximum forecast horizon of 12 hr in Grande de San Miguel river basin in El Salvador. Each of these trained MLP-ANNs models was compared with observed data using visual inspection and goodness-of-fit functions (presented above) and they were also compared with a naïve reference model.

Figure 8.6 presents the results between the observed and simulated discharge for the verification data set and 1 hr of lead time. These results show an excellent representation of both high and low discharge in La Canoa station and confirm the fact that the MLP-ANN is capable of predicting the discharge at 1 hr ahead.

It is also important to explain how the MLP-ANN forecast model is weighting its inputs. In Figure 8.7 is shown the relation between the input variables for the 1 hr of lead time and the resulting weights for each hidden nodes or neurons in the MLP-ANN model. The figure shows that the discharge overweighs the rainfall in the network since the past discharge has a major effect on the forecast model.

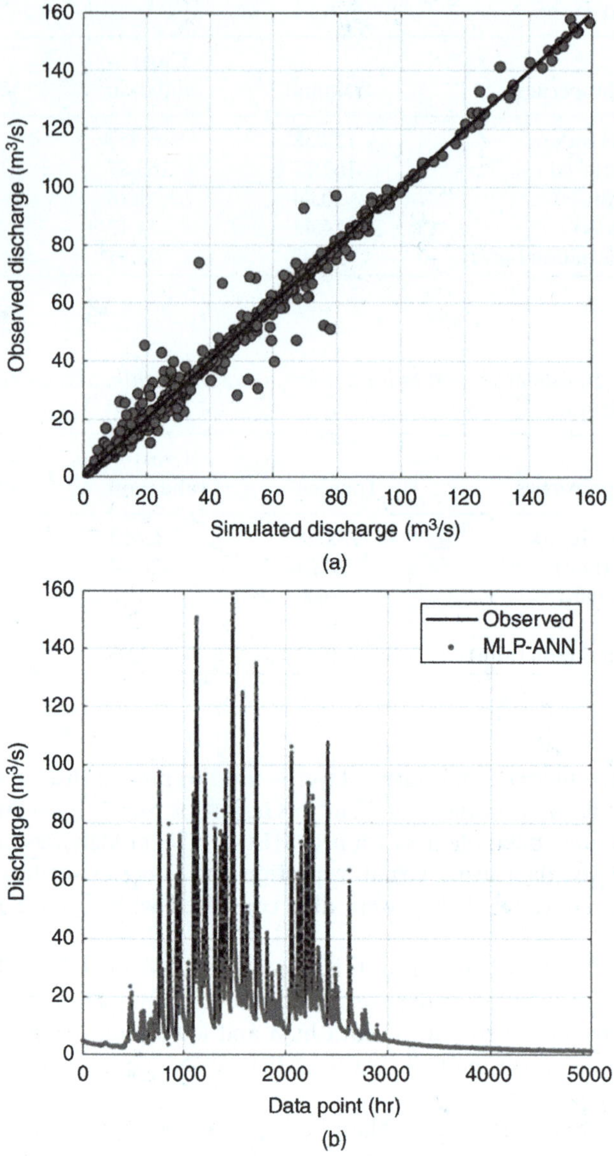

Figure 8.6 (a) Scatter plot and (b) hydrograph comparison between observed and simulated (MLPANN) for a lead time of 1 hr in the verification data set.

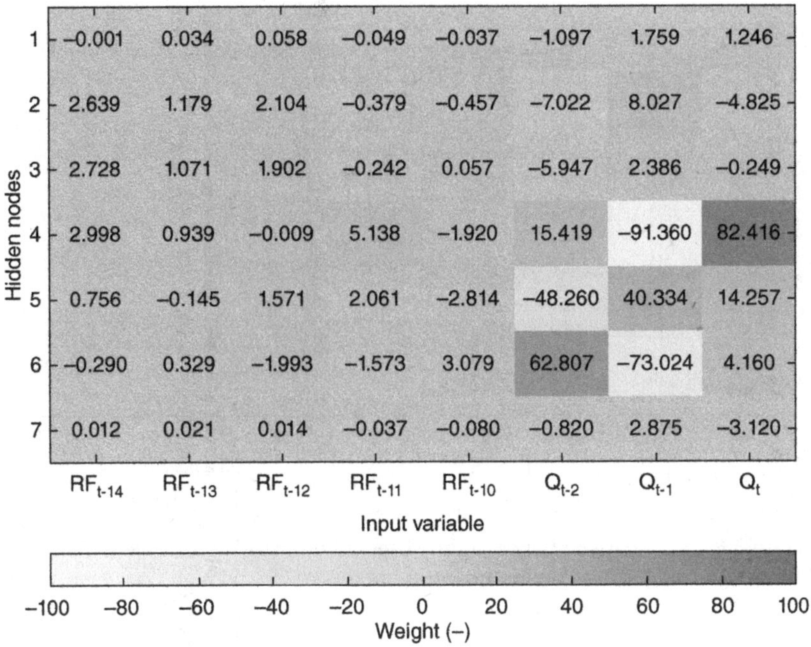

Figure 8.7 Relation between the model input and the resulting weights for each nodes or neurons in the MLP-ANN forecast model for a lead time of 1 hr.

In comparison with the naïve reference model, the results show that both models are capable of reproducing the predicting discharge at 1 hr ahead. It can be seen in Figure 8.8a that the MLP-ANN and the naïve model are capable of reproducing high and low discharge in the verification period. Moreover, Figure 8.8b presents the result for a single high-flow event in which both models reproduce the peak discharge in La Canoa station, but the naïve model exhibit a delay in the time of peak of the discharge.

Table 8.5 and Figure 8.9 present the performance metrics and scatter plots between the predicted and observed hydrograph for different forecast lead times (3, 6, 9, and 12 hr). These results show a diminish in the model performance as the lead time increases, especially for the RMSE, the NRMSE, NSE, and KGE values. The Mean Error (ME) exhibits better results for the naïve reference model than the MLP-ANN model for lead time larger than 2 hr. These results basically means that the average model error in the naïve reference model is lower than the MLP-ANN models, but the magnitude of the error is not measured in the mean error formula. It should also be noted that with a forecast lead time of 12 hr,

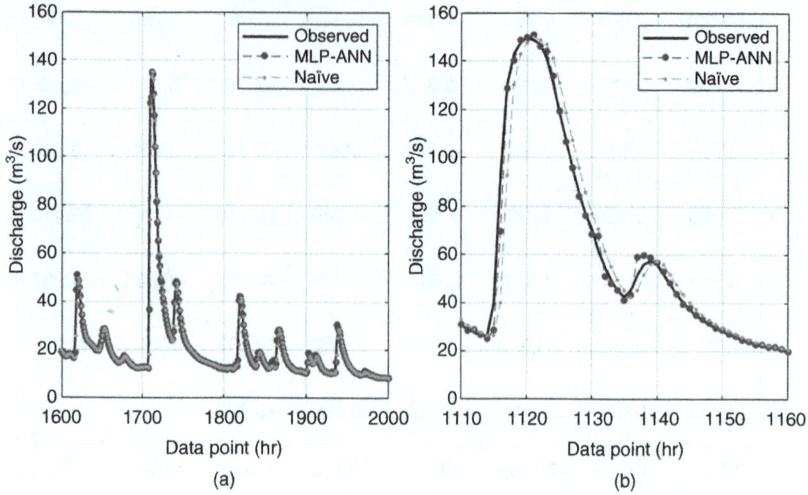

Figure 8.8 Comparison between the observed discharge, the naïve model, and MLP-ANN for Qt + 1in continuous period of record (a) and high flow event (b) in the verification data set.

Table 8.5 Summary of goodness-of-fit function for MLP-ANN and Naïve forecast model for the verification data set.

	Models	ME (m^3/s)	NSE (–)	RMSE (m^3/s)	NRMSE (–)	KGE (–)
Q_{t+1}	Naïve	0.001	0.971	3.282	0.171	0.985
	MLP-ANN	0.033	0.993	1.578	0.082	0.996
Q_{t+2}	Naïve	0.001	0.896	6.164	0.322	0.948
	MLP-ANN	0.071	0.957	3.957	0.207	0.977
Q_{t+3}	Naïve	0.002	0.799	8.588	0.448	0.899
	MLP-ANN	–0.015	0.901	6.021	0.314	0.943
Q_{t+4}	Naïve	0.003	0.693	10.616	0.554	0.846
	MLP-ANN	0.056	0.843	7.596	0.397	0.914
Q_{t+5}	Naïve	0.003	0.586	12.318	0.643	0.793
	MLP-ANN	–0.154	0.787	8.382	0.461	0.86
Q_{t+6}	Naïve	0.004	0.484	13.757	0.718	0.742
	MLP-ANN	–0.137	0.766	9.263	0.484	0.836
Q_{t+9}	Naïve	0.006	0.22	16.91	0.883	0.61
	MLP-ANN	–0.482	0.679	10.855	0.567	0.761
Q_{t+12}	Naïve	0.008	0.042	18.749	0.979	0.521
	MLP-ANN	–0.487	0.576	12.467	0.651	0.682

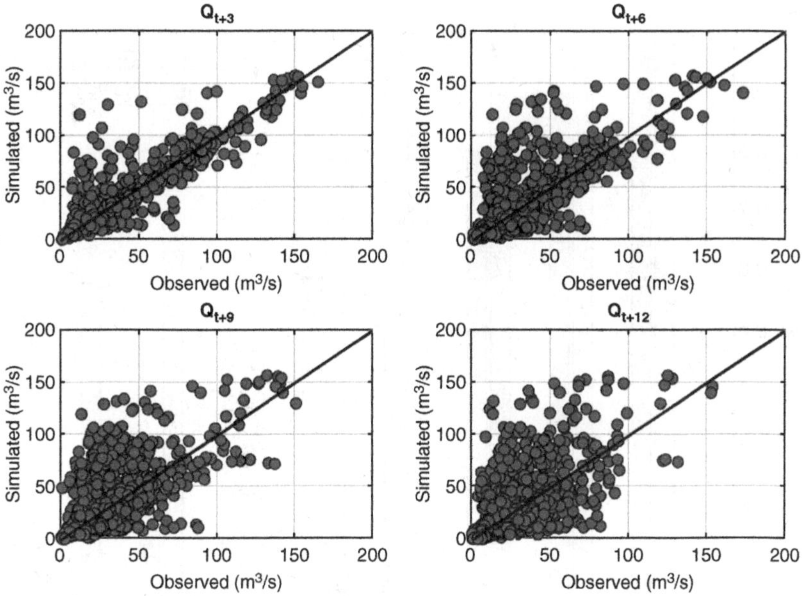

Figure 8.9 Scatter plot between the observed and simulated MLP-ANN discharge for lead time of 3, 6, 9, and 12 hr ahead.

the naïve reference model performance is very poor (NSE equal to 0.042). The MLP-ANN hydrological model is producing better results but still is not satisfactory (NSE equals 0.576) since there are some high-flow events where the MLP-ANN model underestimated the results (see Fig. 8.10). This poor representation of the observed discharge could be due to lack of recent information in model structure that allows the model to better learn the rainfall-runoff relationship. More investigation should be conducted on how to improve the MLP-ANN model performance for a lead time of 12 hr in Grande de San Miguel catchment.

Overall, the model results show excellent performance for lead time no larger than 4 hr since the Nash-Sutcliffe efficiency (NSE) is above 0.80. In addition, the forecast model for lead time between 5 and 9 hr provides good results since the NSE is between 0.68 and 0.79 (see Fig. 8.10).

The resulting number of hidden nodes or neurons of the trained MLP-ANNs is shown in Table 8.6. The number of hidden nodes is associated with the ability of the network to map the output and the number of input variables (Solomatine, 2016). To illustrate this, the forecast model Q_{t+4} has eight input variables, and the number of hidden nodes is seven.

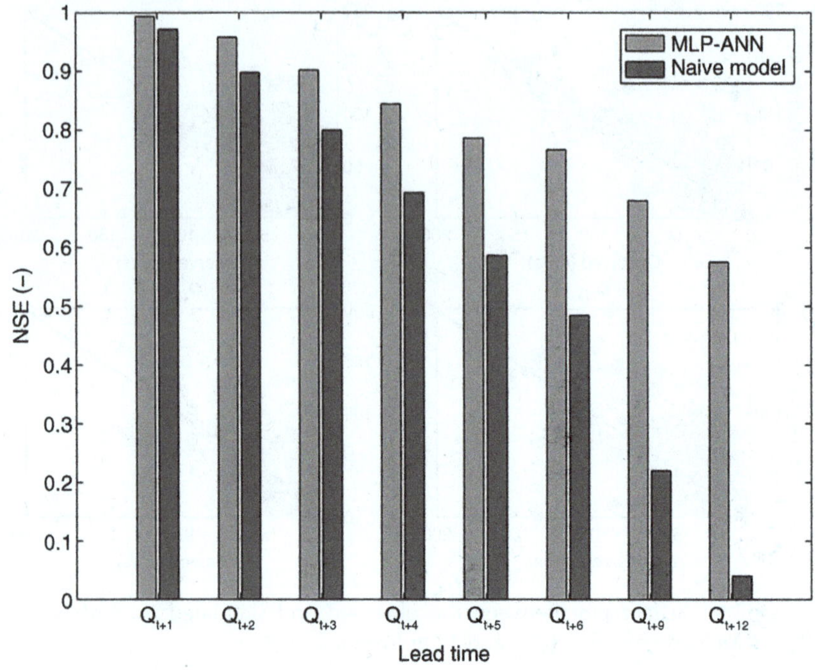

Figure 8.10 Model performance comparison between the naïve and MLP-ANN forecast model for different lead times.

Table 8.6 Number of hidden nodes or neurons in the MLP-ANN forecast models.

Input variable	Output variable	Hidden nodes
RF_{t-14}, RF_{t-13}, RF_{t-12}, RF_{t-11}, RF_{t-10}, Q_{t-2}, Q_{t-1}, Q_t	Q_{t+1}	7
RF_{t-13}, RF_{t-12}, RF_{t-11}, RF_{t-10}, RF_{t-9}, Q_{t-2}, Q_{t-1}, Q_t	Q_{t+2}	7
RF_{t-12}, RF_{t-11}, RF_{t-10}, RF_{t-9}, RF_{t-8}, Q_{t-2}, Q_{t-1}, Q_t	Q_{t+3}	6
RF_{t-11}, RF_{t-10}, RF_{t-9}, RF_{t-8}, RF_{t-7}, Q_{t-2}, Q_{t-1}, Q_t	Q_{t+4}	7
RF_{t-10}, RF_{t-9}, RF_{t-8}, RF_{t-7}, RF_{t-6}, Q_{t-2}, Q_{t-1}, Q_t	Q_{t+5}	7
RF_{t-9}, RF_{t-8}, RF_{t-7}, RF_{t-6}, RF_{t-5}, Q_{t-2}, Q_{t-1}, Q_t	Q_{t+6}	6
RF_{t-6}, RF_{t-5}, RF_{t-4}, RF_{t-3}, RF_{t-2}, Q_{t-2}, Q_{t-1}, Q_t	Q_{t+9}	7
RF_{t-3}, RF_{t-2}, RF_{t-1}, Q_{t-2}, Q_{t-1}, Q_t	Q_{t+12}	7

8.7. Operational MLP-ANN Forecast Model

The trained MLP-ANNs models can be used with real-time observation as guidance for the flood situation in Grande de San Miguel catchment. Many forecast systems can be used as an environmental decision support system in order to organize the complex process of data acquisition in real time and link the model with these hydrometeorological data.

According to Werner et al. (2013) two categories of forecast systems can be defined *a model-centric and a modular approach.* The former basically means that that any change in the model or in the data used may require a redesign and redevelopment of the system. The latter is a more flexible approach, which can adapt to changing needs in modeling requirement and data transformation. After training and validating the MLP-ANN, it was decided to place the developed MLP-ANN models in operational mode using a model-centric approach since the Ministry of Environment and Natural Resources of El Salvador does not have any modular forecast platform (e.g., Delft-FEWS). The conceptual design of the forecast system is shown in Figure 8.11 where all the real-time rainfall and water level in the Grande de San Miguel catchment are imported from MARN Database using a MATLAB routine. Afterward, another MATLAB routine allows one to preprocess the real-time observations to delete erroneous data, detect outliers, transform water level into discharge units, and aggregate hydrometeorological data from 15 min to 1 hr time step. Subsequently, the hourly rainfall and discharge data are used to

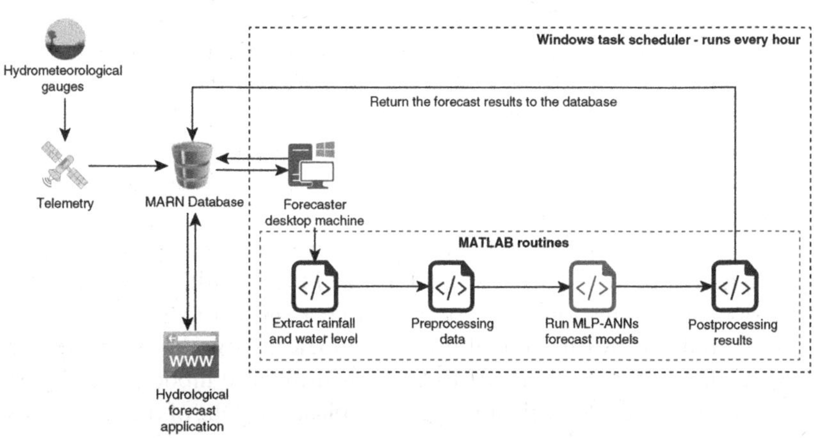

Figure 8.11 Conceptual design of the operational MLP-ANN forecasting in Grande de San Miguel river basin.

feed the trained MLP-ANN forecast models for 1, 2, 3, 4, 5, 6, 9, and 12 hr ahead. Finally, the forecast results are sent to the MARN database to be used in the hydrological forecast Web application, whose results can be seen in the following webpage: http://srt.snet.gob.sv/apps/public/pronosticorgsm. The previous routines are scheduled to run every hour using a Windows Task Scheduler.

8.8. Conclusions and Recommendations

In this study, an operational MLP-ANN is developed to address the short-term hydrological forecast in Grande de San Miguel catchment. The maximum forecast horizon was determined by the physical characteristics of the basin since the mean response time of the catchment is 12 hr. Additionally, the input variables were selected using an analysis of interdependency between hydrometeorological variables. Finally, the results of the trained MLP-ANNs were compared with the observed discharge at La Canoa and a naïve reference model, which allows one to benchmark model performance and to evaluate how much a sophisticated forecast model (MLP-ANN) is improving the results based on a simple autoregressive forecast model.

The model results show excellent performance for lead time no larger than 4 hr since the Nash-Sutcliffe efficiency (NSE) is above 0.80. Moreover, the forecast model for lead time between 5 and 9 hr provides good results since the NSE is between 0.68 and 0.79. The trained MLP-ANN forecast lead time of 12 hr does not provide a satisfactory result since there are some high-flow events where the model underestimates the results. This poor representation of the observed discharge could be due to lack of recent information in model structure that allows the model to better learn the rainfall-runoff relationship. More investigation should be conducted on how to improve the MLP-ANN model performance for a lead time of 12 hr in Grande de San Miguel river basin.

The trained MLP-ANNs forecast models were put in operational mode using a model-centric approach based on MATLAB routines and Windows Task Scheduler. This forecast system approach allows one to import, preprocess, and transform the real time hydrological data. Additionally, the process allows us to run the MLP-ANNs hydrological models and export the results to a hydrological forecast Web application. One of the limitations of the design of the forecast system is that it cannot adapt to changing needs in modeling requirement and data transformation, which would require of a redesign of the forecast system framework.

Acknowledgments

I would like to thank the Ministry of Environment and Natural Resources of El Salvador (MARN) for providing the data used in this study.

References

Abrahart, R. J., & See, L. (2000). Comparing neural network and autoregressive moving average techniques for the provision of continuous river flow forecasts in two contrasting catchments. *Hydrological Processes*, *14*(11–12), 2157–2172.

Abrahart, R. J., & See, L. M. (2007). Neural network modeling of non-linear hydrological relationships. *Hydrology and Earth System Sciences, 11*(5), 1563–1579.

Ahani, A., Shourian, M., & Rad, P. R. (2018). Performance assessment of the linear, nonlinear and nonparametric data driven models in river flow forecasting. *Water Resources Management*, *32*(2), 383–399.

Badrzadeh, H., Sarukkalige, R., & Jayawardena, A. W. (2015). Hourly runoff forecasting for flood risk management: Application of various computational intelligence models. *Journal of Hydrology*, *529*, 1633–1643. https://doi.org/10.1016/j.jhydrol.2015.07.057

Bowden, G. J., Dandy, G. C., & Maier, H. R. (2005). Input determination for neural network models in water resources applications. Part 1, Background and methodology. *Journal of Hydrology*, *301*(1–4), 75–92. https://doi.org/10.1016/j.jhydrol.2004.06.021

Chacon-Hurtado, J. (2012). *Streamflow forecast for hydropower generation-a data assimilation approach* (PhD Thesis). Unesco-IHE.

Corzo, G. A., Solomatine, D. P., de Wit, M., Werner, M., Uhlenbrook, S., & Price, R. K. (2009). Combining semi-distributed process-based and data-driven models in flow simulation: A case study of the Meuse River basin. *Hydrology and Earth System Sciences*, 16.

Corzo Perez, G. A. (2009). *Hybrid models for hydrological forecasting: Integration of data-driven and conceptual modeling techniques*. IHE Delft Institute for Water Education.

Coulibaly, P., Anctil, F., & Bobée, B. (2000). Daily reservoir inflow forecasting using artificial neural networks with stopped training approach. *Journal of Hydrology*, *230* (3–4), 244–257.

Dawson, C. W., & Wilby, R. L. (2001). Hydrological modeling using artificial neural networks. *Progress in Physical Geography: Earth and Environment*, *25*(1), 80–108. https://doi.org/10.1177/030913330102500104

Dayhoff, J. E. (1990). *Neural network architectures: An introduction*. Van Nostrand Reinhold Co.

Fernando, D. A. K., & Jayawardena, A. W. (1998). Runoff forecasting using RBF networks with OLS algorithm. *Journal of Hydrologic Engineering*, *3*(3), 203–209.

Govindaraju, R. S., & Rao, A. R. (2013). *Artificial neural networks in hydrology* (Vol. 36). Springer Science & Business Media.

Gupta, H. V., Kling, H., Yilmaz, K. K., & Martinez, G. F. (2009). Decomposition of the mean squared error and NSE performance criteria: Implications for improving hydrological modeling. *Journal of Hydrology*, *377*(1–2), 80–91. https://doi.org/10.1016/j.jhydrol.2009.08.003

Gupta, H. V., Sorooshian, S., & Yapo, P. O. (1998). Toward improved calibration of hydrologic models: Multiple and noncommensurable measures of information. *Water Resources Research*, *34*(4), 751–763. https://doi.org/10.1029/97WR03495

Häggström, M., & Lindström, G. (1990). *Application of the HBV model for flood forecasting in six Central American rivers*. SMHI. http://urn.kb.se/resolve?urn=urn:nbn:se:smhi:diva-2454

Hsu, K., Gupta, H. V., & Sorooshian, S. (1995). Artificial neural network modeling of the rainfall-runoff process. *Water Resources Research*, *31*(10), 2517–2530.

Hu, T., Wu, F., & Zhang, X. (2007). Rainfall-runoff modeling using principal component analysis and neural network. *Hydrology Research*, *38*(3), 235–248.

Krause, P., Boyle, D. P., & Bäse, F. (2005). Comparison of different efficiency criteria for hydrological model assessment. *Advances in Geosciences*, *5*, 89–97. https://doi.org/10.5194/adgeo-5-89-2005

Lachtermacher, G., & Fuller, J. D. (1994). Backpropagation in hydrological time series forecasting. In *Stochastic and statistical methods in hydrology and environmental engineering* (pp. 229–242). Springer.

Legates, D. R., & McCabe, G. J. (1999). Evaluating the use of "goodness-of-fit" measures in hydrologic and hydroclimatic model validation. *Water Resources Research*, *35*(1), 233–241. https://doi.org/10.1029/1998WR900018

Levenberg, K. (1944). A method for the solution of certain problems in least squares. *Quarterly of Applied Mathematics, 2,* 164–168. https://doi.org/10.1090/qam/10666

Maier, H. R., & Dandy, G. C. (1996). The use of artificial neural networks for the prediction of water quality parameters. *Water Resources Research*, *32*(4), 1013–1022.

Marquardt, D. W. (1963). An algorithm for least-squares estimation of nonlinear parameters. *Journal of the Society for Industrial and Applied Mathematics*, *11*(2), 431–441.

Minns, A. W., & Hall, M. J. (1996). Artificial neural networks as rainfall-runoff models. *Hydrological Sciences Journal*, *41*(3), 399–417.

Nash, J. E., & Sutcliffe, J. V. (1970). River flow forecasting through conceptual models, part I: A discussion of principles. *Journal of Hydrology*, *10*(3), 282–290.

Rajurkar, M. P., Kothyari, U. C., & Chaube, U. C. (2002). Artificial neural networks for daily rainfall-runoff modeling. *Hydrological Sciences Journal*, *47*(6), 865–877.

Riverside Technology (2012). *Development and implementation of a regional Flash Flood Guidance and Early Warning System: Lake Ilopango/Rio Jiboa, El Salvador*. Technical Report, Riverside Technology, Colorado, USA.

Servicio Hidrologico Nacional (2004). *Sistema de alerta temprana por inundaciones. Experiencia en El Salvador*. Technical Report, Servicio Nacional de Estudios Territoriales (SNET), San Salvador, El Salvador.

Shamseldin, A. Y. (1997). Application of a neural network technique to rainfall-runoff modeling. *Journal of Hydrology, 199* (3–4), 272–294.

Shannon, C. E. (1948). A mathematical theory of communication. *The Bell System Technical Journal, 27*(3), 379–423.

SNET (2005). *Balance hídrico integrado y dinamico en El Salvador*. Technical Report (p. 118), Servicio Nacional de Estudios Territoriales, San Salvador, El Salvador

Solomatine, D. (2016). *Data-driven modeling: Machine learning, data mining and knowledge discovery*. Lecture Notes, UNESCO-IHE, Delft, the Netherlands

Solomatine, D., & Wagener, T. (2011). 2.16 Hydrological modeling, 24. In P. Wilderer (Ed.), *Treatise on water science* (vol. 2, pp. 435–457). Oxford: Academic Press.

Solomatine, D. P., & Dulal, K. N. (2003). Model trees as an alternative to neural networks in rainfall-runoff modeling. *Hydrological Sciences Journal, 48*(3), 399–411. https://doi.org/10.1623/hysj.48.3.399.45291

Solomatine, D. P., & Siek, M. B. (2006). Modular learning models in forecasting natural phenomena. *Neural Networks, 19*(2), 215–224.

Solomatine, D. P., Maskey, M., & Shrestha, D. L. (2008). Instance-based learning compared to other data-driven methods in hydrological forecasting. *Hydrological Processes: An International Journal, 22*(2), 275–287.

Sudheer, K. P., & Jain, S. K. (2003). Radial basis function neural network for modeling rating curves. *Journal of Hydrologic Engineering, 8*(3), 161–164.

Wang, W., Van Gelder, P. H., Vrijling, J. K., & Ma, J. (2006). Forecasting daily streamflow using hybrid ANN models. *Journal of Hydrology, 324*(1–4), 383–399.

Werner, M., Schellekens, J., Gijsbers, P., van Dijk, M., van den Akker, O., & Heynert, K. (2013). The Delft-FEWS flow forecasting system. *Environmental Modeling & Software, 40*, 65–77. https://doi.org/10.1016/j.envsoft.2012.07.010

WMO (2018). *Multi-hazard early warning systems: A checklist*. Outcome of the First Multi-Hazard Early Warning Conference (p. 20). WMO.

Wu, C. L., & Chau, K.-W. (2010). Data-driven models for monthly streamflow time series prediction. *Engineering Applications of Artificial Intelligence, 23*(8), 1350–1367.

9

NOISE FILTER WITH WAVELET ANALYSIS IN ARTIFICIAL NEURAL NETWORKS (NOWANN) FOR FLOW TIME SERIES PREDICTION

Daniel A. Vázquez[1], Gerald A. Corzo Perez[1], and Dimitri P. Solomatine[1,2,3]

[1] *IHE Delft Institute for Water Education, Delft, The Netherlands*
[2] *Water Resources Section, Delft University of Technology, Delft, The Netherlands*
[3] *Water Problems Institute of the Russian Academy of Sciences, Moscow, Russia*

A data preprocessing technique, known as Wavelet Analysis from the field of signal processing, was used to enhance an artificial neural network (ANN) model by addressing its primary drawbacks, notably nonstationarity and noise. Wavelet transforms are well-suited for handling time series contaminated by nonstationary traits and can extract relevant information at various time series frequencies. To improve the model's performance, the ANN calibration weighting scheme assigns more weight to representative subseries and reduces the impact of noisy data on the rest. Noise filtering techniques (soft/hard single threshold) where applied to manipulate the subseries, further enhancing the model's predictive skill. This approach resulted in the creation of a Noise Filter with Wavelet Analysis in Artificial Neural Network (NOWANN) scheme. To identify the best model, a model-based approach was used for input variable selection, k-fold cross-validation to determine the

Advanced Hydroinformatics: Machine Learning and Optimization for Water Resources, Special Publications 78, First Edition. Edited by Gerald A. Corzo Perez and Dimitri P. Solomantine.

DOI: 10.1002/9781119639268.ch9

optimal training split, early stopping to prevent overfitting, and an exhaustive search in the variable space (hidden nodes versus decomposition levels) to find the most suitable ANN structure. Other data-driven methods (DDM) testing the effectiveness of wavelets as data preprocessors were briefly discussed, including least median squares, model tree (M5P), pace regression, and linear regressions. For wavelet coefficients time-invariance, the a-trous wavelet transform was used. The results indicate that the NOWANN schemes outperformed all previous models, confirming the effectiveness of this approach.

9.1. Introduction

Artificial Neural Network (ANN) is an alternative to traditional approaches such as concept and process-based methods and belongs to the data-driven group. Abrahart et al. (2004) identified key points, advantages, and disadvantages on the use of ANN for hydrological modeling, and they are briefly summarized as follows: (1) Complex processes with high non-linearity could be modeled by ANN, sometimes necessary in hydrological processes; (2) simple tool, with low data demand and fast execution, ideal for operational modeling; (3) the model can be dynamic, in the sense that it can be updated as more data become available, continually strengthening itself; (4) it can solve very complex problems where other modeling tools are questionable, such as sediment transport, groundwater modeling, and so on.

However, like any model, it has its drawbacks: ANN is an input-output model, therefore prone to problems due to inappropriate use of data and poor model development; it does not adequately handle noise, trend, aperiodicity, or other forms of nonstationarity; and it does not provide additional information on the physical processes that occur in the basin.

In this field, there is an extensive literature with different applications: Dawson and Wilby (2001) have provided in their work a valuable guide to help the construction of ANN models of rainfall-runoff. The following points stand out from the literature consulted. To start with, May et al. (2011) considered that the selection of input variables (IVS) is fundamental and crucial in the identification of the optimal ANN model. In their work, the taxonomy of IVS strategies and algorithms is explored in detail. Bowden et al. (2005) presented a methodology to determine an appropriate set of model inputs, exploring methods such as partial mutual information to measure input-output dependency, self-organizing maps to eliminate redundant inputs, and genetic algorithm to determine the most influential inputs.

Subsequently, the appropriate class of models must be identified. The most common are the backpropagation neural network (BPNN), the radial basis function network (RBFN), and the self-organizing feature map (SOFM). Regarding BPNN (the most common), Shamseldin et al. (1999) have evaluated the importance of choosing the type of transfer function to forecast flow in rivers. In his experience, the logistics function indicated superiority among those tested. Kisi and Cigizoglu (2005) have compared all these techniques to predict river flow and found the RBF to be superior based on performance criteria.

The next point to consider is the selection of suitable data sets to avoid local minima or loss of generalization capabilities during the calibration process. Arlot and Celisse (2010) provide guidelines for choosing the best cross-validation procedure based on the problem trait in question. Another important factor in the model-building process is during the network training step. Abrahart et al. (2004) have mentioned techniques that prevent overfitting during calibration. For example, the early stopping method states that (Coulibaly et al., 2000), if in a predefined number of epochs the validation check fails to decrease its performance, the training stops. A second alternative is weight decay that penalizes large weights to produce a smoother input-output mapping.

As mentioned before, the literature is extensive in the subject and also broad in the range of applications for hydrology. To name a few, Shrestha et al. (2009) emulated Monte Carlo (MC) simulations with ANNs to speed up the assessment of the model parametric uncertainty; Zhang et al. (2009) forecasted daily reservoir inflows by using quantitative precipitation forecasting into ANNs model. In addition, Coulibaly et al. (2000) used an early stop training approach to avoid model overfitting for real-time reservoir inflow forecast. Bhattacharya et al. (2007) modeled the complex mechanism of sediment transport and reported better results compared with the traditional models at hand.

However, in the present study, all the previous points were addressed to some extent and special attention was paid to data preprocessing, considered of utmost importance in the development of ANN. An emerging technique taken from the field of signal analysis is put to the test. Wavelet transforms (WT) is one of the many mathematical transformation methods that was used in its early days in seismic signal analysis and is now widely used in many other fields. This work uses WT due to its ability to transform signals of any type (rainfall and discharges time series of our interest) to obtain time-frequency information from it, which was not available in its raw condition. This technique complements data-driven models due to its ability to analyze signals contaminated with noise, trending, aperiodicity, or other forms of nonstationarity, which DDMs cannot handle well.

Articles about wavelet with ANN are available to the reader (Adamowski & Sun, 2010; Adamowski, 2008; Cannas et al., 2005; Khalil et al., 2012; Okkan, 2012; Okkan & Serbes, 2013; Partal, 2008). Recommended articles are efforts to couple a modeling tool with a highly flexible data preprocessing technique, with different purposes such as Himalayan runoff forecast, short-term river flow for melt-caused floods, river flow predictions, short-term total dissolved solids forecast, and inflow of reservoirs.

The knowledge obtained in other fields such as economics and finance, where this approach is much better established, has contributed a lot to develop concepts and capture limitations of the approach. Some algorithms used for WT analysis may have unwanted behaviors that are not easily detected. One drawback is the use of information brought from the future but located in the present time step; this violates the restrictions on the use of information not yet seen by the model. This can create false expectations of seemingly very promising results. Another problem is the time-varying properties of some algorithms, as well as errors and assumptions at the limits of the signal. This problem is addressed in the next section. Table 9.1 shows the work done in the field, focusing on the use of wavelet neural networks. Table 9.1 is a relevant bibliographic summary for this study, where the types of models and processes used (linked to the wavelet method) are identified. Most of these works have been taken into account for the formulation of the methodology that is followed here.

9.2. Materials

9.2.1. Wavelet Transform

The basic idea behind Wavelet Analysis is to use wavelike functions that, when convolved with the signals, can extract important information and express it in more useful forms. These functions are operated in two ways: they can be translated at all-discrete locations of the signal, and they can be stretched or squeezed so that different frequencies of the signal are studied. As a result, the signal has two representations (approximation and detail). In turn, each element of approximation can be decomposed back to the coarsest resolution (multiresolution representation). After this interrogation, the wavelet transform space indicates how well the signal is represented by the adopted wavelet function. Now, the ANN model decides which information is most relevant. The following concepts have

Table 9.1 Bibliographic reference summary used for this work.

Authors	Models	Predictand variable	Data pre-processing				Data sets splitting procedure	Stopping criteria	Error
			IVS	Decomp algorithm	Mother wavelet	Number of decom-positions			
Adamowski & Sun, 2010	ANN; WANN	(A B) 1, 2, and 3 days ahead (flow prediction)	Model-based: flow from 1 day up to 15 days lag were tested	A-trous WT	Haar	8 levels (no explanation) All subseries were used as inputs (all subseries contain important information)	For all cases mean, max, min, and std were calculated for each data set (A) 80% calib, 20% val (B) 80% calib, 20% val	No information	R2; RMSE

(*continued*)

Table 9.1 *(continued)*

			Data pre-processing						
Authors	Models	Predictand variable	IVS	Decomp algorithm	Mother wavelet	Number of decompositions	Data sets splitting procedure	Stopping criteria	Error
Maheswaran & Khosa, 2012	Wavelet Volterra	(B) 1 month ahead (flow prediction) (C) 1 day ahead (flow prediction)	Not explained	A- trous wavelet transform	Haar, Db2, Db3, Db4, Sym4, Spline-B3	Visual inspection (e.g., Perform decompositions until elimination of the residual component in the detail representation)	(B) 70% cal, 30% val (C) 3 years calibration, 2 years validation	No information	MAE; RMSE; NSC

Okkan & Serbes, 2013	ANN; WANN-MLR; WMLR	monthly (stream flow pre-diction) at time t	Model-free: cross-, auto-, and partial-auto-correlation	DWT- Mallat algorithm (trape-zoidal rule)	Haar		Best 3 levels (by trial and error) Ineffective subseries were eliminated by using all regression methods and evaluating the Mallows Cp coef. to prevent collinearity	50% cali-bration, 50% val-idation	Best training results	MSE; RMSE; R^2; Matt-Whitney test

(*continued*)

Table 9.1 *(continued)*

Authors	Models	Predictand variable	Data pre-processing					Stopping criteria	Error
			IVS	Decomp algorithm	Mother wavelet	Number of decompositions	Data sets splitting procedure		
Kisi & Cimen, 2011	ANN; MLR; AR; WANN	1 month ahead (stream flow prediction)	Not explained	DWT	Not mentioned	5 resolution levels (no explanation) Correlation values between inputs (subseries) and outputs provided information for determination of effective subseries to be	Mean, standard deviation, skewness, and lag 1, 2, and 3 autocorrelation (A) 80% cal, 20% val (B) 80% cal, 20% val	100 epochs, error converged a certain value	MSE; MARE; MAE; R; ANOVA t-test

Paresh Chandra Deka et al	ANN; WANN	Multiple lead times, up to 4 days (flow prediction)	Model-based: input combinations consist of lagged data up to past 4 days	DWT	Haar Db2 Coif1	Trial and error (from 1 up to 6 levels), best RMSE of the testing set	Statistical behavior was analyzed for each data set (mean, std, Cxx, min and max) 2/3 for calibration, 1/3 for validation	Early stopping	RMSE; CE; MRE
Partal, 2008	ANN (3 types) WANN (3 types)	1 month ahead (streamflow prediction)	Model-based: Many combinations of input data were evaluated separately with lag time up to 5 months	DWT- Mallat algorithm (trapezoidal rule)	not mentioned	8 levels (no explanation) Ineffective sub-series were determined by autocorrelation and later on removed from the input neuron	Case A) 80% calibration, 20% validation Case B) 80% calibration, 20% validation		MSE; R; RMSE

(continued)

Table 9.1 *(continued)*

Authors	Models	Predictand variable	Data pre-processing: IVS	Decomp algorithm	Mother wavelet	Number of decom-positions	Data sets splitting procedure	Stopping criteria	Error
Adamowski, 2008	ANN; WANN MARS;	1 day ahead stream flow prediction	Model-based	A-trous WT	not mentioned	4 levels (until the lower resolution)	10-fold cross-validation	No information	R^2; RMSE; SEE
Venkata et al.	ANN; WANN AR	1 month ahead rainfall prediction	Model based	DWT	Db; Morlet	3 levels (exhaustive search)	The first 60% calibration and 40% data for validation	Cross-validation was performed	RMSE; R; COE
Cannas et al., 2005	ANN; WANN1 WANN2	1 month ahead runoff prediction	Heuristic procedure	CWT, DWT	DB4	1 level (exhaustive search)	The first 40 yr for cal, the second 9 yr for val and the last 20 yr for testing	Cross-validation was used as stop criterion	NSE; RMSE; MAE; M4E

been adopted from the work of Addison (2010), who provides an excellent description of the Wavelet Transform.

9.2.2. The Continuous Wavelet Transform (CWT)

The wavelet transform of a continuous signal for the wavelet function is defined as

$$T(a,b) = w(a)\int_{-\infty}^{\infty} x(t)\,\psi^*\left(\frac{t-b}{a}\right)dt \text{ Addison (2010)}. \tag{9.1}$$

Usually w(a) (weighting function) is normalized to $\frac{1}{\sqrt{a}}$, is in the sense of the wavelet energy, and the convolution is between the signal x(t) and the wavelet function ψ. In the wavelet, the dilation and contraction are governed by the dilation parameter a, and the movement is governed by the translation parameter b. The shifted and dilated version of the mother wavelet is denoted $\psi\left[\frac{t-b}{a}\right]$. The normalized wavelet function is often written more compactly as $\psi_{a,b}(t) = \frac{1}{\sqrt{a}}\psi\left(\frac{t-b}{a}\right)$; so the wavelet transform can be written as $T(a,b) = \int_{-\infty}^{\infty} x(t)\,\psi^*_{a,b}(t)\,dt$; $T(a,b) = \langle x, \psi_{a,b}\rangle$. The product of the wavelet and the signal integrated over the window length of the wavelike function is called convolution.

9.2.3. The discrete wavelet transform (DWT)

The a and b parameter can be sampled with a logarithmic discretization of the a scale, then to move in discrete steps b by making b proportional to a, $\psi_{m,n}(t) = \frac{1}{\sqrt{a_0^m}}\psi\left(\frac{t-nb_0a_0^m}{a_0^m}\right)$. When the discrete wavelet parameters a_0 and b_0 are chosen to be 2 and 1, respectively, the arrangement is known as the dyadic grid, $\psi_{m,n}(t) = \frac{1}{\sqrt{a_0^m}}\psi\left(\frac{t-nb_0a_0^m}{a_0^m}\right) = \frac{1}{\sqrt{2^m}}\psi\left(\frac{t-n2^m}{2^m}\right)$. The discrete wavelet transform values, given on a scale location grid of index m, n, are known as detail coefficients and they are expressed as follows $T_{m,n} = \int_{-\infty}^{\infty}\frac{1}{\sqrt{2^m}}\psi\left(\frac{t-n2^m}{2^m}\right)x(t)\,dt$. The scaling function is associated with the smoothing of the signal and it has the same form as the wavelet, given by $\varphi_{m,n}(t) = \frac{1}{\sqrt{2^m}}\varphi\left(\frac{t-n2^m}{2^m}\right)$. When the scaling function is convolved with the signal, the approximation coefficients are produced; $S_{m,n} = \int_{-\infty}^{\infty}\frac{1}{\sqrt{2^m}}\varphi\left(\frac{t-n2^m}{2^m}\right)x(t)\,dt$. We can represent the original signal x(t) by doing the inverse process, through a combined series expansion using the approximation and details coefficients:

$$x(t) = \sum_{n=-\infty}^{\infty} S_{m_{0,n}} \varphi_{m_{0,n}}(t) + \sum_{m=-\infty}^{m_0} \sum_{n=-\infty}^{\infty} T_{m,n} \psi_{m,n}(t), \tag{9.2}$$

Addison (2010), where the signal detail is defined as $d_m(t) = \sum_{m=-\infty}^{\infty} T_{m,n}\psi_{m,n}(t)$. If we add the signal detail at a certain scale to the signal approximation, we get the signal approximation at an increased resolution; this is known as multiresolution representation.

However, is also possible to reconstruct the approximations and details themselves from their coefficients vectors. By making use of the multiresolution decomposition algorithm that allows generating all coefficients at any scale just by knowing the coefficients at previous scales.

$$S_{m+1,n} = \frac{1}{\sqrt{2}} \sum_k c_k S_{m,2n+k} = \frac{1}{\sqrt{2}} \sum_k c_{k-2n} S_{m,k} \text{ Addison (2010)}, \tag{9.3}$$

$$T_{m+1,n} = \frac{1}{\sqrt{2}} \sum_k b_k S_{m,2n+k} = \frac{1}{\sqrt{2}} \sum_k b_{k-2n} S_{m,k}. \tag{9.4}$$

For the derivation of these formulas, the reader should consult Addison (2010). Many of the concepts explained here were taken from the aforementioned work.

Now in the two-channel subband encoder algorithm, there is a decomposition step that consists of retaining one coefficient of two (downsampling) known as decimation (no information repeats). The aforementioned algorithm is suitable for compression purposes, however, lack of redundancy in the coefficients creates a system whose outputs depend explicitly on time (shift-variant system), an undesirable property for forecasting problems. Another problem arises when choosing algorithms, the well-known cone of the influence region. Errors in the limit regions (especially in the right boundary) propagate within the decomposition (approximation and details coefficients) proportionally to the chosen scale within an area of the cone. This represents a problem for forecasting purposes, where at each time step a new entry is placed on the right-hand side of the boundary. To visualize the problem, in Figure 9.1 a segment of the signal is analyzed. When the wavelet function interrogates the signal along the time axis, at locations and scales (dyadic system), and when reaching the end of the signal, the problem arises. The assumptions are necessary to close the analysis of the band-pass filter (for the high-pass filters and also for the low- pass ones). To overcome the problem, future points must be interpolated. If we use future values of the signal, we are violating the test phase of our model (future data are unknown at that

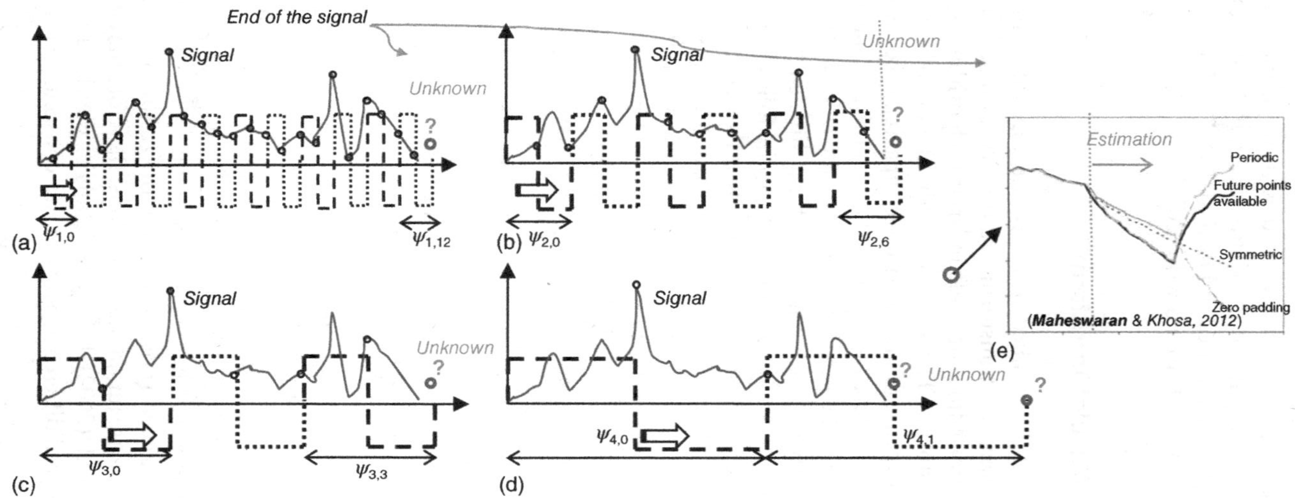

Figure 9.1 Boundary limits and traditional extension methods. (a) Wavelet locations for the first scale (dilation 2^1); (b) wavelet locations for the second scale (dilation 2^2); (c) wavelet locations for the third scale (dilations 2^3); (d) wavelet locations for the fourth scale (dilation 2^4); and (e) extension methods to extend the right or left-hand side of the signal.

time) and if interpolations are performed, the wavelet transform and approximation coefficients are contaminated with impurities brought in by the extension method applied at the right-side limit. Errors commonly found in hydrology can be consulted in Du et al. (2017), Maheswaran and Khosa (2012), and Quilty and Adamowski (2018).

9.2.4. The A-trous Wavelet Transform

The a-trous algorithm is called in many ways by different authors, namely, redundant wavelet transform, stationary wavelet transform, a-trous algorithm, translation or shift-invariant algorithm, or undecimated wavelet transform. Through the redundancy offered by the algorithm and at the expense of greater storage capacity, time invariance is accomplished. Another feature of the method that motivates the use of this algorithm is the fact that the wavelet coefficients are calculated from data only obtained previously in time, appropriate for a directed (time-varying) data stream. The reader is encouraged to refer to the following references to obtain greater details: Aussem et al. (1998), Kozakevicius and Schmidt (2013), Renaud et al. (2002), Renaud et al. (2003), and Shensa (1992).

Let's proceed and briefly show the a-trous algorithm using the redundant Haar. The first approximation resolution level, $\varphi(1)$, is obtained from $\varphi(0)$ (original signal) by using the following equation, where $c_k = \left(\frac{1}{2}, \frac{1}{2}\right)$ is the scaling coefficients that satisfy the following dilation equation:

$$\varphi(t) = \sum_k c_k \varphi(2t - k), \tag{9.5}$$

and, for the first resolution, we never use future values. This is valid for a directed (time-varying) data stream, see below:

$$S_{m+1,n} = \sum_k c_k S_{m,n-2^k}, \tag{9.6}$$

leads to $S_{1,}1 = c_1 S_{0,1-2^0} + c_1 S_{0,1-2^1} = c_1 S_{0,}0 + c_2 S_{0,-1} = \frac{1}{2}\left(S_{0,}0 + S_{0,-1}\right)$. Observe the difference with the decimated form $S_{m+1,n} = \frac{1}{\sqrt{2}} \sum_k c_k S_{m,2n+k}$ (DWT algorithm), where the filter was chosen to be c = (1, 1), in this case, future values are considered (not allowed for the testing phase), and at each decomposition, we are jumping 2n locations, therefore the decimation takes place. Renaud et al. (2002) define the details or wavelet transform, in this case just as the signal difference between two consecutive resolutions,

$$T_{m,n} = S_{m-1,n} - S_{m,n}. \tag{9.7}$$

Alternatively, the wavelet function can be calculated with this relation:

$$\frac{1}{2}\psi\left(\frac{t}{2}\right) = \varphi(t) - \frac{1}{2}\varphi\left(\frac{t}{2}\right). \tag{9.8}$$

After the convolution is done $T_{m,n} = \left\langle x(t), \psi\left(\frac{t-b}{a}\right)\right\rangle$, the same results are expected. Due to the link between decompositions at a different level, the multiresolution algorithm can be constructed $S_{0,n} = S_{m,n} + \sum_{m=1}^{M} T_{m,n}$.

9.2.5. Artificial Neural Networks

The artificial neural network is a well-established method. It had its beginnings with the threshold logic model (McCulloch & Pitts, 1943) and the model paved the way for the now known ANN. It was developed in the field of machine learning, quickly accepted, and established in applications where it is not possible to have a physically based model. ANNs are generally presented as systems of interconnected "neurons" that learn from experience as humans do. Zijderveld (2003) presents comprehensive documentation that readers not so familiar with ANN can review to delve into details. The mentioned author explains neural networks as a directed graph consisting of nodes with synaptic links and interconnection activation characterized by four properties:

1. Each neuron is presented by a set of linear synaptic links and externally applied bias.
2. The synaptic links of a neuron weight their respective input signals.
3. The weighted sum of the input signals defines the induced local field of the neuron in question.
4. The activation link squashes the induced local field of a neuron to produce an output.

The components of neural networks are weighting factors, summation, transfer, scaling, output, error, and learning functions. Once the network is ready and prepared for a particular task, the following step is the learning process. In this process, there are two approaches (supervised and unsupervised). In the first (of our interest) both the input series and the output series are presented to the model and the learning process is carried out by comparison. Optimization is carried out by taking the objective function as the difference between the modeled and the observed value. In this process, the back-propagation algorithm is in charge of adjusting the weights between nodes to find the best approximation for the objective function.

9.2.6. Other Data-Driven Models: Least Median Square, Linear Regressions, Pace Regression, and Model Tree

The same idea will be tested with other data-driven models, some of which offer advantages over an ANN modeling tool. One drawback of ANNs is that no additional information can be extracted due to the (black box) nature of the model. Also, if a simpler model can do the job, there is no need for a complex one. The present study does not seek to give an exhaustive analysis of wavelets with other DDMS, but to show that the same procedure could also be valid with other modeling techniques. Models used in this work are (1) LeastMedSq: Robust regression using the median rather than the mean; (2) LinearRegression: Standard linear regression; (3) PaceRegression: Linear regression models using Pace regression; and (4) M5P: Model tree. Readers can turn to Witten and Frank (2006) for details on the methods cited above; Weka software was used to develop the models in this section (http://www.cs.waikato.ac.nz/ml/weka/).

9.3. Methodology

9.3.1. Sequence of Experiments

The numbers of experiments were many and they are captured in the following sequence:

Input variable selection (IVS)

Model-based: Forward selection

model1 = $P_{(t)}$, $Q_{(t)}$

model2 = $P_{(t)}$, $P_{(t-1)}$, $Q_{(t)}$

model3 = $P_{(t)}$, $P_{(t-1)}$, $P_{(t-2)}$, $Q_{(t)}$

model4 = $P_{(t)}$, $P_{(t-1)}$, $P_{(t-2)}$, $P_{(t-3)}$, $Q_{(t)}$

Model-free: Correlation and average mutual information

model5 = model2 = $P_{(t)}$, $P_{(t-1)}$, $Q_{(t)}$

Data splitting of best IVS (model2)

K-fold cross-validation; k =10

Cross-validation (first 85% of the data), the last 15% to test the model

Develop both models: ANN & WANN

For the ANN approach (model1a(1-20)-model4a(1-20)), in total 80 (20x4) models

Original time series is being used

An exhaustive search for the number of hidden nodes [1,20]

For the WANN approach (model1b(1-220)-model4b(1-220)) in total 880 (220x4) models

Perform a-trous wavelet transform to the original time series

An exhaustive search for the number of hidden nodes [1,20] and number of decompositions [1,11]

Evaluate models performance (model1a-model4a and model1b-model4b)

Error metrics RMSE, Cor, NSE, PERS, and others

Select the best fold for the ANN and the WANN model among the model1, model2, model3, and model4 models

Once the best model is selected, de-noising by threshold is applied to the best WANN model, via optimization

Individual threshold values were found for all the detail coefficients via optimization, the objective function was set to the minimum RMSE

9.3.2. Input Variable Selection (IVS)

Forward selection, correlation, and average mutual information were implemented to select the most appropriate variables. The first consists of a linear increment search strategy in which variables are incremented one at a time until performance does not increase. The method is considered slow because the execution of a model is required. The model-free approach takes some dependency measures (linear and nonlinear between input and target) to find the right variables, attractive in terms of computational efforts because no model needs to be run. Model 2 (equivalent to model 5 suggested by correlation and AMI) described in section 9.3.1 was chosen, since it shows the best performance values. For the pragmatic purposes of the present study, the authors do not elaborate on the detailed development of this process, since the particular conditions of the case analyzed (rapid reaction basin and daily data) generate a reduced space for the selection of variables, and this prevents an objective assessment of the strengths or weaknesses of the IVS methods analyzed. For more details on the methods available for the selection of input variables, it is suggested to refer to May et al. (2011).

9.3.3. K-Fold Cross-Validation

The basic idea behind k-fold cross-validation is to reduce the bias associated with the chosen training set and the remaining samples. To do this, k subsets of equal size are selected (randomly or continuously). In each model, the partition consists of k-1 subsets for training and a single subset for validation. The validation set is rotated k times so that the k subsets are used exactly once; in consequence, k different models (folds) are built. For this study, the first 85% of the data was used for the 10-fold cross-validation, and the rest for testing. For each model, an early stopping criterion was used to avoid overfitting.

9.3.4. Wavelet Decompositions as an Input Preprocessor of an Artificial Neural Network

The a-trous wavelet transform algorithm was applied to the input data set; the resulting time series replaces the original Figure 9.2, and Figure 9.3 shows the traditional ANN and wavelet-ANN schematization. In the last, the simple procedure of decomposing signals already brings benefits to the model performance.

The ANN model assigns bigger weights to subseries with more relevant information for the modeling purpose. This is done by assigning different weights to each subcomponent during the training process.

9.3.5. Levels of Decompositions and ANN Structure as Decision Variables in an Optimization Routine

Since decomposition analysis can be performed using the multiresolution algorithm in an iterative process, it can be performed at any desired level of decomposition. However, for practical reasons, you can continue only until one detail value covers the entire sample. Furthermore, the modeler can select an appropriate number of levels according to the nature of the signal or a criterion such as entropy. If little knowledge is available to perform the above actions, this difficulty can be overcome by optimizing the number of decomposition levels and hidden nodes with an intensive search (see section 9.5.2). This is feasible due to the limited space of decision variables (levels of decomposition x hidden nodes).

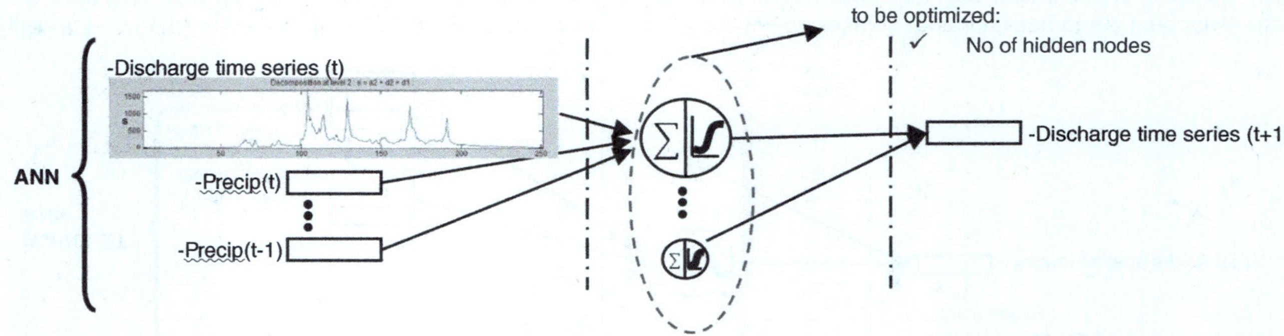

Figure 9.2 Traditional ANN approach.

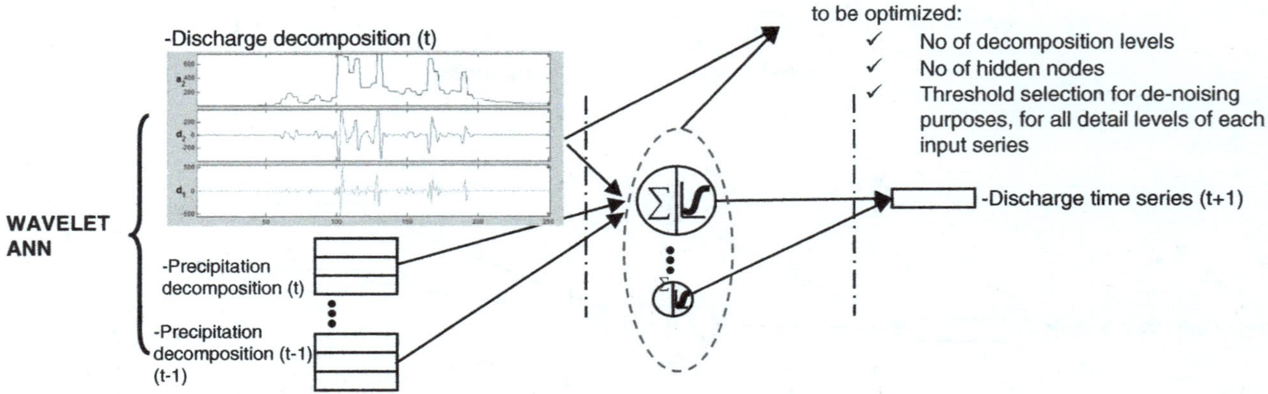

Figure 9.3 WANN approach. In order to create this model, we use the multiresolution representation of the time series signal; each input data set is decomposed in its smooth (average) version known as approximation (first series); and in its details, in this schematization, a second level decomposition with Haar wavelet as a mother wavelet (last two series) is shown.

9.3.6. Wavelet Decompositions as an Input Preprocessor of Other Data-Driven Models

Here the same procedure as explained above is followed, but now with different DDMs. To proceed with the simulations, it was considered convenient to make use of the data mining software called Weka, developed by the University of Waikato (http://www.cs.waikato.ac.nz/ml/weka/). With this software, a collection of machine learning techniques is available, which can be executed in Matlab, thanks to developments by Corzo Perez.

9.3.7. The Noise Filtering Process in Wavelets Analysis

An alternative option is to de-noise the signal. Once the decomposition of the signal is performed, it is possible to alter these resultant coefficients according to our needs. Several methods are recommended in literature such as Universal Threshold, Visu Shrink, Sure Shrink, Bayes Shrink, Normal Shrink, and so forth. For details, the reader can refer to Ruikar and Doye (2011) and Donoho and Johnstone (1994). In the present study, the threshold is found via optimization, and soft thresholding is applied for each decomposition level and again for each input series in the ANN model, so a number of *No dec level* × *ANN Inputs* threshold coefficients were calculated.

Then the calculated thresholds can be treated as follows in the detail representation. In soft thresholding, all coefficients below a certain threshold are set to zero, and those whose magnitude are greater than the threshold are shrunk by that amount. This method recognizes that the coefficients contain both signal and noise, and attempts to isolate them.

$$W_i^{soft} = \begin{cases} 0 & |W_i| < \lambda \\ sign(W_i)(|W_i| - \lambda) & |W_i| > \lambda \end{cases} \tag{9.9}$$

In comparison, hard thresholding retains only the values above the threshold and is considered to correspond to the coherent part of the signal. The rest corresponds to the random or noisy part of the signal (Addison, 2010). Graphic illustrations of the filtering effect are shown in Figure 9.4.

$$W_i^{hard} = \begin{cases} 0 & |W_i| < \lambda \\ W_i & |W_i| > \lambda \end{cases}. \tag{9.10}$$

Some authors (Okkan, 2012; Okkan & Serbes, 2013) affirmed that some subseries are ineffective and must be removed to improve the generalization capability of the model. On the contrary, Adamowski and Sun (2010) have used the complete subseries as inputs to the ANN model,

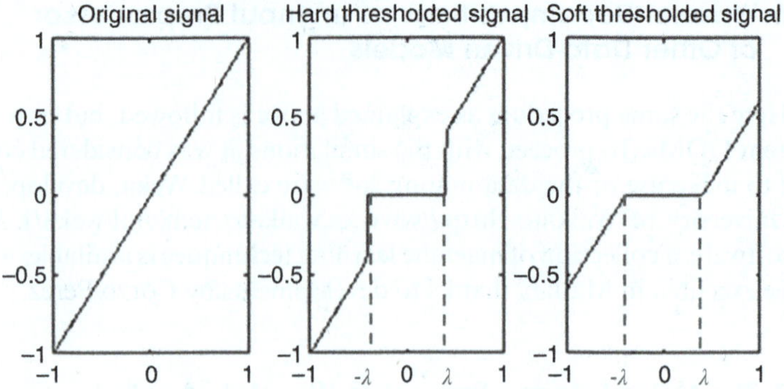

Figure 9.4 Soft and hard threshold for de-noising the signal source (http://nl.mathworks.com/help/wavelit/ref/wthresh.html).

stating that averaging or optimizing the selection of only certain subseries could generate undesired results. The latter stated that all subseries are equally important and contain some information about the original signal. For the present work, and according to the way ANN operates (by assigning different weights to each input series during the training process), it can be said that ANN by itself is selecting which subseries are more significant than others. In the case of doing it with the detail coefficients, to some extent, a noise elimination process is being carried out. For the present study, the statements made previously by the different authors are thought of as partial truths. Every subseries contains some important information about the signal and relevant to the model (no subseries were discarded). However, many factors can introduce errors in our measurements (errors in the instruments, human error, etc.). Even after quality checks, there are still some errors in the data, so this noise can also be eliminated.

To illustrate how threshold de-noising works, it will be briefly introduced below. Let us start from the basis that the noise is presented on fine scales (detailed subseries), contrary to the trends or the main body (coherent part) of a time series, which can be seen in the approximation subseries. The filtering is only done on the detailed subseries. The threshold limits were intentionally chosen as large, to illustrate how the signal transforms (see Fig. 9.5). In practice, and for the present study these values are much smaller in magnitude, therefore, when reconstructing the signal (inverse wavelet transform), the resulting time series would not be so different from the original (let's remember that this is obtained via optimization). Following the key idea that it will not remove important information during the process.

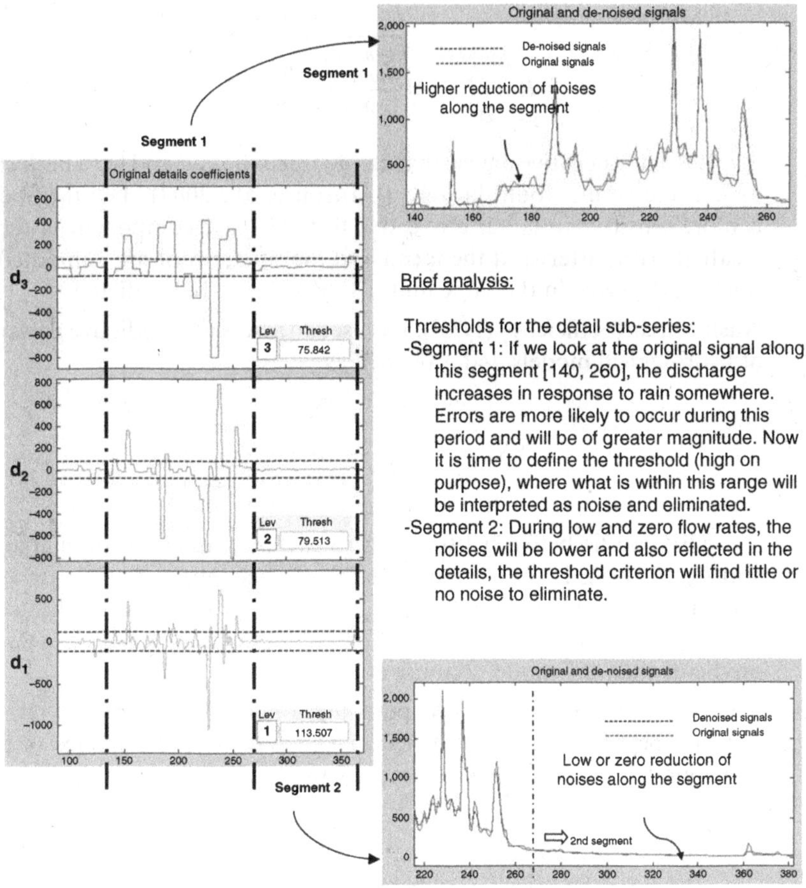

Figure 9.5 De-noising process in the details of the signal.

9.3.8. Error Metrics

There is no one best error metric for evaluating model performance. Some are better suited for certain circumstances than others. In our case, it is considered important to capture peaks and high flows, so RMSE was used to penalize a large variance. Nevertheless, other error metrics are also calculated and evaluated and are listed below.

1. Root mean square error (RMSE): RMSE is the standard deviation of the residual errors (the difference between prediction and observation)

in all samples

$$RMSE = \sqrt{\frac{\sum_{i=1}^{n}\left(O_i - P_i\right)^2}{n}}.$$

It is a nonnegative metric that has no upper bound and for a perfect model, the result would be zero (Dawson et al., 2007). The method is more sensitive to larger errors due to the square component in the residual error, afterward the mean and the square root are computed to bring the error in the same units as the value to be predicted.

2. Nash-Sutcliffe efficiency (NSE): It is used to assess the predictive power of hydrological models and is defined as

$$NSE = 1 - \frac{\sum_{i=1}^{n}\left(O_i - P_i\right)^2}{\sum_{i=1}^{n}\left(O_i - \bar{O}\right)^2}.$$

NSE can lie between 1.0 (a perfect match of modeled discharge) and $-\infty$. If NSE is zero, it indicates that the model predictions are as accurate as the mean observed data. When lower than zero, it indicates that the mean value of the observed time series would have been a better predictor than the model (Krause et al., 2005).

3. Persistence Index (PERS): The persistence index focuses on the relationship between the model performance and the performance of a naïve model, a model that assumes as the forecasted value the same current value.

$$PERS = 1 - \frac{SSE}{SSE_{naive}};\text{ where } SSE_{naive} = \sum_{i=1}^{n}\left(O_{t+L} - O_t\right)^2.$$

Other error metrics were also computed, such as mean absolute error (MAE), normal root mean square error (NRMSE), correlation coefficients (Cor), StdT, StdP, MuT, MuP, RMSEN, NRMSEN, MARE.

9.4. Case Study

In the present study, the Bagmati catchment is being analyzed. A main river with a drainage area of about 3,640 km^2, located in central Nepal. Springs and monsoon rainfall contribute to the streamflow. The catchment originates in the north of Kathmandu Valley and flows downstream to the Gangetic plain (see Table 9.2).

Table 9.2 Description of the case study analyzed.

Basin name	Bagmati	
Location	Nepal	
Topography	Mixed	
River length (km)	170	
Area (km^3)	3,500	
Data set	3-Jan-1988 until 30-Dec-1995	
Cross validation	3-Jan-1988 until 19-Oct-1994	85%
Test	19-Oct-1994 until 30-Dec-1995	15%
Time step	Daily	
Output variable	(t+1) 1 day ahead	
Input variables	Pt, Pt-1, Pt-2, Qt, Qt-1	

9.5. Results and Discussions

9.5.1. ANN Stand-Alone Model

For the ANN stand-alone approach, a total of 200 models was developed. These are the results of the exhaustive search in the variable space (number of hidden nodes (HN) and k folds). Figure 9.6 shows a pseudocode of the steps done to find the best model and the sequence of results is shown in Figures 9.7 and 9.8.

To recap, we divided the subsets into three: calibration, validation (cross-validation), and test set. The latter is done to ensure independence in the test partition, therefore, avoiding the selection bias problem.

```
cv = cvpartition (Data_ANN,'kfold',k); %Divide data set in k folds k=1,..,10. Keep test set
separated, (cvpartition is a built in function in MATLAB that divides the data in k folds)
  for HN=1:20 %number of hidden nodes tried
        for k=1:10 %10 folds
          Run ANN → f (HN, k)%use all subset but k for the training set and
                             k for the validation set
          Store RMSE{HN, k} for train, val & test sets
          bar (1:10, [RMSE_train{HN, k} RMSE_val{HN, k} RMSE_test{HN, k}]
        % the figure is only showing HN=20
        end
      ANN_BestFoldperHN{HN}=min((RMSE_train {HN,:} + RMSE_val{HN,:})/2)
                %best fold per hidden node
         plot (HN, ANN_BestFoldperHN);
end
ANN_Bestfold=min(BestFoldperHN{HN}) %best fold among all hidden nodes
```

Figure 9.6 Pseudocode for the ANN exhaustive.

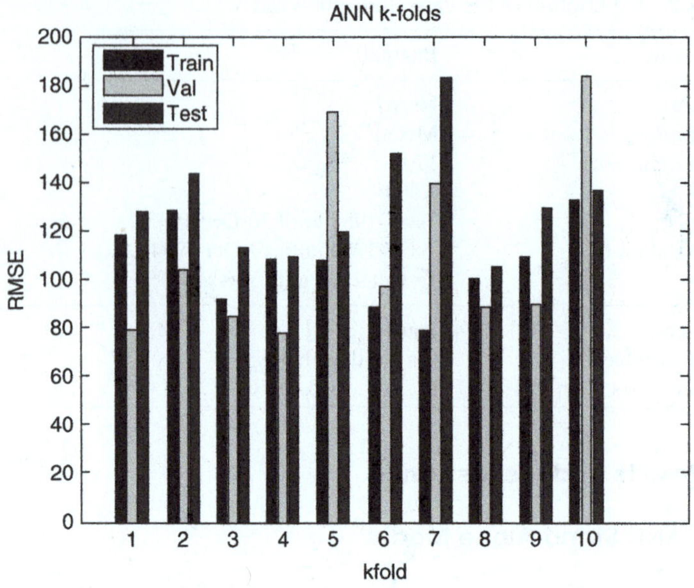

Figure 9.7 All 10-fold's RMSE results for HN = 20.

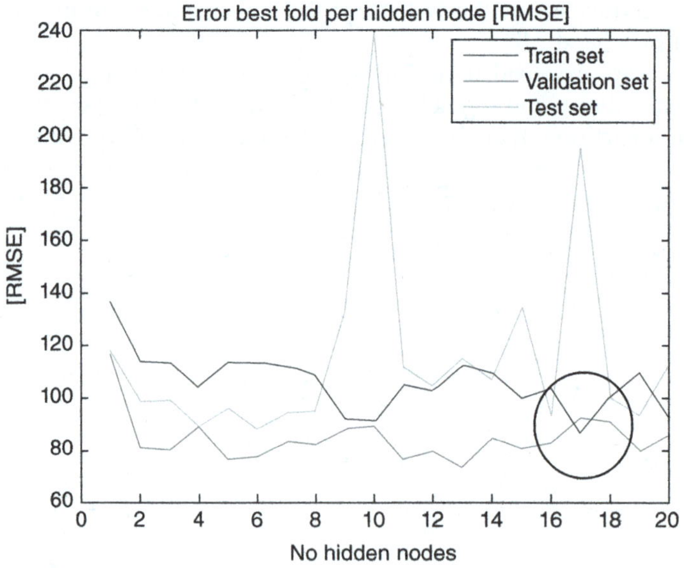

Figure 9.8 Best fold's RMSE results per HN.

For cross-validation between all k-folds, we used the early stop method to avoid overfitting. A common practice is to average the performance across the entire validation set for all the folds used to select the best structure from the ANN model (in this case, HN). This is done to emulate randomness and therefore represent statistical independence. The average here, by default, would decrease the possibility of having poor results in future tests, due to the average effect. Only it would also reduce the chances of getting the best results. However, in the example in this study, the test set was never submitted to the process, nor was it considered to guarantee the best results. The best fold was selected and, subsequently, we evaluated the parameters obtained in the test partition. To keep the same conditions for the following experiments (WANN and WANN without noise), only the third fold (the best fold in the ANN structure) was used. By doing this, the expected improvement would not be a reflection of other factors, such as better matching of subsets. Therefore, the improvement could be attributed to the chosen modeling method.

Figure 9.7 shows how the behavior differs according to the particular data selection or fold during the training process, highlighting the importance of the k-fold cross-validation procedure. The same chart was built for the remaining 19 HNs, however only the best result is shown here (for HN = 20). Although Figure 9.8 highlights that nodes 9–11, 15, and 17 provide poor results for the test set, the modeler is not allowed to use this set to choose the best model. The best model is chosen via the best cross-validation set. Table 9.3 shows a summary of statistics and error measures belonging to the best ANN model (third fold of HN = 20), in section 9.5.3. The training set gave an RMSE value of 91.81 m^3/s considered as good for the case study; the validation set gave an even better result of 84.79 m^3/s. In the test set, a mildly acceptable result was obtained (112.44 m^3/s). A high correlation was found between the observed and simulated discharge for all data sets, being in all cases greater than 0.9.

9.5.2. Wavelet ANN Model

Results in Figures 9.10, 9.11, and 9.12 depict the individual best folds per decomposition level and number of hidden nodes. Figure 9.9 shows the pseudocode used for the WANN structure. No obvious pattern is observed in the variable space for the model performance, however after spotting the best combination through cross-validation, a clear improvement is observed in performance. In doing so, the training data set RMSE was reduced in 16.1%, from 91.8114 m^3/s (ANN) to 77.0193 m^3/s (WANN). The validation set showed a slight drop in performance, 1.6%

```
cv = cvpartition (Data_ANN,'kfold', k);%Divide data set in k folds k=1,10. Keep test
set separated, (cvpartition is a built-in function in MATLAB that divides the data in k
folds)
for HN=1:20 %number of hidden nodes tried
    for DL=1:11 %number of decomposition levels tried
        for K=1:10 %10 folds – these was only done for visualization purposes
                Run ANN → f (HN, DL, k) %use all subsets but k for the training set and
                                                 k for the validation set
              Store RMSE{HN, DL}(k) for train, val & test sets

        end
              ANN_BestFoldperHNperDL{HN, DL}=Same fold as Figure 8
    Identify the indexes for ANN_BestFoldperHNperDL{HN, DL} % so as to keep track
                                                                      of the best fold
      end
  end
Use the identified indexes to generate the RMSE contour for the train, val and test
ANN_Bestfold=min(RMSE{HN,k})       %best fold among all hidden nodes and
                                          decomposition levels
```

Figure 9.9 Pseudocode for the WANN exhaustive.

Table 9.3 Best fold among all HNs: ANN stand-alone model; best fold among all HNs and DL: WANN model; best fold optimized: de-noised WANN model.

	Train	Validation	Test
RMSE	91.81	84.79	112.45
NSE	0.90	0.83	0.77
Cor	0.95	0.92	0.91
NRMSE	32.38	40.75	48.12
MAE	39.01	41.85	54.64
StdT	283.58	208.09	233.68
StdP	266.88	202.20	251.02
MuT	157.21	137.03	121.12
MuP	157.33	137.46	156.09
PERS	0.78	0.79	0.72
RMSEN	196.49	184.42	211.34
NRMSEN	69.28	88.50	90.35
MARE	0.31	0.42	1.04
K-fold=3			

	Train	Validation	Test
RMSE	77.02	86.50	90.89
NSE	0.93	0.83	0.85
Cor	0.96	0.91	0.93
NRMSE	27.16	41.57	38.89
MAE	36.96	43.36	41.51
StdT	283.58	208.09	233.68
StdP	267.28	194.05	245.26
MuT	157.21	137.03	121.12
MuP	157.98	130.57	135.45
PERS	0.85	0.78	0.81
RMSEN	196.49	184.42	211.34
NRMSEN	69.28	88.50	90.35
MARE	0.34	0.43	0.36
K-fold=3			

	Train	Validation	Test
RMSE	76.99	81.16	82.12
NSE	0.93	0.85	0.88
Cor	0.96	0.92	0.94
NRMSE	27.15	39.00	35.14
MAE	40.92	43.46	43.08

(*Continued*)

Table 9.3 (*continued*)

	Train	Validation	Test
StdT	283.58	208.09	233.68
StdP	267.78	196.31	239.86
MuT	157.21	137.03	121.12
MuP	157.24	133.46	132.83
PERS	0.85	0.81	0.85
RMSEN	196.49	184.42	211.34
NRMSEN	69.28	88.50	90.35
MARE	0.53	0.59	0.59
K-fold=3			

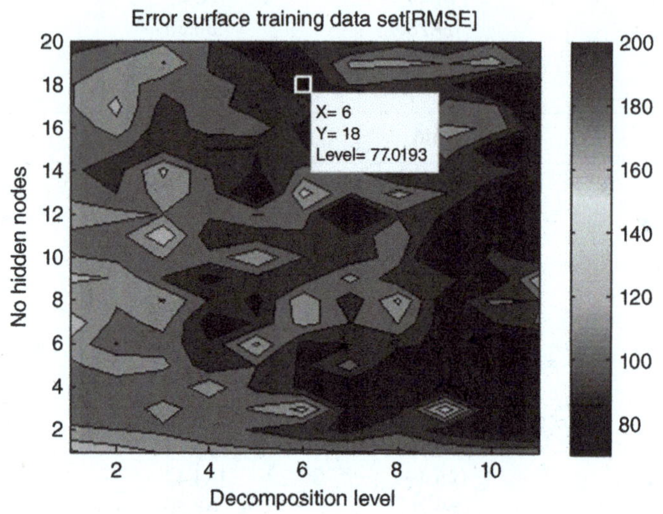

Figure 9.10 Exhaustive search RMSE-train.

from 84.7939 m^3/s to 86.4985 m^3/s, now the test set improved in 21.64% from 112.44 m^3/s (ANN) to 90.88 m^3/s (WANN). In terms of correlation, values of 0.96, 0.91, 0.93 were obtained for the training, validation, and testing sets, respectively. These results are in agreement with the research of Adamowski and Sun (2010), where it was stated that use of the full subseries without omitting or averaging subseries as inputs should improve the ANN model performance.

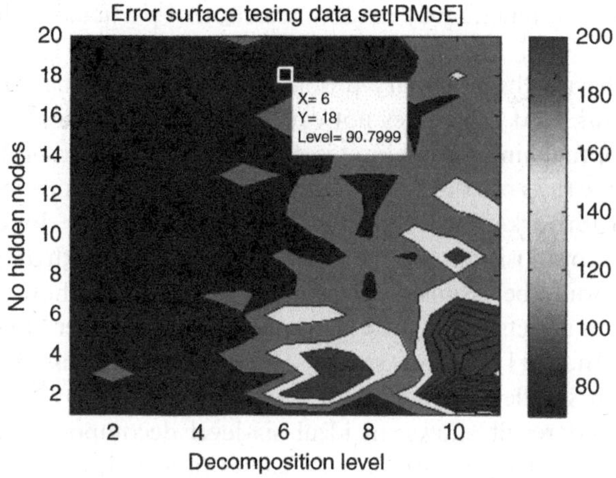

Figure 9.11 Exhaustive search RMSE-test.

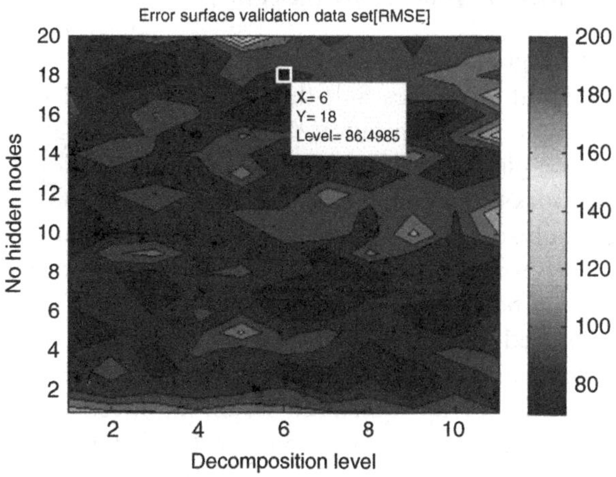

Figure 9.12 Exhaustive search RMSE-val.

Obs: To make a fair comparison, in terms of potential improvement in the models, it has been decided to keep the calibration, validation, and test subsets fixed. In other words, the best fold found previously was used to later test it using the Wavelet-ANN combination and also for the model with noise filtering. By doing this, the expected improvement would not be a reflection of other factors, such as better matching of subsets or better matching by random selection of particular values within subsets.

To deal with the boundary problems on the right, the a-trous wave algorithm was used. This does not cause distortions in this limit; on the contrary, approaching the left limit is different. There are several extension methods. Like the ones shown in Figure 9.1e, we use the symw method (see https://de.mathworks.com/help/wavelet/ref/wextend.html). In effect, the larger the decomposition, the 2 ^ x would be the bandwidth and therefore more data should be assumed. In Figure 9.11, we can see how the boundary distortion affects the highest decompositions; however, it does not in Figures 9.9 and 9.11 (deserves study). This could be because the subset of tests is much smaller (only 15% for tests and 85% for cross-validation). The optimized result shows an ideal six-level decomposition, which is reflected in 2 ^ 6 (bandwith), only 64 values (affected) out of more than three thousand available records.

9.5.3. De-Noised Wavelet ANN Model

The same model showed just before is improved even further by de-noising the input time series. Soft-threshold was applied and the threshold value was found via optimization. Godlike (a generalization of various population-based global optimization scheme) was used for this purpose (http://nl.mathworks.com/matlabcentral/fileexchange/24838-godlike-a-robust-single-multi-objective-optimizer). The objective function was set as RMSE in the cross-validation process; however, any other statistical o error measure can be set as an objective function as well. The training set minimally improved in 0.03% (77.01 m^3/s to 76.98 m^3/s) compared with the WANN model and 16.1% (91.81 m^3/s to 81.1604 m^3/s) with the ANN stand-alone model. For the validation set an improvement of 6.2% and 4.3% were accomplished respectively to the WANN and ANN models. As in the case of the test set, performance increased by 9.6% compared with the WANN model and increased by about 27% to the ANN stand-alone model. Results are considered highly accurate for the case study, however, if more effort is invested in keeping tuning the parameters even major accuracies can be accomplished. In this exercise, the number of hidden nodes and decomposition levels was set as fixed, based on the best result of the WANN model and the only variable to optimize was the threshold value. Besides, the optimizer was run with a small population and a modest number of iterations, therefore if bigger computational resources are at disposal is thought that excellent results could be accomplished.

In Figure 9.13 visual inspection of time series shows consistent best estimates of flow in this order: 1st- de-noised wavelet ANN (light), 2nd- wavelet ANN (grey) and 3rd ANN stand-alone (dark). In the residual error graph, it is clear that the WANN model can capture peak flows more accurately than the ANN model, however the de-noised WANN is even better. Even for the other flow regimes (medium and low flows), the WANN model outperforms the ANN model, where the grey spikes in the residual are almost always smaller than dark in absolute magnitude. In the case of the WANN and de-noised WANN, these other regimes present similar behavior.

9.5.4. Other Data-Driven Models: Least Median Square, Model Tree (M5P), Pace Regression, and Linear Regressions

Four different DDM were tested, these being (1) least median square, (2) model tree (M5P), (3) pace regression, and (4) linear regressions. Results can be visualized in Figures 9.14 and 9.15. The charts are divided into the training, validation, and testing sets for all decomposition levels. It can be seen that for different DDMs, data preprocessing with wavelet analysis is always improving the model skills, however, only for certain decomposition levels. This is an indicator that wavelet subseries can play an important role in the modeling process, reflected by the improved results that must be cautiously found in the decomposition space at the expense of greater computational resources. Among models, the coupled wavelet-M5P (DDM 2) seems to consistently provide the best results and is quite comparable with the performance of the coupled WANN model done in the previous section. In second are the methods (3) and (4) with wavelet as preprocessor, presenting similar performances in all data sets. Finally, method (1) had the lowest performance within the studied models, clearly showing that the problem to solve cannot be assumed with linear behavior and is negligibly improved by the wavelet subseries.

Figure 9.16 depicts the simulated discharges during the testing phase for all the proposed DDMs. As mentioned before, the wavelet model tree (WM5P) outperformed others in all regimes (low, medium, and high flow). Nevertheless, the method manages to overcome (or close) to the 100 m^3/s RMSE barrier (considered as good) but is still not better than the WANN model (Table 9.4).

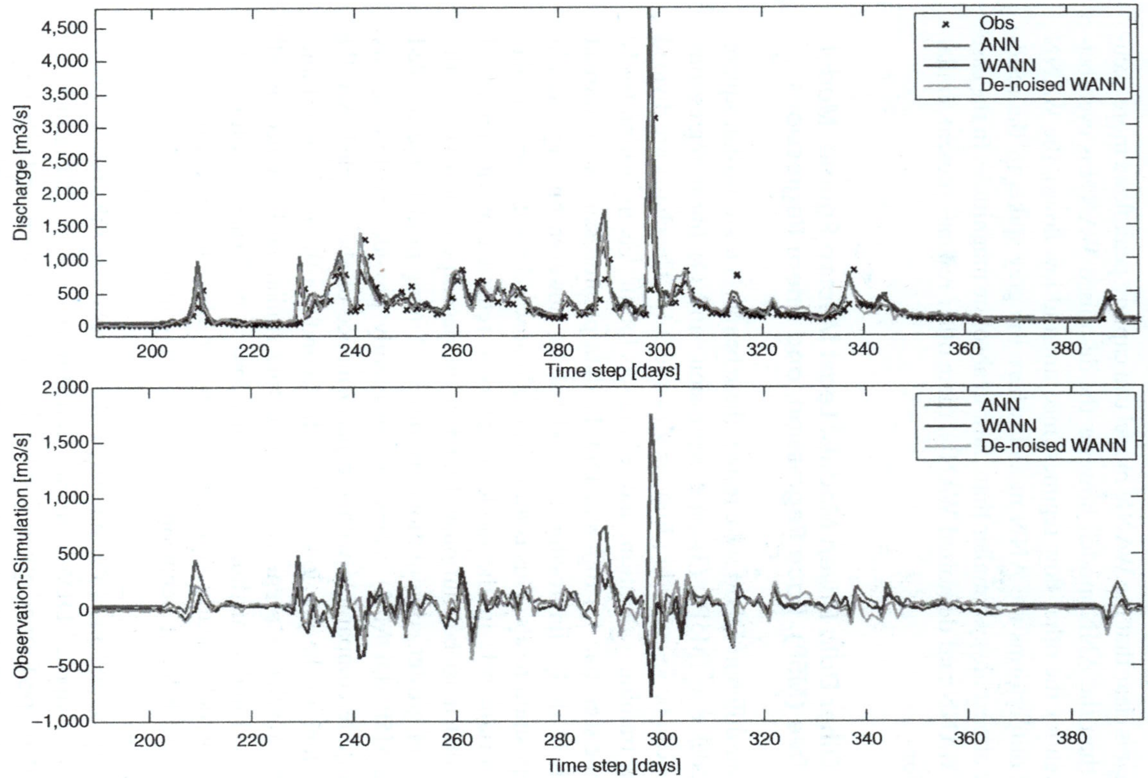

Figure 9.13 Testing data set: Simulated (upper figure), residual errors (lower figure) for the ANN, WANN, and de-noised WANN.

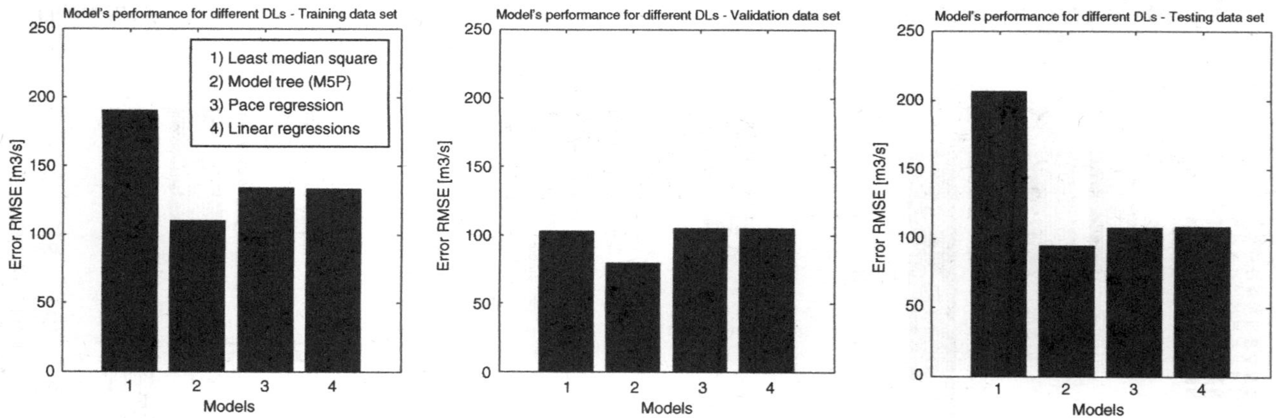

Figure 9.14 Data-driven model with raw time series: Training data set, validation data set, testing data set.

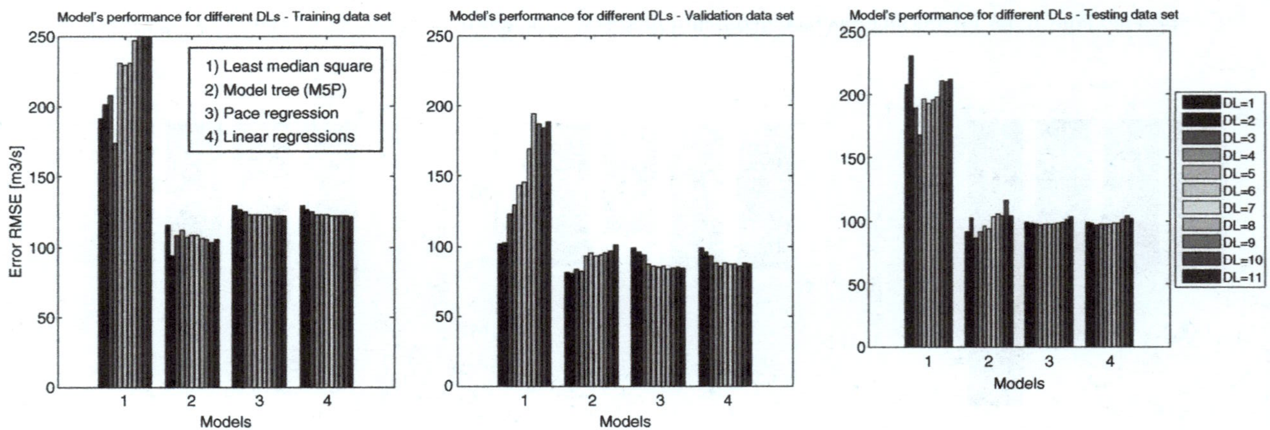

Figure 9.15 Data-driven models with wavelets as preprocessor: Training data set, validation data set, testing data set.

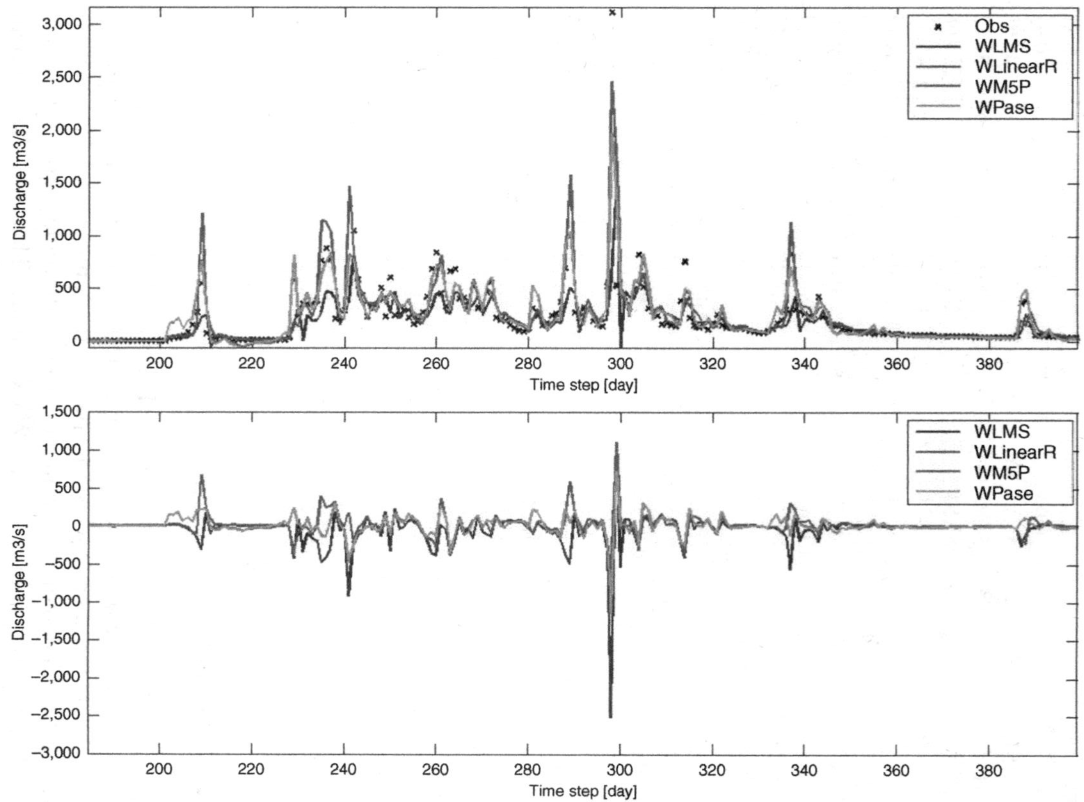

Figure 9.16 Testing data set: Simulated discharge (upper) residual errors(lower) for the least median square, model treee (M5P), pace regression and linear regressions all with wavelets as preprocessor.

Table 9.4 WM5P model results; WANN model results.

	Train	Validation	Test
RMSE	93.97	80.60	102.75
NSE	0.89	0.85	0.81
Cor	0.94	0.93	0.91
NRMSE	33.14	38.74	43.97
MAE	39.97	38.46	37.13
StdT	283.58	208.09	233.68
StdP	259.31	171.49	245.43
MuT	157.21	137.03	121.12
MuP	156.72	125.92	132.37
PERS	0.77	0.81	0.76
RMSEN	196.49	184.42	211.34
NRMSEN	69.28	88.47	90.35
MARE	0.28	0.32	0.25
	Train	Validation	Test
RMSE	77.02	86.50	90.89
NSE	0.93	0.83	0.85
Cor	0.96	0.91	0.93
NRMSE	27.16	41.57	38.89
MAE	36.96	41.36	41.51
StdT	283.58	208.09	233.68
StdP	267.28	194.05	245.26
MuT	157.21	137.03	121.12
MuP	157.98	130.57	135.45
PERS	0.85	0.78	0.81
RMSEN	196.49	184.42	211.34
NRMSEN	69.28	88.50	90.35
MARE	0.34	0.43	0.36

9.6. Conclusions

Wavelets and ANN coupled together proved to be of great value and easy to apply, the best of all the models tested. So far, many researchers have come to the same conclusion; however, the contribution of this work is the insertion of de-noising techniques using optimization tools. The results ended up being superior to the best WANN model. This was to be expected, since filtering the noise in the time series should reduce the spread of these errors to the result. Related jobs in the future may test these procedures under different catchment conditions. Other suggestions to explore could be the following:

1. Some improvements can be made due to the features that wavelet analysis offers. For instance, de-noising can be done to both input and output. Consequently, the complexity of the problem to be modeled must be reduced.
2. Due to time constraints, only one noise filtering technique was tested. Other authors are encouraged to explore more advanced techniques and combine them with optimization processes. Some methods to test can be Universal Threshold, Visu Shrink, Sure Shrink, Bayes Shrink, Normal Shrink, and so forth. For details, the reader can refer to Ruikar and Doye (2011) and Donoho and Johnstone (1994).
3. In this study, only the Redundant Haar Wavelet was used. However, some authors (Adamowski & Sun, 2010; Maheswaran & Khosa, 2012) state that this algorithm outperforms others only when the signal has sharp changes distributed in small lengths; but time constraints did not allow testing these results.
4. After the filtering process, the signal can be recomposed and inspected for different flow regimes. The underlying idea is that higher errors (higher noise) occur for maximum or highly changing flows, while lower errors occur for low or invariant flows. The clean and recomposed signal must correspond to the previous statement.

It is worth mentioning that the proposed method, in addition to the benefits it offers, also has limitations. Better predictions were obtained at the expense of higher computational requirements. Also, attention must be taken when selecting different bandpass families and decomposition algorithms. Some algorithms can implicitly go against certain rules, such as the unknown state of future values, providing false expectations.

References

Abrahart, R., et al. (2004). *Neural networks in hydrology*. Rotterdam: AA Balkema.

Adamowski, J., & Sun, K. (2010). Development of a coupled wavelet transform and neural network method for flow forecasting of nonperennial rivers in semi-arid watersheds. *Journal of Hydrology, 390*(1–2), 85–91. https://doi.org/10.1016/j.jhydrol.2010.06.033

Adamowski, J. F. (2008). Development of a short-term river flood forecasting method for snowmelt driven floods based on wavelet and cross-wavelet analysis. *Journal of Hydrology, 353*(3), 247–266.

Addison, P. S. (2010). *The illustrated wavelet transform handbook: Introductory theory and applications in science, engineering, medicine and finance*. CRC Press.

Arlot, S., & Celisse, A. (2010). A survey of cross-validation procedures for model selection. *Statistics Surveys, 4*, 40–79.

Aussem, A., et al. (1998). Wavelet-based feature extraction and decomposition strategies for financial forecasting. *Journal of Computational Intelligence in Finance, 6*(2), 5–12.

Bhattacharya, B., et al. (2007). Machine learning approach to modeling sediment transport. *ASCE Journal of Hydraulic Engineering, 133*(4), 440–450.

Bowden, G. J., et al. (2005). Input determination for neural network models in water resources applications. Part 2, Case study: Forecasting salinity in a river. *Journal of Hydrology, 301*(1–4), 93–107. https://doi.org/10.1016/j.jhydrol.2004.06.020

Cannas, B., et al. (2005). *River flow forecasting using neural networks and wavelet analysis*. Paper presented at the Geophysical Research Abstracts.

Coulibaly, P., et al. (2000). Daily reservoir inflow forecasting using artificial neural networks with stopped training approach. *Journal of Hydrology, 230*(3–4), 244–257. https://doi.org/10.1016/S0022-1694(00)00214-6

Dawson, C. W., & Wilby, R. L. (2001). Hydrological modeling using artificial neural networks. *Progress in Physical Geography, 25*(1), 80–108. https://doi.org/10.1177/030913330102500104

Dawson, C. W., et al. (2007). HydroTest: A Web-based toolbox of evaluation metrics for the standardized assessment of hydrological forecasts. *Environmental Modeling & Software, 22*(7), 1034–1052.

Donoho, D. L., & Johnstone, I. M. (1994). Ideal denoising in an orthonormal basis chosen from a library of bases. *Comptes Rendus de l'Academie des Sciences-Serie I-Mathematique, 319*(12), 1317–1322.

Du, K., Zhao, Y., & Lei, J. (2017). The incorrect usage of singular spectral analysis and discrete wavelet transform in hybrid models to predict hydrological time series. *Journal of Hydrology, 552*, 44–51. https://doi.org/10.1016/j.jhydrol.2017.06.019

Khalil, B., et al. (2012). Development of a wavelet neural network hybrid model for Total Dissolved Solids short-term forecasting in the Nile Delta drainage system. https://www.researchgate.net/publication/230600536_Development_of_a_wavelet_neural_network_hybrid_model_for_Total_Dissolved_Solids_short-term_forecasting_in_the_Nile_Delta_drainage_system https://www.academia.edu/24910680/Trous_Transform_Using_Wavelet_Extrapolation; https://www.academia.edu/19214235/Kalman_type_Filtering_using_the_Wavelet_Transform

Kisi, O., & Cimen, M. (2011). *A wavelet-support vector machine conjunction model for monthly streamflow forecasting*.

Kozakevicius, A., & Schmidt. (2013) A Trous transform using wavelet extrapolation. https://www.researchgate.net/publication/230600536_Development_of_a_wavelet_neural_network_hybrid_model_for_Total_Dissolved_Solids_short-term_forecasting_in_the_Nile_Delta_drainage_system https://www.academia.edu/24910680/Trous_Transform_Using_Wavelet_Extrapolation; https://www.academia.edu/19214235/Kalman_type_Filtering_using_the_Wavelet_Transform; https://doi.org/10.1016/j.jhydrol.2010.12.041

Krause, P., et al. (2005). Comparison of different efficiency criteria for hydrological model assessment. *Advances in Geosciences, 5*, 89–97.

Maheswaran, R., & Khosa, R. (2012). Comparative study of different wavelets for hydrologic forecasting. *Computers & Geosciences, 46*(0), 284–295. https://doi.org/10.1016/j.cageo.2011.12.015

May, R., et al. (2011). *Review of input variable selection methods for artificial neural networks*. INTECH Open Access Publisher.

McCulloch, W. S., & Pitts, W. (1943). A logical calculus of the ideas immanent in nervous activity. *The Bulletin of Mathematical Biophysics, 5*(4), 115–133.

Okkan, U. (2012). Wavelet neural network model for reservoir inflow prediction. *Scientia Iranica, 19*(6), 1445–1455. https://doi.org/10.1016/j.scient.2012.10.009

Okkan, U., & Ali Serbes, Z. (2013). The combined use of wavelet transform and black box models in reservoir inflow modeling. *Journal of Hydrology and Hydromechanics, 61*, 112.

Quilty, J., & Adamowski, J. (2018). Addressing the incorrect usage of wavelet-based hydrological and water resources forecasting models for real-world applications with best practices and a new forecasting framework. *Journal of Hydrology, 563*, 336–353. https://doi.org/10.1016/j.jhydrol.2018.05.003

Partal, T. (2008). River flow forecasting using different artificial neural network algorithms and wavelet transform. *Canadian Journal of Civil Engineering, 36*(1), 26–38.

Renaud, O., et al. (2002). *Wavelet-based forecasting of short and long memory time series*: Université de Genève/Faculté des sciences économiques et sociales.

Renaud, O., et al. (2003). Kalman-type filtering using the Wavelet Transform. https://www.researchgate.net/publication/230600536_Development_of_a_wavelet_neural_network_hybrid_model_for_Total_Dissolved_Solids_short-term_forecasting_in_the_Nile_Delta_drainage_system https://www.academia.edu/24910680/Trous_Transform_Using_Wavelet_Extrapolation; https://www.academia.edu/19214235/Kalman_type_Filtering_using_the_Wavelet_Transform

Ruikar, S. D., & Doye, D. D. (2011). Wavelet based image denoising technique. *IJACSA, International Journal of Advanced Computer Science and Applications, 2*(3).

Shamseldin, A. Y., et al. (1999). Comparison of different forms of the Multi-layer Feed-Forward Neural Network method used for river flow forecasting. *Hydrology and Earth System Sciences, 6*(4), 671–684. https://doi.org/10.5194/hess-6-671-2002

Shensa, M. (1992). The discrete wavelet transform: Wedding the a trous and Mallat algorithms. *IEEE Transactions on Signal Processing, 40*(10), 2464–2482.

Shrestha, D., et al. (2009). A novel approach to parameter uncertainty analysis of hydrological models using neural networks. *Hydrology and Earth System Sciences, 13*(7), 1235–1248.

Zhang, J., et al. (2009). Daily reservoir inflow forecasting combining QPF into ANNs model. *Hydrology and Earth System Sciences Discussions, 6*(1), 121–150.

Zijderveld, A. (2003). *Neural network design strategies and modeling in hydroinformatics*. Doctoral Thesis, TU Delft, Delft University of Technology. http://resolver.tudelft.nl/uuid:0c44f94c-58ad-4634-b5f2-13049a636577

[illegible]

[illegible] Wavelet neural network model [illegible]

[illegible]

Part III

Knowledge Discovery and Optimization

10

APPLICATION OF NATURAL LANGUAGE PROCESSING TO IDENTIFY EXTREME HYDROMETEOROLOGICAL EVENTS IN DIGITAL NEWS MEDIA: CASE OF THE MAGDALENA RIVER BASIN, COLOMBIA

Santiago Duarte[1,2], Gerald A. Corzo Perez[1], Germán Santos[3], and Dimitri P. Solomatine[1,2,4]

[1] *IHE Delft Institute for Water Education, Delft, The Netherlands*
[2] *Delft University of Technology, Delft, The Netherlands*
[3] *Colombian School of Engineering Julio Garavito, Bogotá, Colombia*
[4] *Water Problems Institute of the Russian Academy of Sciences, Moscow, Russia*

Human behavior and decision making are dynamically influenced by digital media, making digital news a vast data source of events and points of view. Meanwhile, artificial intelligence now has advanced capacity for natural language processing (NLP) through tools such as sentiment analysis and topic identification. This research uses machine learning algorithms to prove the hypothesis that it is possible to find correlations between news information and water resource problems. The 207 water bodies in the Magdalena River Basin, Colombia, were analyzed alongside 19,490 news articles published between 2016 and 2020 in 42 online newspapers. A platform for the visualization of spatiotemporal information was developed as a proof of concept. There has been a noticeable increase in digital

Advanced Hydroinformatics: Machine Learning and Optimization for Water Resources, Special Publications 78, First Edition. Edited by Gerald A. Corzo Perez and Dimitri P. Solomantine.

DOI: 10.1002/9781119639268.ch10

news about the Magdalena River Basin in recent years and some correlation with extreme events, but the comparison between the sentiments and measured physical variables was inconclusive. This is likely due to the difficulty in filtering the topic from obtained news, as well as in precisely identifying the spatial location and temporal range of events. In the future, NLP techniques to analyze news articles could be used to provide additional information for decision making at the basin level.

10.1. Introduction

Integrated water resources management (IWRM) aims at making informed decisions with the involvement and information from all stakeholders. Over the last decade, digital media has implicitly increased data-based information on most societal activities carried out by humans (Nielsen, 2015). Water resources, extreme conditions, and hydrometeorological events are not exceptions. For example, flood-related digital media information has been collected in the last decades, from tweets (Murthy & Longwell, 2013) to social media (Rossi et al., 2018), and used to provide information toward a smart flood-risk management (L. Smith et al., 2017) and improved policies (A. Smith et al., 2017). However, challenges posed by conflicting interests are not easily solved (Giupponi & Gain, 2017). The problem of decision making in water management using IWRM support from stakeholders is based on several criteria. While multiple actors participate in hydrology, the possibilities to involve citizens' science in hydrology has increase, as presented detailed in Nardi et al. (2020) where a framework to address citizens' science and hydrology problems are approached. It would seem that the relation between the information provided by citizens and extreme events is obvious, but in practice, the technologies require integration, as well as validation of this relation. Citizens' data collected from local observations to crowdsourcing (Le Coz et al., 2016), measured directly or not (Kam et al., 2019), it's nowadays a new database in decision-making process. In this context, the decision-making process in IWRM is based on history, documentation, and key elements of past and current decisions (Pires et al., 2017). In Avellaneda et al. (2020), crowdsourcing information regarding volunteers' calibration data improved hydrological model results. In the same way, analysis of media (digital and analogical) regarding historical hydrological disaster event that affected the Paraná coastal region in Brazil concluded in an improved understanding and characterization of floods and debris flows (Picanço et al., 2017).

One of the main challenges in decision making is the fact that human behavior and decisions are influenced by digital media (Keegan & Rowley, 2017). In general, decision-making results are presented as all kinds of policies and human interventions around water bodies, from forecasting estimations and emergency alerts, to nature-based solutions or artificial structures (Watts, 2016). Moreover, the implications of these actions on digital media behavior is starting to be used to validate decision-making results. Yu et al. (2019) discussed the applications of social media in hydroinformatics, including detection of fake news, analysis of floods management, and optimization of water resource planning, among others. Further, relations between digital media (e.g., tweets) and disaster severity could improve risk planning management when distributing resources in an emergency (Kankanamge et al., 2020). Thus, it is necessary to quantify the short-term and the long-term effects of these decisions, while measuring and looking for physical variables, social equity, environmental impacts, and so on (Badham et al., 2019). Consequently, new tools have been developed and used to study decision-making connections with digital media. For example, Bai et al. (2020) applied a machine learning technique (ML-KNN) to evaluate flood disaster risk based on social media (Weibo).

In this research, we address a relatively new paradigm in relating digital information (digital news) to historical decision making about water resources. We do so by analyzing digital citizens' media information across space and time. To understand these relations in other areas of science, artificial intelligence (AI) has developed some branches of data analysis in the semantics of language and how they can be integrated and quantified, the so-called natural language processing (NLP). This area explores AI tools to understand how natural language can be used for analyzing data (Chowdhury, 2003). Most of the advances focus on machine translation, speech recognition, and speech synthesis (Hirschberg & Manning, 2015). NLP has also been used to build concepts, such as sentiment analysis (Bello, 2020; Cambria et al., 2017). Sentiment analysis is a technique that provides a feeling-like index that relates to the semantics of positive and negative emotions, combined with AI. NLP learning algorithms allow providing a real-time analysis of data sets. The work by Mishev et al. (2020) shows some examples of how this has been implemented in finance problems. Garvey and Maskal (2020) use the NLP Google tool to quantify positive or negative effects on the news about AI. In general, research in this area has shown an important potential. All these advances motivate researchers to study the relation between digital media and water resources management. In this context, in a country such as Colombia, where most of its Gross Domestic Product (GDP) depends on water resources, decisions regarding the Magdalena River Basin are thus quite important, especially in an economic recovery scenario after COVID-19 (Borja-Vega et al., 2020). Moreover, digital media may help identify social patterns related to the environment and water resources. In this research, we evaluate particular cases (e.g., Hidroituango) using NLP, from which results are obtained and conclusions are drawn. This will also be relevant for future researchers who can analyze the effects of decision-making processes as proposed solutions to specific water-related problems and their consequences on human behavior.

10.2. Case Study

The description of this case study is divided into water resources and a digital socioeconomic perspective. The description of the water resources covers all important and general hydrometeorological information about the river basin in the Colombian context. The socioeconomic and digital perspective covers the information about not only the importance of the river basin but also the characteristics of this digital information in terms of quantity, quality, and availability.

The Magdalena River Basin is located in Colombia, South America, as shown in Figure 10.1. The basin extension is 273,459 km^2, and the main river length is 1,558 km. The average discharge flow is 7,293 m^3/s. The annual mean precipitation is 2,050 mm, with wet seasons between March to May and September to November, and dry seasons between December to February and June to August.

In total, this macro basin integrates 724 municipalities of 19 departments of the country, which produce 83% of Colombia's GDP (USD 260 billion in 2018), 70% of its hydropower, and 70% of its agricultural production. In recent years, this basin has been highly subjected to human interventions, involving a wide variety of stakeholders (public and private). Adding the capacities of the projects in operation and under construction, it has an installed capacity of around 9.97 GW, distributed among 35 large- and medium-scale projects, and approximately 0.5 GW in minor hydraulic generation projects. This research has been developed as an entry to the competition "Conectate con el Río Magdalena" under the auspices of the Netherlands' Ministry of Foreign Affairs.

The Hidroituango project's location (Cauca River) is also presented in Figure 10.1. This is one of the most important infrastructure projects in Colombia in recent years, with a power capacity of 2,400 MW. In 2018, the collapse of a diversion tunnel due to a geological fault resulted in a major emergency on the Cauca River. This caused an accumulation in the

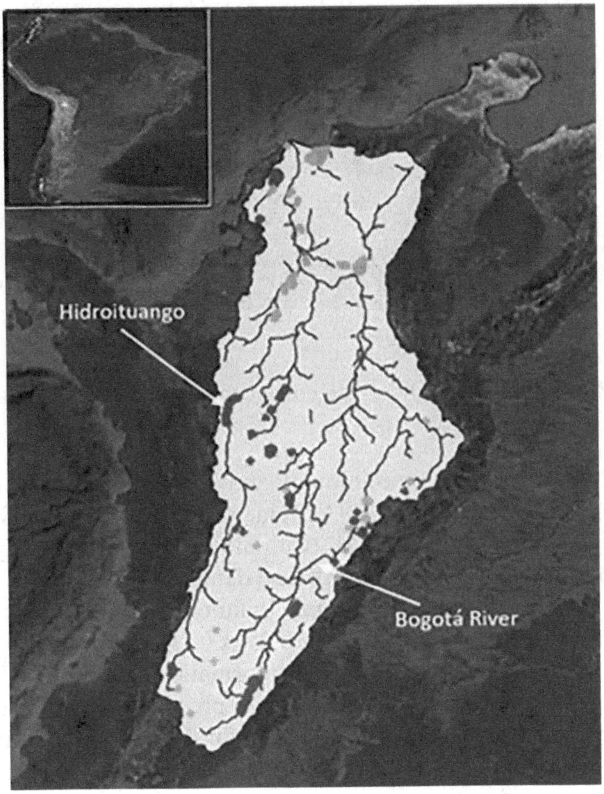

Figure 10.1 Magdalena River Basin scheme: Rivers and water bodies. WGS-84. dams (dark grey), rivers (lines), lakes (white), swamps (light grey).

reservoir and a decrease in the level of the Cauca River. For this reason, the comparison of the sentiment analysis and the measurable physical impact is an objective of our study. Figure 10.2 shows the impact of the emergency.

Three hydroclimatic variables (precipitation, temperature, and river levels) were used in this research. Two river-level stations in situ were selected to evaluate the sentiment index. For both stations, the monthly minimum and maximum river levels were considered for the analysis. The first one was the Puerto Valdivia hydrological station located next to Hidroituango Dam. The average minimum and maximum river-level values were found to be 300.12 and 423.69 cm, respectively. The second station, La Campina in Girardot Municipality, recorded the data from Bogotá River's flows. Both stations presented the results on the flows and the levels (minimum, average, and maximum) at monthly intervals from

Nueva emergencia en Hidroituango mantieneen vilo a 12 municipios

Figure 10.2 Journal article shows the emergency (from Tiempo, 2018).

2016 to 2020. The average minimum and maximum river-level values were found to be 200.40 and 493.82 cm, respectively. This information was obtained from the IDEAM (Colombia's Institute of Hydrology, Meteorology and Environmental Studies) database.

Likewise, ERA5 data on the precipitation and the temperature at monthly intervals, with a resolution of 0.25° x 0.25° for the analysis period, were used to compare the results. The information was obtained through Google Earth Engine. The average medium precipitation in the course of the Magdalena River Basin was 233.59 mm for the period, and the average maximum precipitation was 1,805.38 mm. For the same period, the average medium, maximum, and minimum temperatures in the course of the Magdalena River Basin were 23.65 °C, 30.83 °C, and 9.68 °C, respectively.

Among the multiple sources of digital citizens' information (e.g., Facebook, Twitter, news, text messages), only digital newspapers were selected for this research. According to *Prensa* (2015, Dec 17):

> The penetration of the press in Colombia is 78% and regional newspapers and the popular press play an important role. Although there appear to be no studies on the time Colombians spend reading the newspaper, recent research revealed that the number of newspaper readers in Colombia reached 6.3 million in 2013 versus 6.1 million in 2012. As of 2015, there are only 4 newspapers with national circulation. There are about 90 different regional newspapers between paid, popular and free press.

Additionally, Perdomo (2022) found:

> Colombian media has historically been dominated by publishers and broadcasters with strong ties to the political and business elites. That influence

seems to be transferring to the digital world, but new players and social media are clearly changing consumption trends. "El Tiempo" newspaper website is by far the most used news site (40%), followed by the online versions of "El Espectador" and "Caracol TV" (27%). But some digital native media are growing in importance. (5 Jan, p. 121)

10.3. Methodology

Figure 10.3 shows the main tasks performed to apply, evaluate, and validate the use of NLP in the mapping between water resource problems and digital news. These steps allowed us to extract, detect, and analyze qualitative and quantitative information from newspaper articles related to water bodies.

10.3.1. Step 1 (Digital Citizens' Search for Information Sources)

In this study, digital newspapers were selected as the sources of digital citizens' information. In this sense, the digital news information

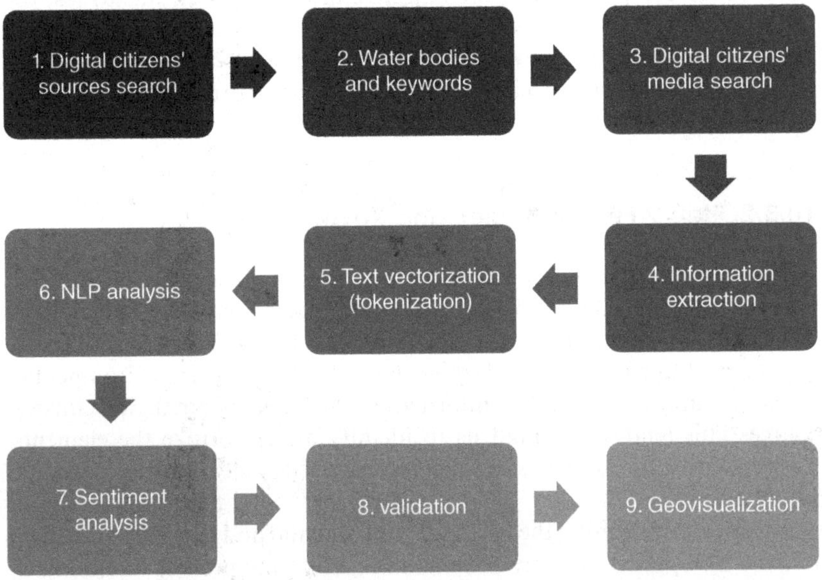

Figure 10.3 Methodology process.

provided text records of water-related events (environmental, extreme climatic events, supply). First, all available open sources of information (digital newspapers) were identified. No journalist blogs, newsletters, or non-water-related newspapers were considered for the analysis. The digital newspaper sources were selected under the following criteria: spatial coverage (regional or national), digital information availability, Web access permission, and number of articles. Nevertheless, because of the exploratory nature of this research, the newspaper selection was limited, and not all available newspapers were used. Only departments in close proximity to the Magdalena River Basin were selected. Two regional newspapers were picked for each department, and four national newspapers were used.

Once the newspaper sources were identified, the uniform resource locator (URL) of each newspaper was obtained. For all selected digital newspapers' websites, the types of available information (articles, photos, videos, and documentaries) were studied. Thus, the news text section was identified for the purpose of obtaining information only from each article's author. No photos or videos were analyzed.

Although the most representative and oldest newspaper in the country has provided digital information since 1990 and even earlier, most of this information is poor and mainly related to sociopolitical problems, not to water resources. In contrast, the objective of our research was to test the hypothesis that it would be possible to relate water resource events and associated problems to news information. Therefore, the time range was reduced; the digital news search included only news articles from 1 January 2016 to 31 May 2020.

10.3.2. Step 2 (Water Bodies and Keywords)

Next, the most important water bodies in the basin were identified (rivers, reservoirs, lakes, and swamps). An analysis of the water bodies in the Magdalena River Basin was then carried out. The analyzed characteristics are (1) size (area or length); (2) importance for the region (environmental, economic, and social); and (3) information availability from digital citizens' sources.This analysis allowed us to identify and prioritize the elements. Consequently, this information allowed us to select local sources of digital information.

Understanding that the basin is a fundamental and transversal element in the development of many communities, a filtering process to omit irrelevant or unrelated news was carried out. Then, a set of keywords to be used in the news search process and filtering related to the main events and

Table 10.1 Keywords defined for the news search.

Climate change and climate variability	Seasonal events	Environment	Related human activities
Cambio climático (Climate change)	*Sequía* (Drought)	*Ecosistema* (Ecosystem)	*Minería* (Mining)
El Niño	*Verano* (Summer)	*Vegetación* (Vegetation)	*Energía* (Energy)
La Niña	*Inundación* (Flood)	*Peces* (Fish)	*Riego* (Irrigation)
	Invierno (Winter)	*Cuenca* (Basin)	*Contaminación* (Pollution)
	Nivel (Water level)		*Navegabilidad* (Navigability)

problems in the basin was defined. The keywords are presented in Spanish in Table 10.1 because of the language used in the case study. The aggregation of these words in groups is also included.

Because of the exploratory context of this research, other keywords that might provide additional results were excluded. This was also intended to reduce the possible redundancy of newspaper articles. For this purpose, words such as *desbordamiento* (overflow) might be included in articles related to the word *inundación* (flood) or *invierno* (winter). Consequently, the newspaper articles related to *desbordamiento* were already included in the search results.

10.3.3. Step 3 (Digital Citizens' Media Search)

We adapted the custom search API (application programming interface) of Google Cloud and a custom search engine (CSE) from Google Search Console for the news search process. For this case, only the results from the selected newspapers' websites were obtained. The search process was performed as the combination of a water body, a keyword, and a date range in a process called *Web crawling*, for example: "Río Bogotá" "Inundación" "01/06/2016 to 30/06/2016".

The specifications of these parameters had these objectives: Specifying the full name of the water body allowed us to locate the spatial region of analysis. For example, if the word *rio* (river) was not included, the information obtained would be related to the City of Bogotá, not the Bogotá River Basin.

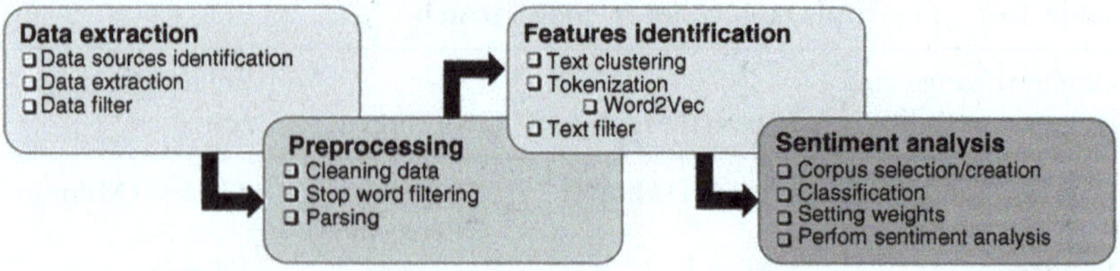

Figure 10.4 Data-mining process.

The keyword filtered information regarding the water resource main events and problems. In this sense, our search excluded any irrelevant or unrelated news, for example, "Soccer match on the banks of the Medellin River." In this case, while the news might mention the water body, the focus of the main article was related to other social topics. The time period selection allowed us to filter information regarding its digital publishing date for further analysis.

Figure 10.4 summarizes the data mining and analysis process developed in Steps 4 to 7.

10.3.4. Step 4 (Data Extraction and Preprocessing)

As a result of Step 3, for every newspaper article related to each water body, the URL was obtained through the Web-crawling process. The consequent step was to perform Web scraping, based on the parameters of each newspaper website obtained in Step 1. In this process, the information was filtered, downloaded, and processed from an HTML format to a comma-separated values (CSV) file as unstructured data. Python scripts with Googleapiclient, BeautifulSoup, and Panda libraries were used to download, extract, and transform the digital news. These scripts are available in a GitHub repository at https://github.com/sduarte09/buscandoelmagdalena.

Next, the unstructured data were transformed into a structured data format, so the text of each article, the title, the date, and the ID of the news websites were extracted. This information was then used to filter the news articles in order to remove duplicates, empty website pages, music/video links, or image galleries, as illustrated in Figure 10.5.

Once the structured data were filtered, the data were cleaned, checking the accent marks, changing uppercase into lowercase letters, and removing advertising in the middle of the text. Likewise, for the implementation of NLP techniques, a stop word filter was needed. In the NLP context,

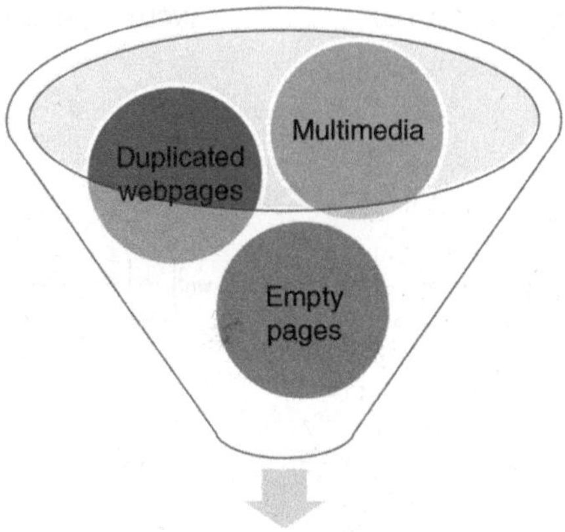

Figure 10.5 News-filtering process.

these are words that do not add much meaning value to a sentence or a document. The bag of stop words can be obtained from the GitHub collective repository at https://github.com/Alir3z4/stop-words. Other stop words were added to the bag, following a trial-and-error procedure on the article text, regarding the dialect.

10.3.5. Step 5 (Text Vectorization-Tokenization)

Subsequently, the feature text identification or tokenization started with the word embedding, applying the Word2Vec function, which provides concepts for the word embedding (Goldberg & Levy, 2014). The Word2Vec method identifies text mapping and then creates a model for reproducing such mappings to generate "word embedding" (given a text corpus), which is the process of mapping words to a vector space to realize statistical and probabilistic operations. The most important objective of this method is to preserve the relation between words, this is minimizing the distance between vectors within a similar context. In example, while through a tokenization, it is possible to represent the word *flood* in a text corpus X as [0,0,0,0,0,0,1,0,0,0,0,0,0] and the word *drought* as [0,0,1,0,0,0,0,0,0,0,0,0,0], where the vector size is equal to the vocabulary in the text corpus, these vectors do not show any relation between the meaning of the words *flood* and *drought*, both extreme event opposites within the hydrological cycle.

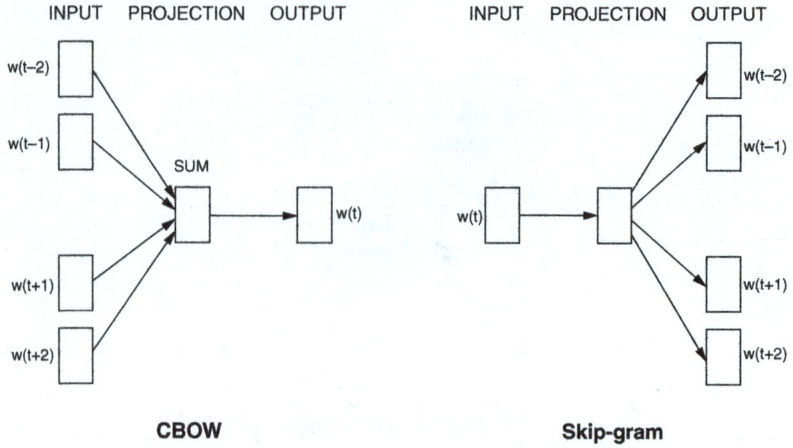

Figure 10.6 Graphical representation of the CBOW model and the Skip-gram model (from Mikolov et al., 2013).

Additionally, large values of data (e.g., big data corpus) could present a challenge to computational calculations (Mikolov et al., 2013). In this sense, the Word2Vec function looks for words that occur in similar contexts to have similar embeddings, that is, represent the meaning of the word, for example:

The flood in the Bogotá River triggered emergency responses from the Colombian government.

The drought in the Bogotá River triggered remedial actions taken by the Colombian government.

The words *flood* and *drought* will have similar word vectors due to a similar context. Two architectures/algorithms are widely used to represent the "meaning" of a word according to a context: CBOW (Continuous Bag of Words) (predict the target word from the bag-of-words context) and Skip-gram (predict the context words, given the target) (Jang et al., 2019). Both implementations use neural networks, as shown in Figure 10.6, where $w(t)$ denotes the analyzed word, and $w(t \pm s)$ represents the word at a distance of s words in the sentence context.

For example:

Sentence: "Several parts in Kuala Terengganu were inundated with flood waters, following continuous rain."

Distance: −8 −7 −6 −5 −4 −3 −2 −1 0 1 2 3 4

If $w(t)$ = flood, then $w(t+1)$ = waters, $w(t-2)$ = inundated, and $w(t+4)$ = rain. In practice, Skip-gram gives better word representations when the monolingual data are small. However, CBOW is faster and more suitable for larger datasets (Mikolov et al., 2013). In the Skip-gram approach, the objective of the neuronal network is to maximize the probability of any context word, given the current central word, mathematically formulated as follows:

$$F.O. = \frac{1}{T}\sum_{t=1}^{T}\left(\sum_{\substack{j=-k \\ j\neq 0}}^{k} \log p\left(w_{t+j}|w_t\right)\right),$$

where k denotes the size of the training window (which can be a function of the central word), and $p(w_{t+j}| w_t)$ represents the probability of the word $w(t+j)$, given $w(t)$:

$$p\left(w_{t+j}|w_t\right) = \frac{\exp\left(u_w{}^T * v_{wj}\right)}{\sum_{n=1}^{N}\exp\left(u_n{}^T * v_{wn}\right)},$$

where N signifies the number of words in the vocabulary.

10.3.6. Step 6 (NLP Analysis)

Once the text had been tokenized and the Word2Vec function had been applied to embed the words in vectors, NLP tools were used to create a word cloud for each water body, so the main words linked to each article could be represented in a graph. This allowed us to obtain a better picture of the most representative elements (social, economic, environmental, and climatic) related to each water body. The Natural Language ToolKit (NLTK), Scipy, and Word Cloud libraries were used in this process.

Counting the number of news items for each time step and each water body, as well as identifying the main topics about each water body, were also part of this study. Through this information, patterns between extreme events and digital news could be explored. Later, this information was aggregated on monthly and annual levels. However, most of the NLP free libraries work only for English-language texts or have a paragraph word-count limitation for non-English languages (because of an implicit third-party translation process). This makes the application of NLP difficult in non-English regions. Nevertheless, some public packages and libraries allow NLP analysis in Spanish-language texts.

10.3.7. Step 7 (Sentiment Analysis)

An important analysis that could be derived from NLP tools is sentiment analysis. Sentiment analysis shows the perception of an author when writing the article, interpreting this sentiment with a number that ranges between 0 (negative sentiment) and 1 (positive sentiment). For this purpose, a neuronal network is built to classify the writer's sentiment through a numerical value; a summary of the involved processes is shown below. This requires statistical data sampling, data normalization, input variable selection, training, and validation.

Algorithm 10.1 Sentiment analysis pseudo code

Input: $x=[x_1, \ldots, x_i]$, $x_t \varepsilon R^m$ and $s=[s_1, \ldots, s_n]$, x_t and $s_t \varepsilon [0,1]$
Initialize parameters: $w=[w_1, \ldots, w_j]$, $w_t \varepsilon R^n$
for t=1, ... , n do
  Calculate $y=w \bullet x$
  Update w
  Calculate error(y,s)
end for
Output: $s_{n+1} \varepsilon [0,1]$

In this research, we used the *spanish_sentiment_analysis* library (Hofman, 2017/2021), which was already trained and tested in its open source version. The library used almost one million reviews of users of the following webpages: *Trip Advisor, PedidosYa, Apestan, QuejasOnline, MercadoLibre, SensaCine, OpenCine, TASS,* and *Twitter* (Spanish-language version) for its calibration and validation. TensorFlow and Marisa-trie libraries were used for building the sentiment neuronal network. In this process, the model developed a regression between the users' reviews of the websites and their corresponding scores on a scale from 0 to 5 (websites' 5-star reviews). Then, this information was represented as a numerical value between 0 and 1, where a 5-star review represented a sentiment value of 1. This calibrated model was used to predict the news sentiment over the Magdalena River Basin.

The *spanish_sentiment_analysis* package is an open-source library built on scikit-learn and NLTK in Python. The model works as a simple pipeline that includes a vectorizer (TfidfVectorizer), a feature selector (SelectKBest), and a classifier (MultinomialNB). The feature selector will reduce the number of features generated by the vectorizer. The feature selector function selected was SelectKBest from scikit-learn, performed by checking whether a feature was relevant or irrelevant. The classifier model used was a multinomial naive Bayes with best parameter alpha = 0.03. The parameters and hyperparameters of this pipeline were found by using a GridSearch K cross-validation, with K = 10. Additionally, the Marisa-trie library was used for a static memory-efficient structure. This model was applied to the digital news information of each article. The results were compiled on monthly and annual aggregation levels. Moreover, the model was tested with reviews of users of the pages *eltenedor*, *decathlon*, *filmaffinity,* and *ebay* in their Spanish language for two data sets (positive and negative) of 50,418 reviews each one using the Mean Absolute Error (MAE).

10.3.8. Step 8 (Validation)

The results obtained from the NLP analysis were compared with measured physical variables (e.g., river-level flows) and observed events (social and environmental emergencies), as part of the validation process. This comparison was necessary to verify the relations between events and digital citizens' media. For this purpose, different parameters were validated, as listed below.

1. Sentiment index versus monthly river levels' range next to the Hidroituango Dam and the Bogotá River (Girardot). The monthly river-level range was used as a simple representation of a sudden change in the flow of the river (e.g., Hidroituango emergency).
2. Correlation of monthly climatic variables and sentiment index. The correlation was performed using the monthly sentiment index for the water bodies with information for all the time ranges (at least one article per month). Furthermore, the climatic events were extracted as the maximum and the minimum values of precipitation and temperature for the spatial area of the water body.
3. Confusion matrix of extreme events identified by climatic variables (precipitation and temperature) versus the number of articles over time. Extreme events were identified as monthly values smaller than 5% of the selected range values of the climatic variables. Likewise, months with values that exceeded 95% of the selected time series were classified as extremes. For this analysis, annual time ranges were selected.

10.3.9. Step 9 (Geovisualization)

Finally, these results were incorporated in interactive maps using the Google Earth search engine, in which the largest amount of the information obtained was represented. This allowed a second verification through visualization of the georeferenced data and an analysis of the transitions in the water bodies in the region that had occurred over time.

10.4. Results

10.4.1. Step 1

Initially, 81 digital newspapers were identified after conducting a search of newspapers published in every department of the country. However, only 42 newspapers were selected. A list of the newspaper sources is shown in Table 10.2.

Table 10.2 Selected newspaper sources grouped by region.

Department	Newspapers
All	*El Tiempo - El Espectador - Publimetro - La República - Semana*
Antioquia	*El Colombiano - El Mundo*
Atlántico	*Al día - El Heraldo*
Bolívar	*El Universal - El Bolivarense*
Boyacá	*El Diario - Siete días Boyacá*
Caldas	*La Patria - Eje21*
Caquetá	*Florencianos - Diario del Caquetá*
Cauca	*El Nuevo Liberal - Diario del Cauca*
Cesar	*El Pilón - Diario del Cesar*
Córdoba	*El Meridiano de Córdoba - Tu Prensa Digital*
Cundinamarca	*Extra - Periodismo Público*
Huila	*La Nación - Opanoticias*
La Guajira	*Diario del norte*
Magdalena	*Hoy diario del Magdalena - Seguimiento*
Norte de Santander	*La Opinión - Así es Cúcuta*
Quindío	*La crónica del Quindío - El Quindiano*
Risaralda	*El Diario - El Expreso*
Santander	*Vanguardia Liberal*
Sucre	*El Meridiano de Sucre*
Tolima	*El Nuevo Día - El Olfato*
Valle del Cauca	*El País - Diario Occidente*

10.4.2. Step 2

In total, 207 water bodies were identified and distributed, as shown in Figure 10.7. Because of the scale and the area covered by each regional newspaper source, little or no information was found for small basins within the expected time search.

An examination between the newspaper circulation size (region covered) and the extension of the water body was carried out. The news articles about the largest water bodies reported more information from more sources, while the news items about small water bodies presented the most accurate information at the local level. It was also not possible to access some newspapers because of website restrictions or property law restrictions.

10.4.3. Step 3

As a result of the digital citizens' media search, 19,490 newspaper articles from January 2016 to July 2020 were found. The largest number of articles regarding the analyzed type of water body featured the Magdalena River (4,405 news articles), followed by Hidroituango Dam (2,600 news articles), Cienaga Grande de Santa Marta (swamp, 621 news articles), and Fúquene Lake (94 new articles).

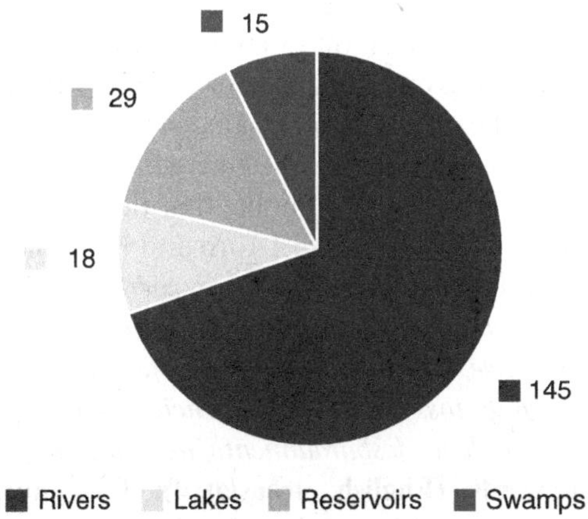

Figure 10.7 Distribution of 207 water bodies.

10.4.4. Step 4

A summary of the Web-scraping process is shown in Figure 10.8, transforming a website into an HTML format, and then into a CSV file. An example of the information extracted from a news article is illustrated in Figure 10.9.

As the main result, the distribution of articles featuring the Magdalena River Basin (by type of water body) is shown in Figure 10.10. The rivers are the main water bodies covered in digital news.

10.4.5. Step 5 (Tokenization)

A news article from Arroyo Alférez is shown as an example of the tokenization performed. The selected article was published in the newspaper *El Universal* on 19 November 2016, with the title, "Inundaciones en San Cayetano y deslizamiento de tierra en la Troncal." Below is the first paragraph of the article and its tokenization (including the removal of stop words):

- **Text (Spanish, original)**: "*Cien casas del corregimiento de san cayetano amanecieron este domingo inundadas por las fuertes lluvias que no le han dado tregua al departamento de bolívar desde la noche del sábado. Las casas afectadas están situadas en los sectores san martín los manguitos y bocatá. la emergencia es atendida por la defensa civil que informó que la situación se debe al desbordamiento de los arroyos el toro y naranjal.*"
- **Text (English, translated):** "The residents of 100 houses in the San Cayetano township woke up this Sunday, [their area] flooded by heavy rains that have not given the department of Bolívar a respite since Saturday night. The affected houses are located in the San Martín los Manguitos sectors and Bocatá. The emergency is handled by the civil defense that has informed [the public] that the situation is caused by the overflow of the river streams El Toro and Naranjal."
- **Tokenization words (Spanish):** *Cien, casas, corregimiento, san, cayetano, amanecieron, domingo, inundadas, fuertes, lluvias, dado, tregua, departamento, bolívar, desde, noche, sábado, casas, afectadas, situadas, sectores, san, martín, manguitos, bocatá, emergencia, atendida, defensa, civil, informó, situación, debe, desbordamiento, arroyos, toro, naranjal*
- **Tokenization words (English, translated):** One hundred, houses, San, Cayetano, township, woke, Sunday, flooded, heavy, rains, given, department, Bolívar, respite, since, Saturday, night, affected, houses, located, San, Martin, Manguitos, sectors, Bocatá, emergency,

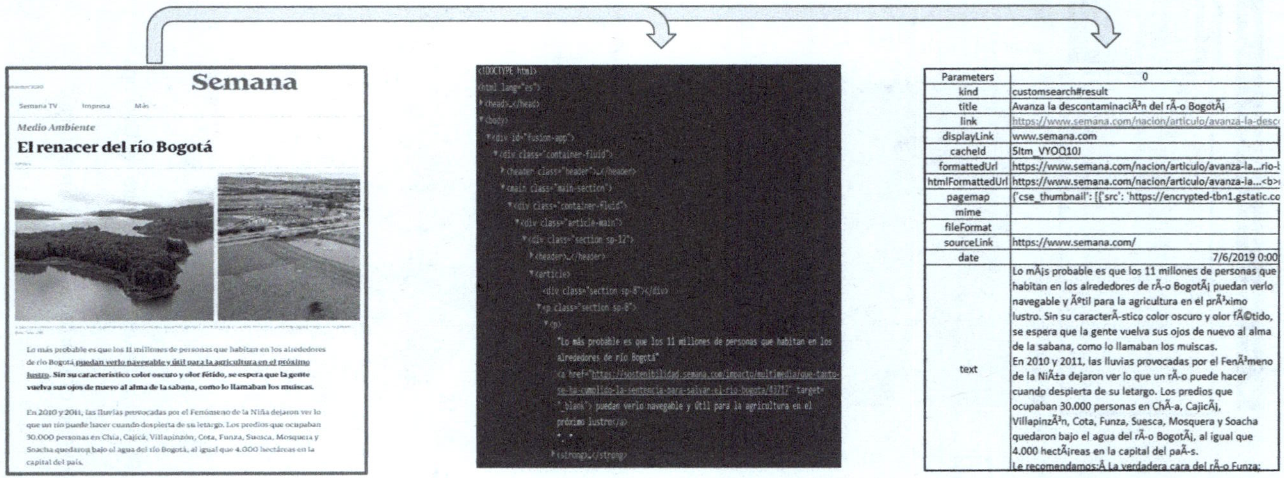

Figure 10.8 Web-scraping simplified process.

Figure 10.9 Example of information extracted from a newspaper website (https://www.elespectador.com).

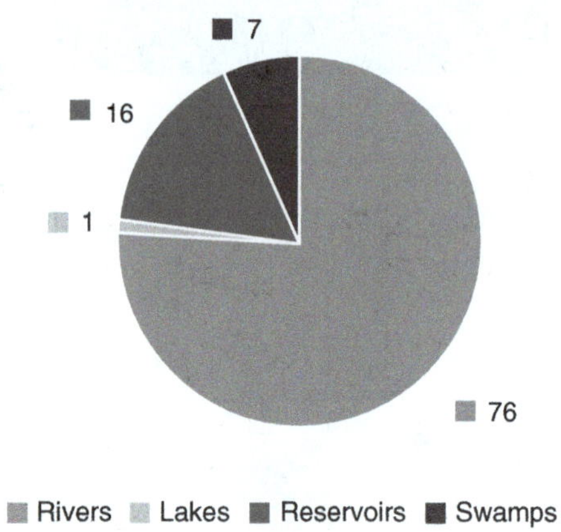

Figure 10.10 Distribution of journal articles (percentage).

attended, civil, defense, informed, situation, generated, overflow, river, streams, Toro, Naranjal

- ***Tokenization vector (word: "overflow"). This result is calculated for each word given the paragraph as corpus:*** [1.61e-03, -1.59e-03, -4.26e-03, -4.87e-03, -1.46e-03, 1.09e-03,-3.36e-03, -1.99e-03, 8.40e-04, -2.99e-03, 2.48e-03, -3.71e-03, 2.06e-03, 2.68e-03, -1.53e-03, 4.51e-03, 4.11e-03, 2.23e-03, -2.47e-03, 1.03e-03, -4.58e-03, 3.82e-03, -4.65e-03, -9.20e-04, 2.32e-03, -2.91e-03, 4.91e-03, -4.60e-04, -4.42e-03, 3.40e-03, -3.38e-03, -2.47e-03, -3.05e-03, 2.19e-03, 2.48e-03, -2.50e-03, -3.09e-03, 2.98e-03, 2.23e-03, -1.60e-03, -3.25e-03, -2.60e-04, 6.70e-04, -4.09e-03, -6.20e-04, 4.08e-03, -3.78e-03, 1.33e-03, 4.60e-04, 2.55e-03, 1.01e-03, -8.60e-04, 4.79e-03, 3.01e-03, 3.02e-03, -1.76e-03, 9.10e-04, 1.79e-03, -3.66e-03, 3.73e-03, 9.90e-04, 2.00e-05, 1.19e-03, -4.63e-03, -1.30e-04, -1.81e-03, -2.02e-03, 3.05e-03, 1.33e-03, 2.33e-03, 2.49e-03, -1.82e-03, 7.40e-04, 1.96e-03, -3.34e-03, 1.90e-04, 1.42e-03, 1.86e-03, 4.70e-04, 3.95e-03, -2.77e-03, 1.17e-03, 1.88e-03, 4.19e-03, 2.70e-03, -3.21e-03, -4.90e-04, -4.94e-03, -4.49e-03, -3.90e-04, 2.70e-03, 4.70e-03, 4.92e-03, -3.30e-03, 3.32e-03, 3.54e-03, -1.27e-03, -3.43e-03, 3.16e-03, -3.77e-03]
- ***Top three words similar to "overflow," given the vector distance (minimum distance):***
    1. affected - distance: 0.275
    2. township - distance: 0.183
    3. rains - distance: 0.135

10.4.6. Step 6 (NLP Analysis)

The main words about each water body over time were found and represented as word cloud images. Figure 10.11 shows the word cloud image for Quebrada el Carmen, with the word *Ecopetrol*, which is the largest and primary petroleum company in Colombia. In the word cloud, we also identified the word *derrame*, which means "oil spill," illustrating the region's main problem related to the river.

An increase in newspaper articles published in the last years of the study period (Fig. 10.12), tripling the number of articles written between 2016 and 2019, was also found.

10.4.7. Step 7 (Sentiment Analysis)

Applying sentiment analysis allowed us to understand the social perception about the water bodies. For the case, the *spanish_sentiment_analysis*

Figure 10.11 Word cloud image for Quebrada el Carmen representing the news articles about the water bodies (rivers) composing the Magdalena River Basin.

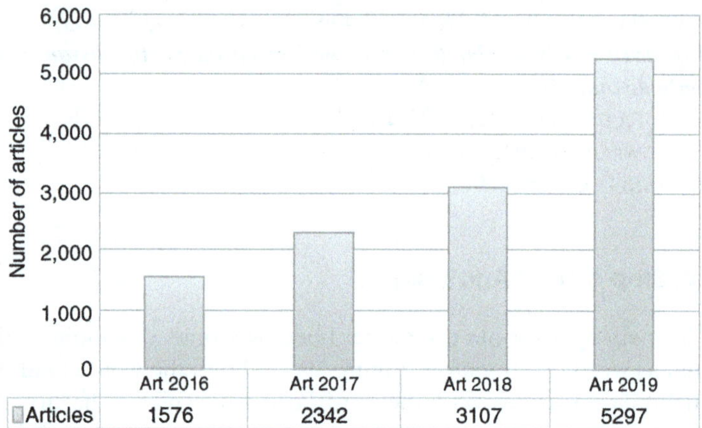

Figure 10.12 Number of newspaper articles about the rivers composing the Magdalena River Basin.

trained model was tested and MAE values of 0.068 and 0.198 were found regarding positive and negative reviews respectively. These values indicate the improved performance of the model regarding positive reviews. This is a logical consequence of the data sources of calibration, as positive reviews are more frequent in these websites. For example, we analyzed the sentiment toward the water body (Hidroituango Reservoir) over time. This sentiment shows the news writer's opinion, decreasing after the emergency,

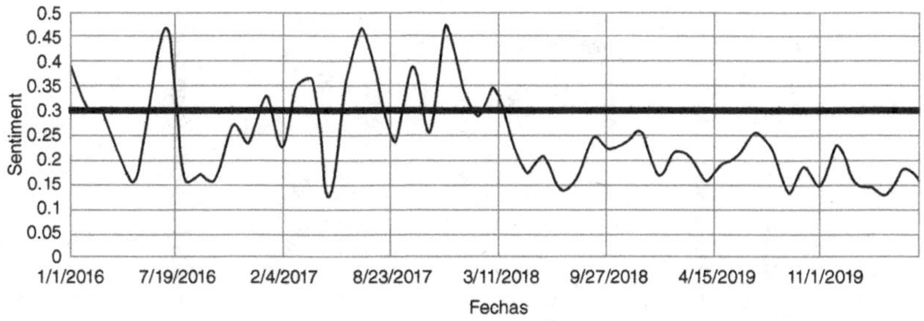

Figure 10.13 Sentiment change over time: Hidroituango.

in comparison with the previous months, as presented in Figure 10.13. This result indicates a negative mainstream opinion about the project due to the emergency. The y-scale values in Figure 10.13 are limited to those founded in Hidroituango Reservoir, while the horizontal line, intend to represent clearly the changes in sentiment before and after the emergency.

10.4.8. Step 8 (Validation)

1. *Sentiment index versus monthly river-level ranges next to Hidroituango Dam and Bogotá River*. The graphical results of the comparison between the sentiment index and the monthly river-level ranges of the Cauca River stations (located next to the dam) related to the Hidroituango Dam emergency in April 2018 are shown in Figure 10.14. Likewise, the Bogotá River sentiment results over time were validated through a river-level station at the river mouth, as shown in Figure 10.15.

   In both cases, larger monthly level ranges are related to decreases in the sentiment toward the water body (dark dashed lines, Figs. 10.14 and 10.15). Nevertheless, due to the characteristics of each water body, its associated problems, and main physical variables, it is not possible to infer a direct relation between sentiments toward water bodies and flow/level changes. This suggests detailed research on each water body to understand the nexus between communities and water bodies.

2. *Correlation of climatic events and sentiment index*. Due to the lack of available news covering the complete time range, only Bogotá River, Cauca River, and Magdalena River were used in this analysis. The results of the correlation of climatic variables (precipitation and temperature) and the sentiment index are shown in Table 10.3.

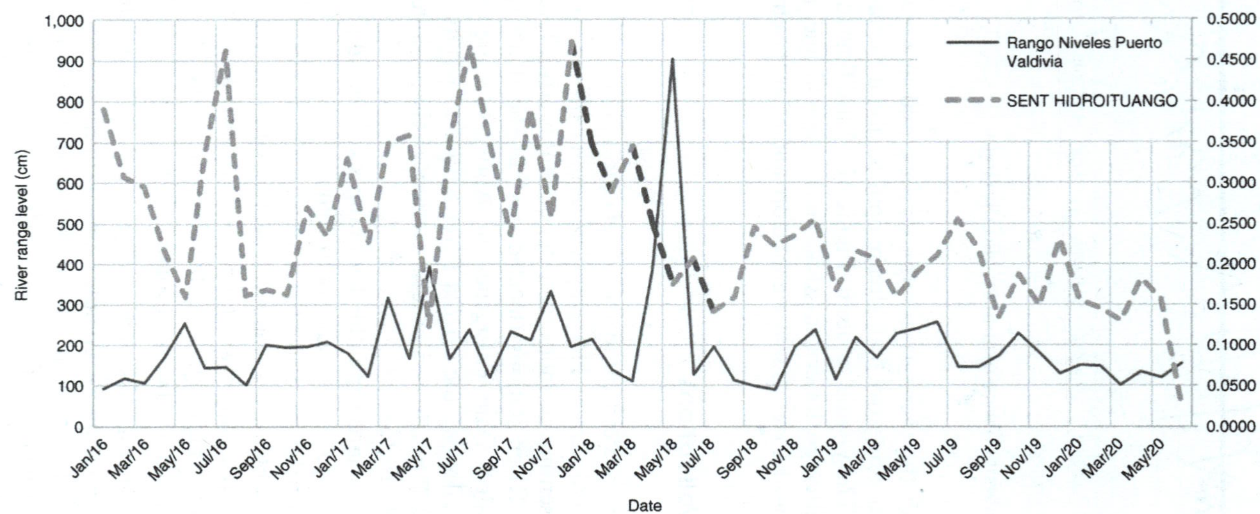

Figure 10.14 Monthly river-level range, Puerto Valdivia station in the Cauca River (Continuous line) versus Hidroituango sentiment (Dash line).

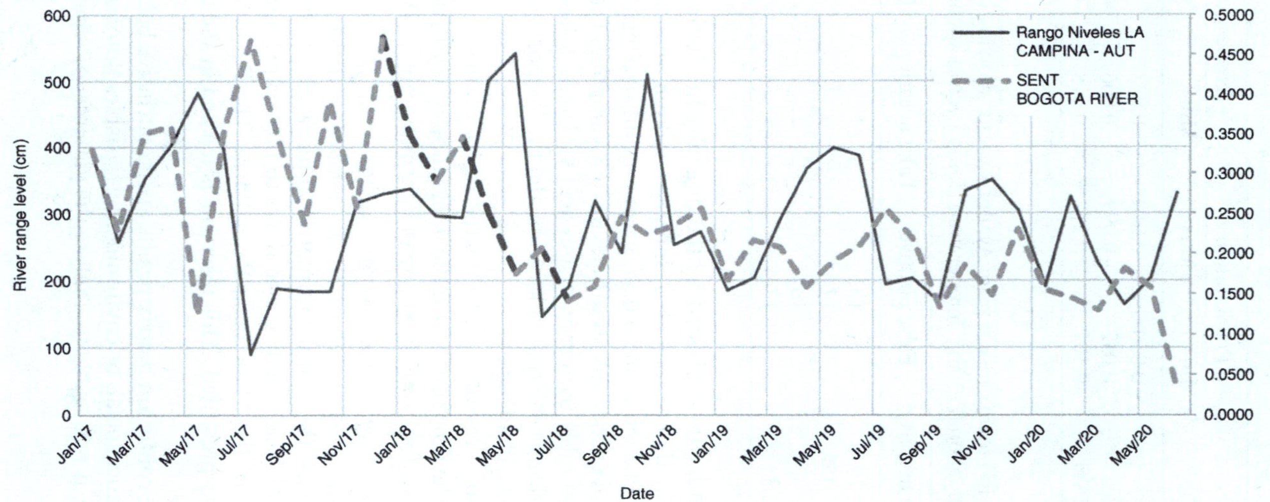

Figure 10.15 Monthly river-level range, La Campina station (Continuous line) versus Bogotá River sentiment (Dash line).

Table 10.3 Correlation of monthly sentiment index and monthly climatic data.

	Precipitation		Temperature	
Element	Min	Max	Min	Max
Bogotá River	−0.050	0.010	0.038	0.023
Cauca River	0.032	−0.099	−0.006	−0.130
Magdalena River	−0.101	−0.217	0.018	−0.197

Table 10.4 Confusion matrix results from Ayapel Swamp, showing high and low values of Precipitation (Pr.) and Temperature (Temp.) versus sentiment index.

	True negative	False positive	False negative	True positive
Pr. min	34	15	1	2
Temp. min	35	17	0	0
Pr. max	33	16	2	1
Temp. max	35	17	0	0

These results are explained by the difficulty in filtering the topic from the obtained news related to climatic conditions (e.g., higher temperatures and precipitation). This is an important statement, considering that most of the newspaper articles include components that cannot be easily measurable by means of physical variables. Likewise, a precise identification of the spatial location and the temporal range of the events remains a challenge.

These results reveal that the spatial distribution, the amount of information, the time range, and the integration of multiple topics are key components for using digital news information in an IWRM.

3. *Confusion matrix of extreme events identified by climatic variables (precipitation and temperature) versus number of articles over time.* Two examples of the confusion matrix application are presented in Tables 10.4 and 10.5. The first one refers to the Ayapel Swamp in the northern area of the Magdalena River Basin. The Betania Reservoir results are also presented. This reservoir is located in the upper river region.

   While the confusion matrix results are inconclusive because of the multiple climatic, social, or environmental variables that can influence a news article, it is possible to analyze the patterns for which the True

Table 10.5 Confusion matrix results from Betania Reservoir, showing high and low values of Precipitation (Pr.) and Temperature (Temp.) versus sentiment index.

	True negative	False positive	False negative	True positive
Pr. min	42	7	1	2
Temp. min	40	9	3	0
Pr. max	40	9	3	0
Temp. max	42	7	1	2

Positive area is presented. In the first case, the low and the high monthly precipitation values imply the relation and the dependence between the Ayapel Swamp operations and the precipitation variable; in comparison, the temperature variable is not that relevant. In contrast, in the second example, low precipitation and high temperatures coincide with the patterns that affect the operation of a reservoir. Low precipitation may be related to water shortage, while high temperature values are related to the water evaporation process. It is also important to note that the increase in the number of articles over time is a very relevant parameter for the confusion matrix. The latest years with most of the articles may not be compared directly with earlier years, so the available information is less. Thus, using the number of articles as a variable in the matrix confusion should be checked in further analysis.

10.4.9. Step 9 (Geovisualization)

Finally, through the Google Earth search engine, it was possible to represent the indexing of the water bodies (swamps, reservoirs, lakes, and rivers) with their respective sentiments over time. The maps can be found at https://santiagoduarte09.users.earthengine.app/. Figure 10.16 shows an example of a sentiment change regarding the river basins between 2016 and 2019.

It was also possible to simultaneously visualize the sentiment variation and the spatial distribution of the ERA5 information (precipitation and temperature) for the same period, as shown in Figure 10.17 for precipitation.

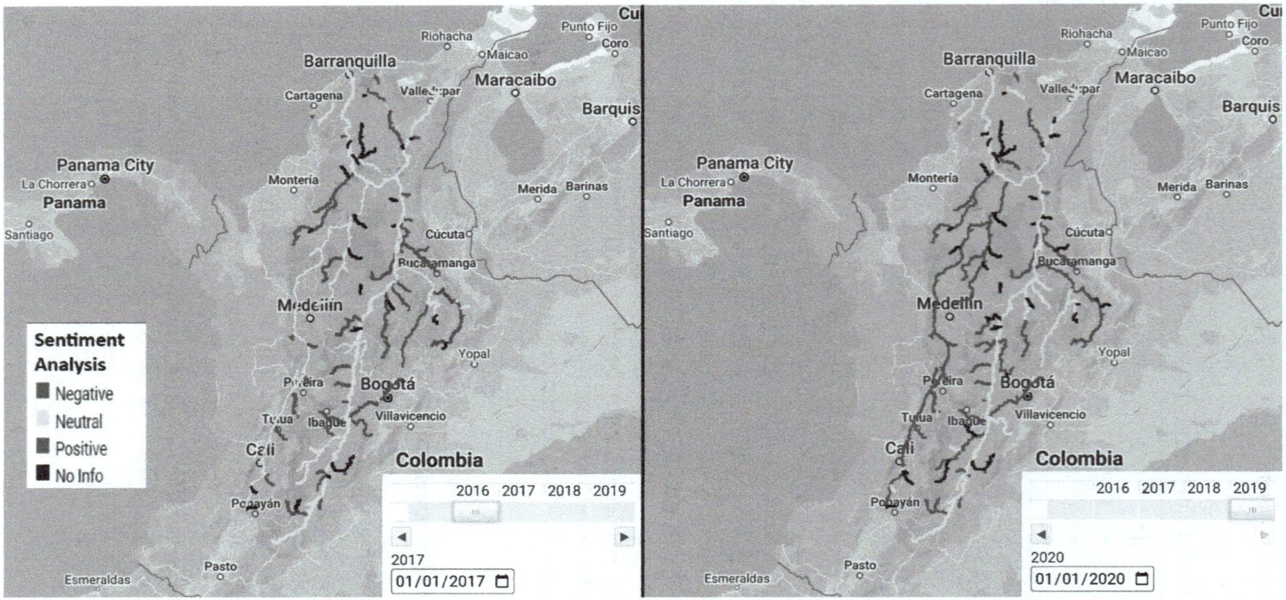

(a) (b)

Figure 10.16 Spatial sentiment changes regarding the Magdalena River Basin, between (a) 2016 and (b) 2019.

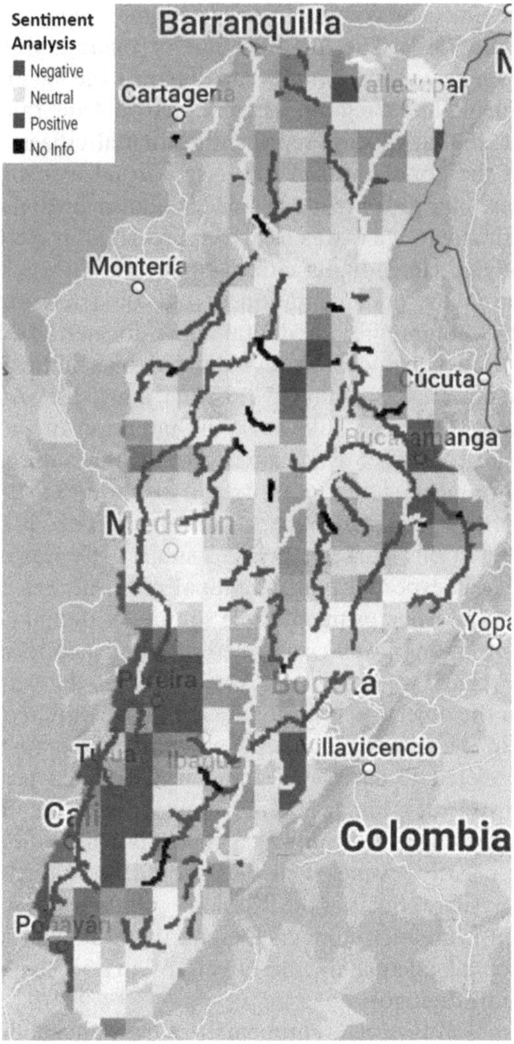

Figure 10.17 Spatial comparison sentiment and climatological variables regarding the Magdalena River Basin, precipitation, 2018 (Dark grey = wet; Light grey = dry).

10.5. Conclusions

Analysis of digital citizens' information allows us to understand the existing relation between communities and water bodies, as well as their cultural appropriation. We found that physical changes in water bodies with significant effects on communities can be analyzed over time through news articles and NLP. The identification of main topics about water bodies included in newspaper articles can illustrate the big picture of water resources problems for a region, as presented in Quebrada el Carmen. In this context, digital news is representing digital citizens' information from a part of the population, in this case journalists, which describe a mainstream view of a water-related topic for a given spatial and temporal range. This mainstream view was clearly appreciated in the Hidroituango emergency analysis. Here, digital citizens' information has been captured indirectly through the documentation of one or a few individuals and shared with the community in digital news format. Additionally, it's also observed and concluded about editorial bias influence on digital news results using NLP. In other words, the journalist writer's political or environmental affinities could twist the intention of the information documented, and consequently it's advised to include subjectivity analysis in further developments. Other authors had also found and analyzed this bias in their works regarding newspapers (Gangula et al., 2019; Valdeavilla & Pulido, 2019). In this work, the writer's subjectivity was not measured, nevertheless, it was expected that individual political and environmental biases were reduced when using multiple sources of digital newspapers, as well as use information of well-established newspaper editorials. Additionally, politically related topic news was filtered within the search news step. Likewise, given the objective and activities of this research, it was deduced (as illustrated in the geovisualization step, within the maps) that most of the news was related to emergencies and water-related problems. These negative stories and big impact events allowed us to identify challenges in IWRM. Furthermore, this implies a coverage bias, as journalists tend to write about negative events. This bias should be studied further when evaluating the positive effects of the decision-making process, and applying NLP outputs in a broader context. Despite these drawbacks, newspaper articles represented adequately the effects of major changes on water bodies (e.g., Hidroituango).

The comparison between sentiments toward water bodies and physical variables is essential for a better understanding of the relations in the water-energy-food nexus. Sentiment index and monthly river-level ranges comparison in Hidroituango Dam and Bogotá River allowed us to satisfactorily identify drastic sentiment changes over. From the correlation between climatic events and sentiment index, two main statements are obtained. First, it is necessary to have an improved geolocation methodology. Spatial location derived from water bodies shows it is not enough when analyzing long rivers, as point-located news are not easily geolocated in the river. Second, it was found that the date of publication of a newspaper article is not representative of the time occurrence of a hydrological event in small water bodies with few data sources. In these cases, the publication date denoted the timestep when the hydrological event was a mainstream topic for the community. Nevertheless, the visualization of georeferenced sentiment analysis results has proven to be a useful tool to understand the spatial changes in water bodies over time. Additionally, the results of the confusion matrix of extreme events provides a new understanding regarding the complexity when linking physical variables with digital media, as external factors could affect the sentiment index regarding hydrological events. In the context of NLP analysis on newspapers, other authors suggest the existence of sets of assumptions that imply a different interpretation of sentiment analysis results regarding the field of study (Puschmann & Powell, 2018), while Baccarella et al. (2018) exhibits the risk of social media and fake news. These statements reinforce the relevance of a digital news filtering process toward a newspaper database of water-resources topics. Consequently, this discussion proposes that a deeper analysis should be carried out when analyzing the perceptions of digital news and our understanding of it. Finally, we conclude by stating that NLP analysis of newspaper articles can provide additional information for decision making at the basin level, based on digital news information about the Magdalena River Basin's water bodies, however, this should be checked against physical variables and improved in spatial and temporal resolution when implementing action plans.

It is foreseen that other sources of information about extreme hydrological events could be explored through NLP (e.g., Twitter). Further, a specialized word-corpus based on water resource management is highly recommended for improved results. Furthermore, additional water cycle components should be incorporated in NLP applications, as aquifers, subsurface flows, and humidity. In future works, a deeper geospatial analysis between digital citizens' information and measured physical variables could provide additional tools for decision-making processes. We also recommend to explore new machine learning approaches for NLP involving water resources, as new advances could improve their results. Moreover, while new machine learning developments could provide new results, new additional challenges arise when trying to implement the work results in a decision-making process for community development and as a new methodology within the broader academic study of water-resource management.

References

Avellaneda, P. M., Ficklin, D. L., Lowry, C. S., Knouft, J. H., & Hall, D. M. (2020). Improving hydrological models with the assimilation of crowdsourced data. *Water Resources Research, 56*(5), e2019WR026325. https://doi.org/10.1029/2019WR026325

Baccarella, C. V., Wagner, T. F., Kietzmann, J. H., & McCarthy, I. P. (2018). Social media? It's serious! Understanding the dark side of social media. *European Management Journal, 36*(4), 431–438. https://doi.org/10.1016/j.emj.2018.07.002

Badham, J., Elsawah, S., Guillaume, J. H. A., Hamilton, S. H., Hunt, R. J., Jakeman, A. J., et al. (2019). Effective modeling for integrated water resource management: A guide to contextual practices by phases and steps and future opportunities. *Environmental Modelling & Software, 116*, 40–56. https://doi.org/10.1016/j.envsoft.2019.02.013

Bai, H., Yu, H., Yu, G., & Huang, X. (2020). A novel emergency situation awareness machine learning approach to assess flood disaster risk based on Chinese Weibo. *Neural Computing and Applications.* https://doi.org/10.1007/s00521-020-05487-1

Bello, H. J. (2020). *Sentiment-analysis-spanish: Sentiment analysis for sentences in Spanish* (Version # 0.0.24) [Python; OS Independent]. Hugo Bello. https://github.com/sentiment-analysis-spanish/sentiment-spanish

Borja-Vega, C., de Groot, K., & Serrano, H. A. (2020). *Colombia, turning the tide: Water security for recovery and sustainable growth.* Snapshot (Spanish). Water Security Diagnostics, Volume 1, Report 152351. Washington, DC : World Bank Group. http://documents.worldbank.org/curated/en/212231615975980795/Snapshot

Cambria, E., Poria, S., Gelbukh, A., & Thelwall, M. (2017). Sentiment analysis is a big suitcase. *IEEE Intelligent Systems, 32*(6), 74–80. https://doi.org/10.1109/MIS.2017.4531228

Chowdhury, G. G. (2003). Natural language processing. *Annual Review of Information Science and Technology, 37*(1), 51–89. https://doi.org/10.1002/aris.1440370103

Gangula, R. R. R., Duggenpudi, S. R., & Mamidi, R. (2019). *Detecting political bias in news articles using headline attention.* Proceedings of the 2019 ACL Workshop BlackboxNLP: Analyzing and Interpreting Neural Networks for NLP, 77–84. https://doi.org/10.18653/v1/W19-4809

Garvey, C., & Maskal, C. (2020). Sentiment analysis of the news media on artificial intelligence does not support claims of negative bias against artificial intelligence. *OMICS: A Journal of Integrative Biology, 24*(5), 286–299. https://doi.org/10.1089/omi.2019.0078

Giupponi, C., & Gain, A. K. (2017). Integrated water resources management (IWRM) for climate change adaptation. *Regional Environmental Change, 17*(7), 1865–1867. https://doi.org/10.1007/s10113-017-1173-x

Goldberg, Y., & Levy, O. (2014). Word2Vec explained: Deriving Mikolov et al.'s negative-sampling word-embedding method. *ArXiv:1402.3722* [Cs, Stat]. http://arxiv.org/abs/1402.3722

Hirschberg, J., & Manning, C. D. (2015). Advances in natural language processing. *Science, 349*(6245), 261–266. https://doi.org/10.1126/science.aaa8685

Hofman, E. (2021). Aylliote/senti-py [Python]. Elliot Hofman. https://github.com/aylliote/senti-py (Original work published 2017).

Jang, B., Kim, I., & Kim, J. W. (2019). Word2Vec convolutional neural networks for classification of news articles and tweets. *PLOS ONE, 14*(8), e0220976. https://doi.org/10.1371/journal.pone.0220976

Kam, J., Stowers, K., & Kim, S. (2019). Monitoring of drought awareness from Google Trends: A case study of the 2011–17 California drought. *Weather, Climate, and Society, 11*(2), 419–429. https://doi.org/10.1175/WCAS-D-18-0085.1

Kankanamge, N., Yigitcanlar, T., Goonetilleke, A., & Kamruzzaman, M. (2020). Determining disaster severity through social media analysis: Testing the methodology with South East Queensland flood tweets. *International Journal of Disaster Risk Reduction, 42,* 101360. https://doi.org/10.1016/j.ijdrr.2019.101360

Keegan, B. J., & Rowley, J. (2017). Evaluation and decision making in social media marketing. *Management Decision, 55*(1), 15–31. https://doi.org/10.1108/MD-10-2015-0450

Le Coz, J., Patalano, A., Collins, D., Guillén, N. F., García, C. M., Smart, G. M., et al. (2016). Crowdsourced data for flood hydrology: Feedback from recent citizen science projects in Argentina, France and New Zealand. *Journal of Hydrology, 541*, 766–777. https://doi.org/10.1016/j.jhydrol.2016.07.036

Mikolov, T., Le, Q. V., & Sutskever, I. (2013). Exploiting similarities among languages for machine translation. *ArXiv:1309.4168* [Cs]. http://arxiv.org/abs/1309.4168

Mishev, K., Gjorgjevikj, A., Vodenska, I., Chitkushev, L., & Trajanov, D. (2020). Evaluation of sentiment analysis in finance: From lexicons to transformers. *IEEE Access.* https://doi.org/10.1109/ACCESS.2020.3009626

Murthy, D., & Longwell, S. A. (2013). Twitter and disasters: The uses of Twitter during the 2010 Pakistan floods. *Information, Communication & Society, 16*(6), 837–855. https://doi.org/10.1080/1369118X.2012.696123

Nardi, F., Cudennec, C., Abrate, T., Allouch, C., Annis, A., Assumpção, T., et al. (2020). Citizens AND HYdrology (CANDHY): Conceptualizing a transdisciplinary framework for citizen science addressing hydrological challenges. *Hydrological Sciences Journal, 0*(0), 1–18. https://doi.org/10.1080/02626667.2020.1849707

Nielsen, R. K. (2015). *Local journalism: The decline of newspapers and the rise of digital media.* Bloomsbury Publishing.

Perdomo, V. (2022). *Colombia. Reuters Institute for the Study of Journalism.* https://reutersinstitute.politics.ox.ac.uk/digital-news-report/2021/colombia

Picanço, J., Mesquita, M. J., & Soares, L. F. (2017). The hydrological disasters through historical survey in the Serra do Mar Range, Southern Brazil. In M. Mikos, B. Tiwari, Y. Yin, & K. Sassa (Eds.), *Advancing culture of living with landslides* (pp. 1017–1026). Springer International Publishing. https://doi.org/10.1007/978-3-319-53498-5_115

Pires, A., Morato, J., Peixoto, H., Botero, V., Zuluaga, L., & Figueroa, A. (2017). Sustainability assessment of indicators for integrated water resources management. *Science of the Total Environment, 578*, 139–147. https://doi.org/10.1016/j.scitotenv.2016.10.217

Prensa (2015, December 17). *De quién son los medios*. https://www.monitoreodemedios.co/panoramaprensa/

Puschmann, C., & Powell, A. (2018). Turning words into consumer preferences: how sentiment analysis is framed in research and the news media. *Social Media + Society, 4*(3), 2056305118797724. https://doi.org/10.1177/2056305118797724

Rossi, C., Acerbo, F. S., Ylinen, K., Juga, I., Nurmi, P., Bosca, A., et al. (2018). Early detection and information extraction for weather-induced floods using social media streams. *International Journal of Disaster Risk Reduction, 30,* 145–157. https://doi.org/10.1016/j.ijdrr.2018.03.002

Smith, A., Porter, J. J., & Upham, P. (2017). "We cannot let this happen again": Reversing UK flood policy in response to the Somerset Levels floods, 2014. *Journal of Environmental Planning and Management, 60*(2), 351–369.

Smith, L., Liang, Q., James, P., & Lin, W. (2017). Assessing the utility of social media as a data source for flood risk management using a real-time modeling framework. *Journal of Flood Risk Management, 10*(3), 370–380. https://doi.org/10.1111/jfr3.12154

Tiempo, C. E. E. (2018, May 17). *Nueva emergencia en Hidroituango mantiene en vilo a 12 municipios. El Tiempo.* https://www.eltiempo.com/colombia/medellin/causas-de-la-nueva-emergencia-en-hidroituango-218440

Valdeavilla, D., & Pulido, M. (2019). *Bias in Filipino newspapers? Newspaper sentiment analysis of the 2017 Battle of Marawi*. Proceedings of the 4th International Conference on Internet of Things, Big Data and Security, 408–413. https://doi.org/10.5220/0007752104080413

Watts, G. (2016). Hydrology with impact: How does hydrological science inform decision-makers? *Hydrology Research, 47*(3), 545–551. Scopus. https://doi.org/10.2166/nh.2015.044

Yu, Y., Zhu, Y., Wan, D., Zhao, Q., Shu, K., & Liu, H. (2019). Applications of social media in hydroinformatics: a survey. *arXiv:1905.03035* [cs]. http://arxiv.org/abs/1905.03035

11

THREE-DIMENSIONAL CLUSTERING IN THE CHARACTERIZATION OF SPATIOTEMPORAL DROUGHT DYNAMICS: CLUSTER SIZE FILTER AND DROUGHT INDICATOR THRESHOLD OPTIMIZATION

Vitali Diaz[1,2], Gerald A. Corzo Perez[1], Henny A. J. Van Lanen[3], and Dimitri P. Solomatine[1,2,4]

[1] *IHE Delft Institute for Water Education, Delft, The Netherlands*
[2] *Delft University of Technology, Delft, The Netherlands*
[3] *Hydrology and Quantitative Water Management Group, Wageningen University, Wageningen, The Netherlands*
[4] *Water Problems Institute of the Russian Academy of Sciences, Moscow, Russia*

In its three-dimensional (3-D) characterization, drought is an event whose spatial extent changes over time. Each drought event has an onset and end time, a location, a magnitude, and a spatial trajectory. These characteristics help to analyze and describe how drought develops in space and time (i.e., drought dynamics). Methodologies for 3-D characterization of drought include a 3-D clustering technique to extract the drought events from the hydrometeorological data. The application of the clustering method yields small artifact droughts. These small clusters are removed from the analysis with the use of a cluster size filter. However, according to the literature, the filter parameters are usually set arbitrarily, so this study concentrated on

Advanced Hydroinformatics: Machine Learning and Optimization for Water Resources, Special Publications 78, First Edition. Edited by Gerald A. Corzo Perez and Dimitri P. Solomantine.

DOI: 10.1002/9781119639268.ch11

a method to calculate the optimal cluster size filter for the 3-D characterization of drought. The effect of different drought indicator thresholds to calculate drought is also analyzed. The approach was tested in South America with data from the Latin American Flood and Drought Monitor for 1950–2017. Analysis of the spatial trajectories and characteristics of the most extreme droughts is also included. Calculated droughts are compared with information reported at a country scale and a reasonably good match is found.

11.1. Introduction

In recent decades, methods for drought calculation have increasingly treated this phenomenon as an event characterized in space and time (Andreadis et al., 2005; Corzo Perez et al., 2011; Diaz et al., 2020a, 2020b; Herrera-Estrada & Diffenbaugh, 2020; Lloyd-Hughes, 2012; Sheffield et al., 2009; van Huijgevoort et al., 2013; Vernieuwe et al., 2020). Each drought event is calculated considering different characteristics, such as duration, spatial extent, and location. A better characterization of drought improves the analysis of its behavior and its possible effects on different economic and environmental sectors (World Meteorological Organization (WMO), 2006). Treating drought as an event with spatial extent has also allowed new methods for its monitoring and prediction to be proposed (Diaz et al., 2018, 2020a).

The approach to calculating drought in three dimensions (longitude, latitude, and time) has followed a gradual process. One of the first works to consider drought with a spatial extent (area) was Yevjevich (1967). He used spatially distributed synthetic precipitation data to define the drought areas. The time series of drought areas allowed him to calculate the onset, end, duration, and magnitude of drought. Another key work that marked a before and after in calculating drought as a phenomenon with spatial extension was the research of Andreadis et al. (2005). They introduced a methodology for the calculation of drought clusters. In their methodology, the drought areas are contiguous in space. Over a given region, for example, a country, one or more areas of drought can be distinguished at the same time. The different areas (2-D clusters) are calculated using a clustering technique. Later, this 2-D clustering-based method was expanded to three dimensions (i.e., latitude, longitude, and time). The works of Corzo Perez et al. (2011) and Lloyd-Hughes (2012) are some of the first examples in which drought is calculated as a 3-D object (i.e., a contiguous group of voxels). A voxel is a cube that represents a value in the 3-D space similar to a pixel that is a cell in the 2-D space (Fig. 11.1). Corzo Perez et al. (2011) and Lloyd-Hughes (2012) include new features in the characterization of drought, such as volume (number of voxels) and location. The location is defined as the centroid of the 3-D cluster. Subsequently, the centroids of the drought were used to build the spatial trajectory (i.e., drought tracking). In works such as those of Diaz et al. (2020a, 2020b), Herrera-Estrada et al. (2017), and Herrera-Estrada and Diffenbaugh (2020), spatial trajectories of drought are calculated, serving to analyze the dynamics of drought. The 3-D characterization of drought opens up other possibilities for drought monitoring and prediction. In addition to predicting drought duration and magnitude, its spatial extent, location, and trajectory could also be predicted (Diaz et al., 2018, 2020a).

Research gaps remain in the topic of 3-D drought characterization. When calculating 3-D drought clusters, hereafter also only clusters, it is common practice to apply a cluster size filter to remove small clusters resulting from the methodology and not from the droughts themselves. In most cases, the cluster size filter is chosen arbitrarily, or such a choice is generally driven by past work or experience. Another parameter commonly chosen arbitrarily is the drought indicator threshold (section 11.2.1) (i.e., the value for which a value of the hydrometeorological variable is considered a drought). Although there is extensive research on the latter, its combined effect with the cluster size filter has not been fully analyzed.

This research proposes a method to calculate the optimal cluster size filter for the 3-D characterization of drought. Droughts are calculated with a 3-D clustering technique using different cluster size filters and drought indicator thresholds. The analysis of the most extreme droughts in the analysis period is also presented. The methodology was tested in South America, using data from the Latin American Flood and Drought Monitor (LAFDM) (Sheffield et al., 2014; Wood et al., 2016) for 1950–2017.

11.2. Methods and Data

11.2.1. Drought Calculation

In this methodology, drought is represented by a 3-D cluster. The data must be organized in a grid system where each voxel is geographically referenced (latitude and longitude). The arrangement of these geospatial data voxels can be schematized as a cube with latitude, longitude, and time as its sides (Fig. 11.1). Each 3-D drought cluster is made up of voxels that indicate drought (Fig. 11.1). The 3-D clustering-based methodology considers three

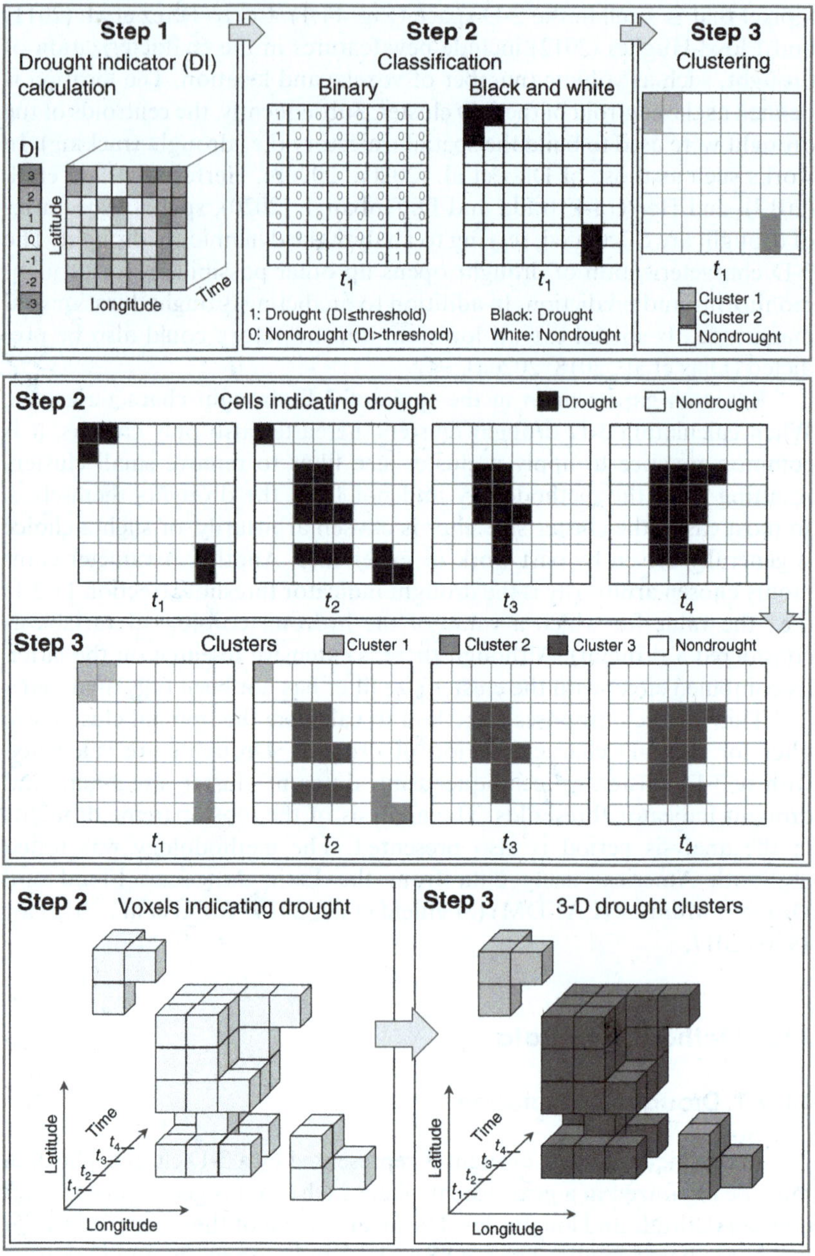

Figure 11.1 Schematic overview of the methodology for 3-D drought clusters calculation: Drought indicator calculation (Step 1), classification of voxels in drought (Step 2), and 3-D clusters calculation (Step 3). The middle and bottom panels provide a 2-D and 3-D view in each time step (t). Note that when we refer to the 2-D space, we use the term *cell*, while in the 3-D space, we use *voxel*.

main steps: (1) calculation of the drought indicator, (2) classification of the voxels in drought, and (3) calculation of the 3-D clusters (Fig. 11.1). After calculating the droughts, their characterization is carried out as described in section 11.2.2.

Step 1. Drought Indicator Calculation

The drought indicator is calculated with the hydrometeorological data. This indicator makes it possible to define anomalies in the data by using a statistic (Fig. 11.1, top panel). The standardized drought indicators, for example, standardize the values of the hydrometeorological variable. Using these standardized values, the drought is identified by means of a threshold associated with a statistical value that defines an anomaly as drought. Values lower than the threshold are considered as drought. Other indicators use a threshold value directly applied to the hydrometeorological variable for the entire analysis period or a moving time window. In this second method, the values that are below the threshold are also considered as drought.

Step 2. Classification of Voxels in Drought

After drought indicator calculation, the indication that a voxel is in drought is carried out with the drought indicator data. To indicate that a voxel is in drought, the binary classification is employed (i.e., the use of 1s and 0s; Fig. 11.1, top panel). In this way, drought is identified with 1s and nondrought with 0s, as indicated in equation (11.1). When the drought indicator (DI) is below a selected threshold (T), a cell is in drought. In equation (11.1), D_s stands for drought state (i.e., drought (1) or nondrought (0)).

$$D_S(t) = \begin{cases} 1 & \text{if } \mathrm{DI}(t) \leq T \\ 0 & \text{if } \mathrm{DI}(t) > T \end{cases}. \tag{11.1}$$

This research also analyzes the effect of the threshold (T) on the calculation of droughts (3-D clusters). The thresholds of 0, −1, −1.5, and −2 are tested. It is noted that, in standardized drought indices, the values equal to or below zero indicate "drought." Equation (11.1) is applied in each voxel in each time step (t).

Step 3. 3-D Clusters Calculation

The third step is the calculation of the 3-D clusters. A clustering method to extract the drought events is applied following Corzo Perez et al. (2011). The drought events correspond to the 3-D clusters calculated with the data of 0s and 1s (Fig. 11.1, middle and bottom panels). The following describes the unsupervised machine-learning-based method to extract the clusters. The method follows the connected-component labeling approach to cluster the voxels in drought (Haralick & Shapiro, 1992). In this method, a two-scan algorithm is applied. First, each cell is numbered. Then, the first run is carried out, in which the binary grid is explored and provisional labels are assigned to connected (contiguous) components (voxels). These labels identify the connection of every voxel with its nearest neighbors. In this first run, the voxel's label does not yet refer to the cluster number but to the voxels with which the given voxel is connected. Finally, the second scan is carried out to find similar voxel connections (i.e., clusters), which are given a unique label. Examination of the grid can be performed by columns, rows, or time. The clustering method is conducted over the whole binary data (Fig. 11.1, middle and bottom panels).

As mentioned, it is a common practice to remove small 3-D clusters that constitute artifact droughts produced by the clustering technique. This task uses a cluster size filter to clean the number of calculated 3-D clusters. This cluster size filter is usually set based on similar studies or experiences, or left at the default value. In this research, we propose a method to calculate the optimal cluster size filter introduced in the following. To test this method, we used the cluster size filters of 0, 4, 9, 16, 25, 36, 49, 64, 81, and 100 cells to remove clusters. The value of 0 indicates no cluster cleaning. Details on the data use are provided in section 11.2.3. The procedure for cluster filtering is as follows. For each time step (t), clustering was carried out to identify 2-D clusters. Afterward, the 2-D clusters below each cluster size filter were removed. Finally, the 3-D clusters were identified for each cleaned sample data. We carried out the procedure in this way because, owing to the large size of this region and the resolution of the data, extremely large events were identified when applying the 3-D clustering. The subsequent filtering of 3-D clusters was not practical (i.e., there were no or few small 3-D clusters to remove). By applying the cluster cleaning to the 2-D clusters in each time step, we removed small isolated areas (2-D clusters) throughout the region, which was found to be a more effective means of cleaning the 3-D cluster data.

For the identification of 3-D drought clusters, we also considered a filter of duration. All the 3-D clusters of 1 month duration were excluded from further analysis. We made sure that none of these clusters was of a large number of voxels, thus avoiding the risk of removing severe events. The 3-D clusters of 1-month duration were removed for each case of cluster size filter and drought indicator threshold.

Drought calculation concluded with the selection of the cluster size filter (Fig. 11.2). The optimal cluster size was defined as the value in the curve

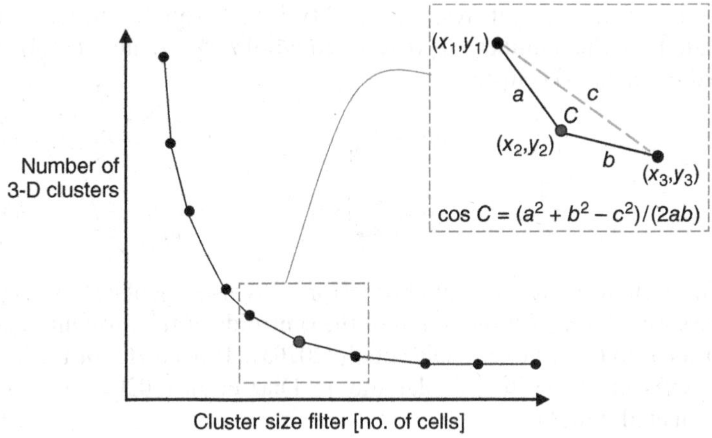

Figure 11.2 Scheme of the method to calculate the optimal cluster size filter. The angle *C* is calculated with the sides *a, b,* and *c* of the triangle formed by the three subsequent points (zoomed-in view).

"cluster size filter vs. the number of 3-D clusters," at which the number of clusters stabilizes (i.e., does not undergo considerable changes for a constant increase of the cluster size filter; Fig. 11.2). This point is defined as the vertex of the curve (Fig. 11.2). To identify this point, we developed a method based on the angle C formed between two continuous segments of the curve "cluster size filter vs. number of 3-D clusters" (Fig. 11.2). We calculated this angle C using the law of cosines (equation (11.2)), which considers the length of the sides of the triangle a, b, and c formed by the coordinates of three subsequent points (i.e., the points (x_1, y_1), (x_2, y_2), and (x_3, y_3); Fig. 11.2). For each triad of points, the angle C was calculated. The vertex corresponds to the smallest angle of all the calculated Cs. The optimal cluster size filter was calculated for each drought indicator threshold.

$$\cos C = (a^2 + b^2 - c^2)/(2ab). \tag{11.2}$$

11.2.2. Drought Characterization

After identifying the 3-D drought clusters, the onset and end in time, duration, and severity (magnitude) were calculated for each drought (3-D cluster) with a similar methodology to Diaz et al. (2020a).

The drought duration (*dd*) and magnitude (*ds*) were obtained with equations (11.3) and (11.4), respectively. The times *ti* and *tf* are the onset and end of each drought, respectively. DV is the drought cluster volume calculated as the number of voxels. Equations (11.3) and (11.4) were applied for each 3-D cluster.

$$dd = ti - tf + 1, \tag{11.3}$$

$$ds = \sum_{t=ti}^{tf} \mathrm{DV}(t). \tag{11.4}$$

For each drought, the spatial trajectory was also calculated. These trajectories were built with the union of the centroids of the drought areas at each time step (t), following Diaz et al. (2020a). Trajectories of the largest events were analyzed in a similar way to Diaz et al. (2020a, 2000b) and Hosseini et al. (2021).

11.2.3. Data

For the identification of droughts, we used data from the Standardized Precipitation Index (SPI) retrieved from the Latin American Flood and Drought Monitor (LAFDM) (https://platform.princetonclimate.com/PCA_Platform/lafdmLanding.html). Within LAFDM, the SPI is calculated with precipitation and numerically indicates the magnitude of the water anomaly, and runs from -3 to 3. Being below zero, SPI shows a dryness condition, whereas above zero, it indicates a wetness condition (McKee et al., 1993). The aggregation period of 6 months in the SPI, denoted by SPI6, is considered a good proxy for monitoring drought condition on the surface (i.e., on runoff and soil moisture), which is more relevant to the potential impact on agricultural activities (WMO, 2012). Thus, SPI6 data were considered for the calculation of droughts. The period of the analysis was 1950–2017 (816 months) on a monthly basis. The spatial resolution was 0.25 deg.

11.3. Results and Discussion

11.3.1. Drought Calculation

Figures 11.3 and 11.4 show the duration, magnitude (number of voxels), and number of 3-D drought clusters (nc) for each drought indicator threshold and cluster size filter. Figure 11.3 presents the results for

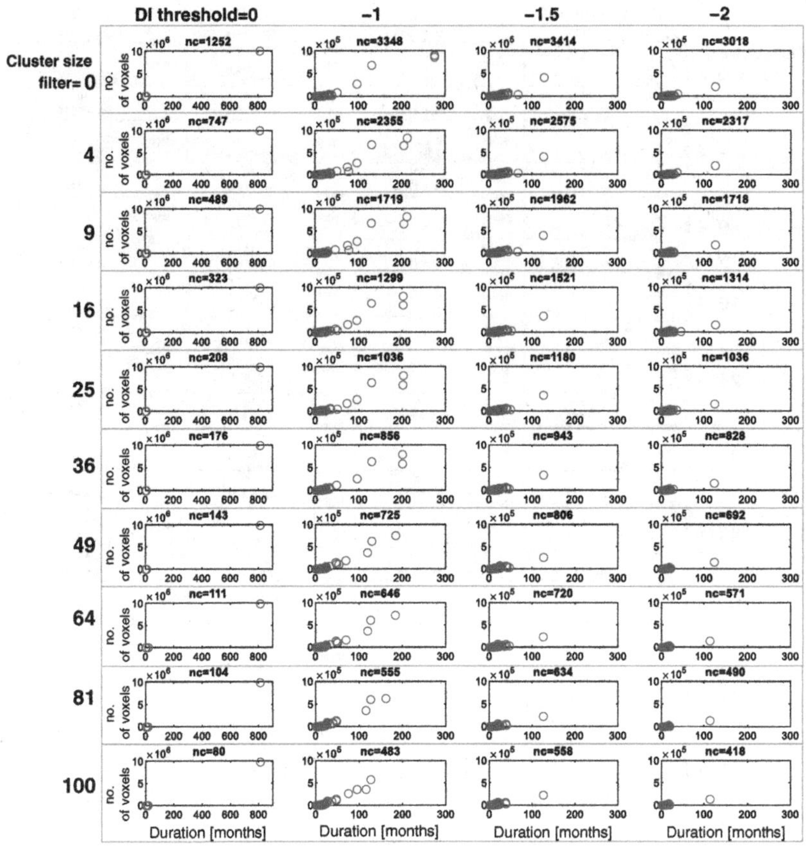

Figure 11.3 Number of 3-D clusters (nc) calculated for different drought indicator (DI) thresholds and cluster size filters. The duration (months) and magnitude (number of voxels) are indicated. Note that the number of voxels for the first threshold is up to 10×10^6, while for the rest it is up to 10×10^5.

all the durations and magnitudes. For better visualization of the results, Figure 11.4 was prepared, which deploys the droughts for durations up to 20 months and 10,000 voxels. Results show that, for the threshold of 0, the least number of droughts is achieved (Figs. 11.3 and 11.4). These clusters are smaller in quantity, but their structure is made up of more voxels than those of the other thresholds, as shown in Figure 11.7. As the drought indicator threshold decreases (from 0 to −2), more clusters are identified,

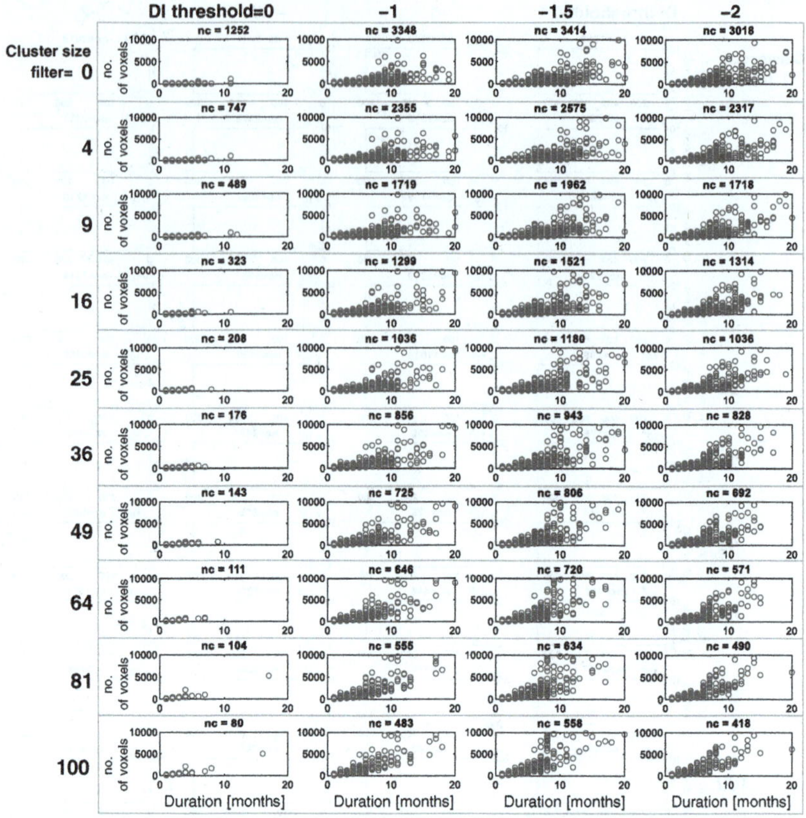

Figure 11.4 As Figure 11.3 but for durations up to 20 months and sizes up to 10,000 voxels.

although they have fewer voxels (Figs. 11.3 and 11.4). This increase in the number of clusters is not constant; there is a decrease when the threshold has a small value, such as drought indicator ≤−2. The latter indicates that the number of clusters with extreme drought (drought indicator ≤−2) tends to be smaller. In the case of the cluster size filter (number of voxels), the decrease in the number of clusters is more evident than in the drought indicator threshold (Figs. 11.3 and 11.4). Results show that when the cluster size filter increases, it separates large clusters into smaller clusters with shorter durations.

Figures 11.3 and 11.4 display the clusters of 1 month duration. Although these clusters have a number of voxels greater than the cluster size filter in each case, they are small compared to the rest of the clusters

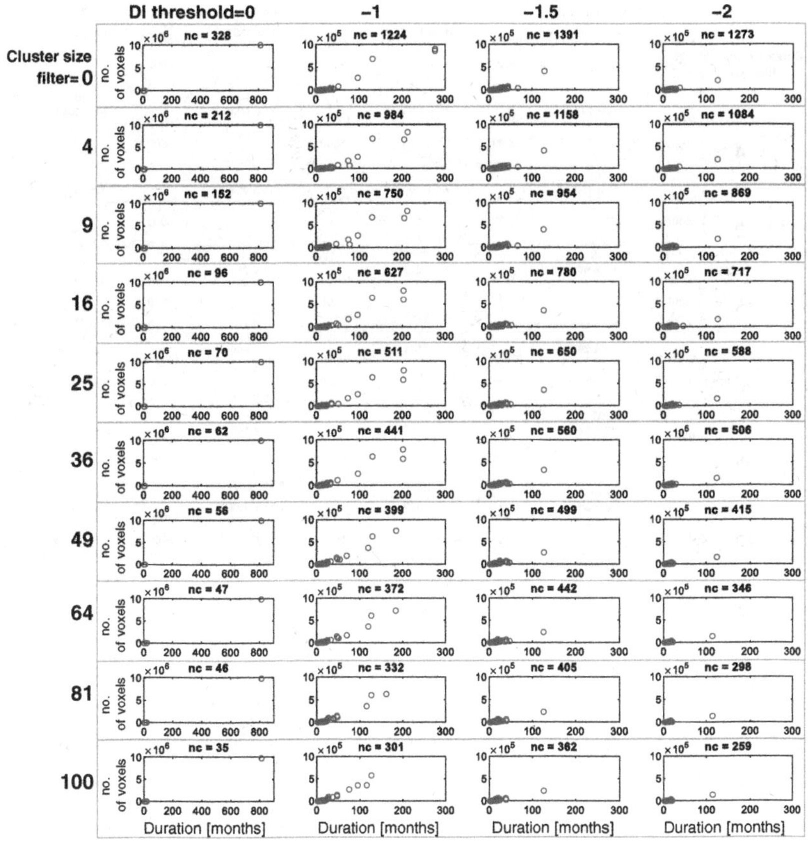

Figure 11.5 Number of 3-D clusters (nc) calculated for different drought indicator (DI) thresholds and cluster size filters. The duration (months) and magnitude (number of voxels) are indicated. Note that the number of voxels for the first threshold is up to 10×10^6, while for the rest it is up to 10×10^5. In these results, the 3-D clusters of 1 month duration were excluded.

in each case (Figs. 11.3 and 11.4); for this reason, they were also removed. Figures 11.5 and 11.6 show the final results after removing the 1-month duration clusters. Figure 11.5 shows the droughts for all durations and magnitudes. In Figure 11.5, scarce events with durations greater than 100 months and a considerable number of voxels are observed; most droughts have durations of less than 20 months. Figure 11.6 shows the results for droughts of up to 20 months and 10,000 voxels. A direct relationship is observed between duration and magnitude.

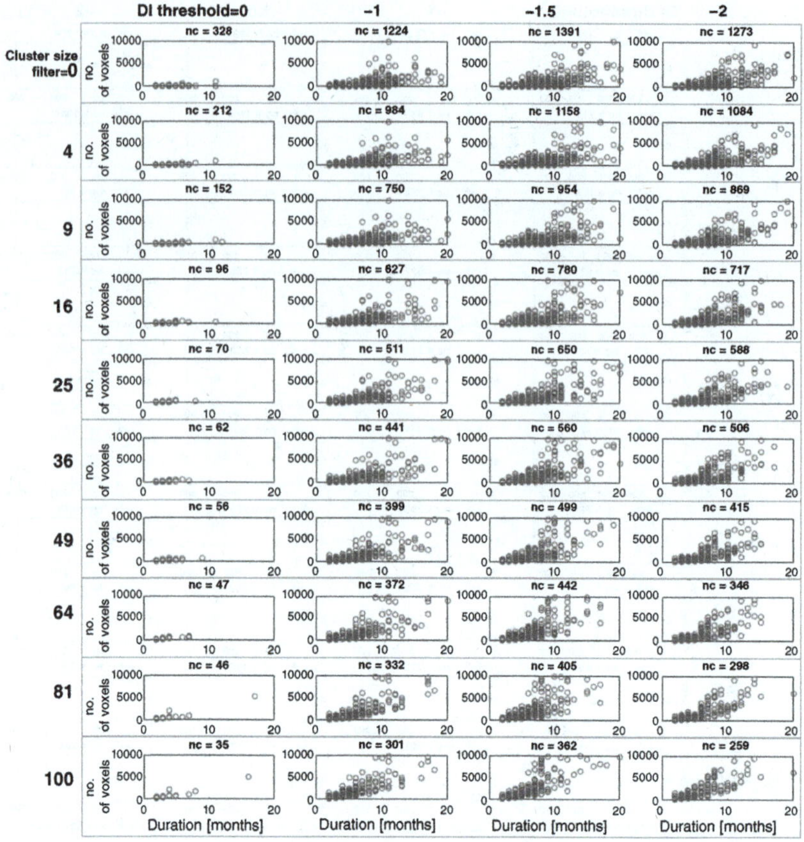

Figure 11.6 As Figure 11.5 but for durations up to 20 months and sizes up to 10,000 voxels. Figure 11.7 shows the percentages of drought areas (PDAs) without considering the 1 month duration clusters. PDAs were calculated for each 3-D cluster as the number of cells of each drought area at each time step divided by the total number of cells of the region (24,877).

This figure helps visually compare the durations and magnitudes, here as the percentage of drought area, of the events for each threshold. Figure 11.7 shows a drought event with a duration almost equal to the analysis period when the threshold is 0. This result shows that, even in this large study area, there are consecutive drought areas in time, with at least 20% of the total study area that are connected to each other and form the long-lasting 3-D cluster. For the threshold of 0, the use of the cluster size filter does not show any significant difference. In the other

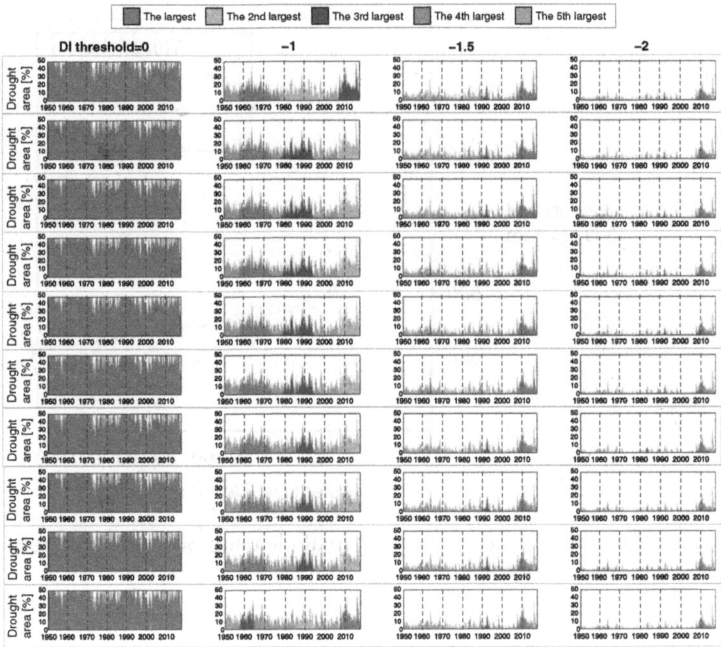

Figure 11.7 Percentage of drought area calculated for each 3-D cluster. Results for each drought indicator (DI) threshold and cluster size filter are presented. Drought areas are shown for the period 1950–2017. The five largest droughts are indicated by shading.

thresholds, more events with shorter durations and less extensive areas are observed. In general, as the threshold decreases, indicating a more severe drought, the events are smaller in magnitude (size) and duration. The results also show that the 2010–2020 decade had more events with considerable magnitudes that indicated severe and extreme drought.

The number of 3-D clusters for each drought indicator threshold and cluster size filter is presented in Figure 11.8. The method developed to find the optimal cluster size filter (section 11.2.1) shows, for instance, that, for the threshold of −1, the cluster size filter is 49 cells. Table 11.1 shows the results for each of the thresholds. In the following sections, the results of the drought characterization are shown for this optimal cluster size filter. For the case of the drought indicator threshold, we focused our analysis on the threshold of −1. In general, the −1.5 and −2 thresholds produced more events but with shorter durations and magnitudes (sizes) (Fig. 11.7). The threshold of 0 produced a long-lasting event and some small events.

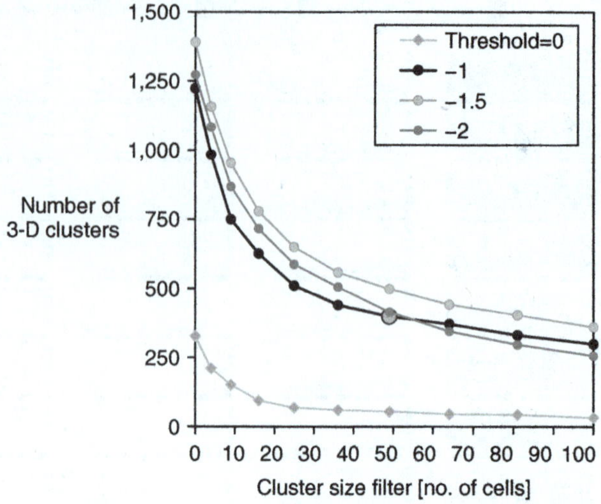

Figure 11.8 Identification of the optimal cleaning filter size. The result for the threshold of −1 is indicated.

Table 11.1 Angle C (Degrees) for each Drought Indicator (DI) threshold and cluster size filter.

Cluster size filter (number of cells)	DI threshold = 0	−1	−1.5	−2
0	–	–	–	–
4	177.2	179.7	179.6	179.9
9	177.6	178.0	179.1	178.7
16	168.0	178.8	178.3	178.6
25	145.1*	175.5	177.0	176.4
36	168.7	171.7	174.9	179.5
49	173.8	**168.1***	177.3	175.9
64	152.4	174.0	170.1*	172.8*
81	153.3	171.5	179.2	173.5
100	–	–	–	–

Note: The lowest *C* angle is indicated with * (asterisk). For the threshold of −1, the result is indicated in bold.

Based on the results, this threshold of 0 is not recommended for drought calculation.

11.3.2. Drought Characterization

A total of 399 drought events were identified. Figure 11.9 shows that the distribution of the duration and magnitude (number of cells) is mainly concentrated in durations of less than 20 months and 10,000 cells. Few events lasting around 24 months or longer are observed. The linear correlation coefficient between the duration and magnitude is $R^2 = 0.89$, indicating an almost linear relationship between these two characteristics.

The centroids of each of the 3-D clusters are shown in Figure 11.10a. The centroids are classified by their duration in four intervals: 2 to 6, 7 to 12, 13 to 24, and 24 or more months. The centroids of the clusters of 2 to 6 months are observed practically throughout the entire study area, as well as the centroids of the 7 to 12-month events. Those of durations of 13 to 24 months are also seen, although with less density. Most of the centroids of 3-D clusters with durations greater than 24 months are located in the central region of the study area, although two centroids are located outside this region, one in the south and one in the north. Figure 11.10b shows the number of 3-D clusters counted in each cell. In general, 14 or more clusters

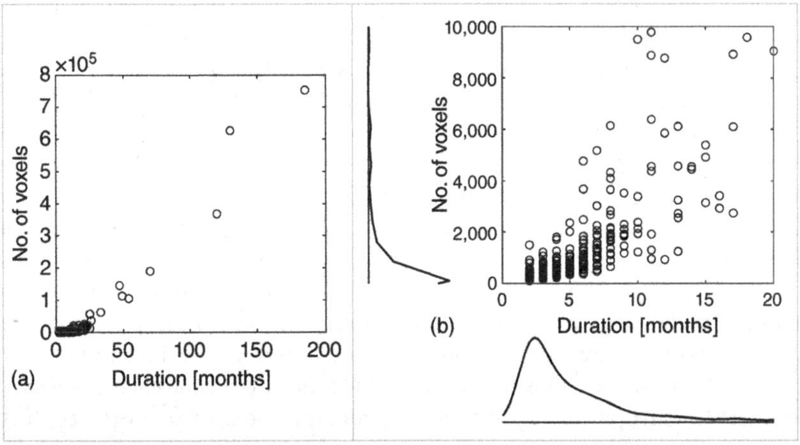

Figure 11.9 (a) Duration (months) and magnitude (number of voxels) of the droughts for the period 1950–2017. (b) Detail for drought durations up to 20 months and 10,000 voxels.

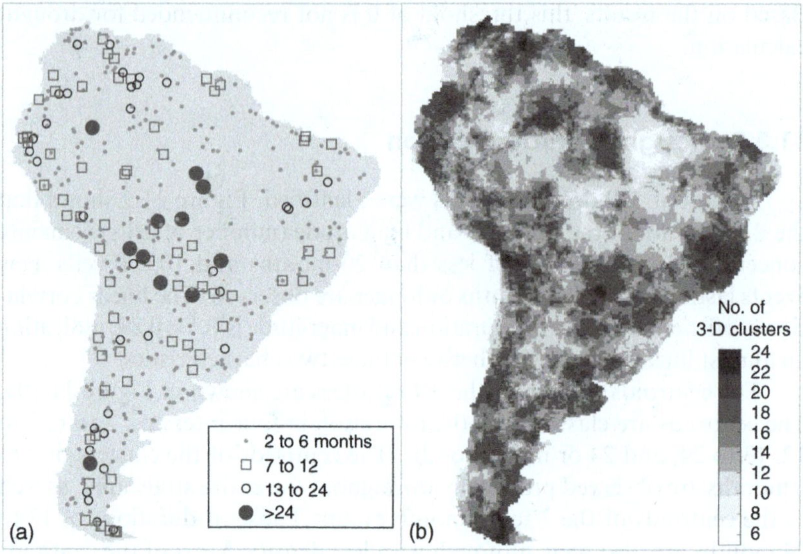

Figure 11.10 (a) Centroids of the 3-D clusters for the period 1950–2017. Centroids are classified by cluster duration (months). (b) Occurrence of the 3-D clusters: Each cell indicates the number of 3-D clusters that took place there in the period 1950–2017.

are observed over the study area, except in some areas of the Amazon basin, the upper Magdalena River basin in Colombia, the north of Uruguay, the northwest of Argentina, and the south of Chile. One of the areas with the least occurrence is the lower Amazon basin, near the discharge.

The following paragraphs present the results of three of the most extreme droughts: those of 1957–1972, 2006–2017, and 1984–1994. The drought with the longest duration is shown in Figure 11.11. The drought lasted from May 1957 to September 1972 (185 months). The 3-D cluster is shown in Figure 11.11a. It is observed that this cluster completely covered the region (Fig. 11.11a,b). However, its spatial distribution over time varied significantly, concentrating mainly in the north and south of the subcontinent. Figure 11.11b shows the distribution of the duration of each cell in the study area. This figure shows the amount of time each cell was in drought for this particular drought. In general, the central, north-central, and south-central regions had the longest amount of time in drought. The spatial trajectories of this drought are shown in Figure 11.11c. It is observed that the centroids of each drought area follow a pattern in which the trajectories go from northeast to south, northwest, and back

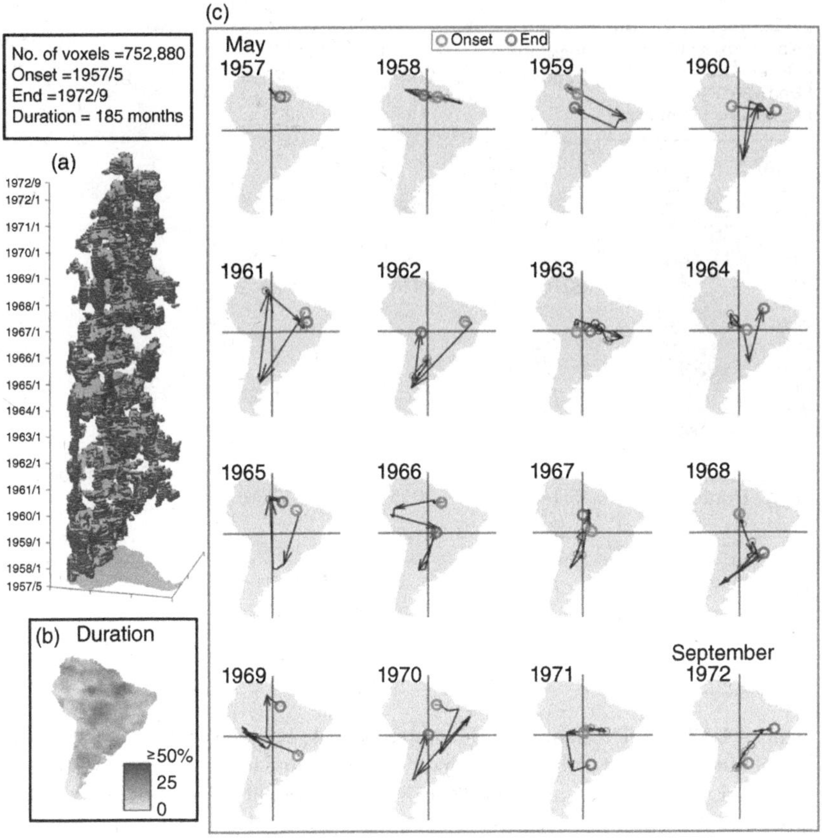

Figure 11.11 (a) Drought from May 1957 to September 1972. (b) Percentage of duration: Each cell shows the percentage of time in drought with respect to the duration of the indicated drought. (c) Monthly drought trajectories per year; the onset and the end of the drought trajectory are indicated for each year by circles.

to northeast, mostly in a clockwise direction, although the direction is counterclockwise in the last 2 yr.

The second-longest drought is shown in Figure 11.12. This drought lasted from December 2006 to September 2017 (130 months). The 3-D cluster is shown in Figure 11.12a. The event was mainly concentrated in the southwestern and southern coast of South America and some areas of Brazil (Fig. 11.12a,b). The trajectories of this drought presented in Figure 11.12c show a different dynamic from the drought shown previously (Fig. 11.11c). The trajectories in the first years are almost diagonal, from

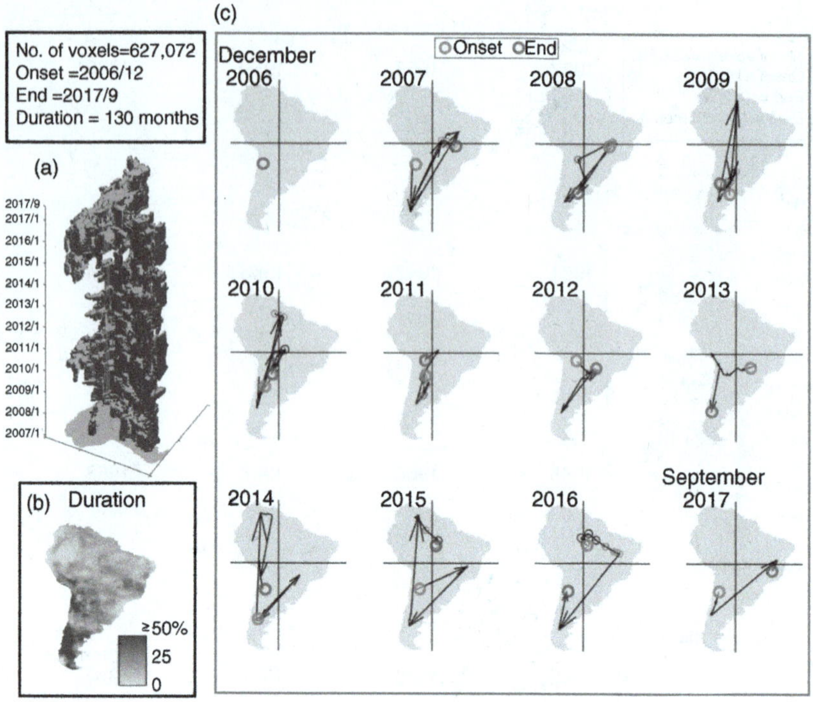

Figure 11.12 (a) Drought from December 2006 to September 2017. (b) Percentage of duration: Each cell shows the percentage of time in drought with respect to the duration of the indicated drought. (c) Monthly drought trajectories per year; the onset and the end of the drought trajectory are indicated for each year by circles.

southwest to northeast. Later, the trajectories are more concentrated in the center and then again extend to the southwest and northeast. At the end of the duration, more trajectories are shown on the coasts of the region.

The third-longest drought is shown in Figure 11.13. The 3-D cluster spans almost the entire study area, although it was mostly concentrated in the northwest and south of the subcontinent (Fig. 11.13a,b). The largest extent was observed in the period from 1988 to 1991 (Fig. 11.13a). The trajectories of this drought show three main patterns (Fig. 11.13c): they run from northeast to south in the first years, they then change from north to south, and finally, they are located in the north.

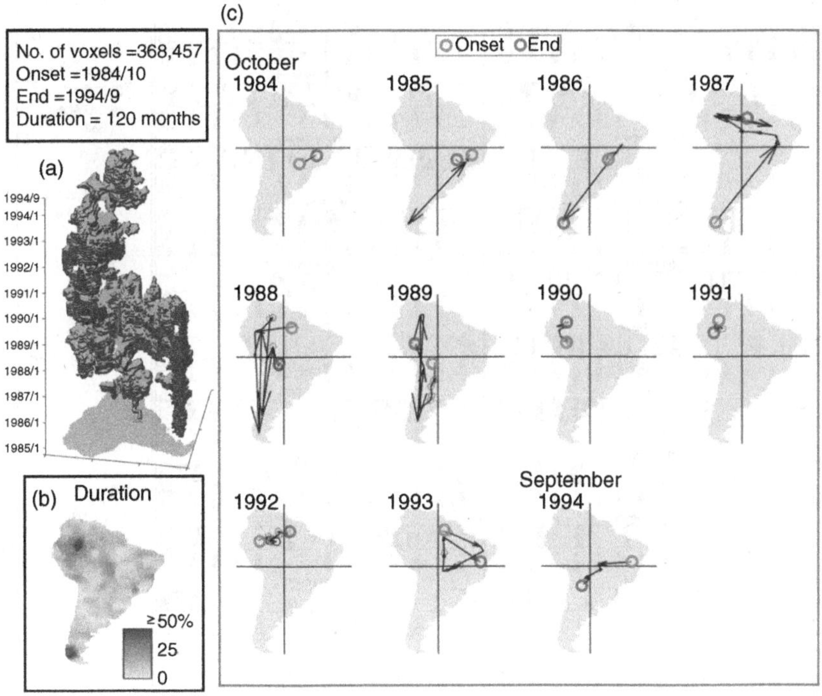

Figure 11.13 (a) Drought from October 1984 to September 1994. (b) Percentage of duration: Each cell shows the percentage of time in drought with respect to the duration of the indicated drought. (c) Monthly drought trajectories per year; the onset and the end of the drought trajectory are indicated for each year by light and dark circles.

11.3.3. Comparison with the Reported Droughts

Figure 11.14 (left) shows the droughts reported in two sources that compile the occurrence and the information on the impacts of droughts in some of the region's countries. The first source is the Emergency Events Database (EMDAT) (Guha-Sapir, 2019). The second source is the *Atlas of Droughts in Latin America and the Caribbean*, which was carried out by the Regional Water Center for Arid and Semi-arid Zones of Latin America and the Caribbean (CAZALAC) (UNESCO and CAZALAC, 2018). In both documents, droughts are not reported for Suriname and French Guiana and are therefore not included in this analysis.

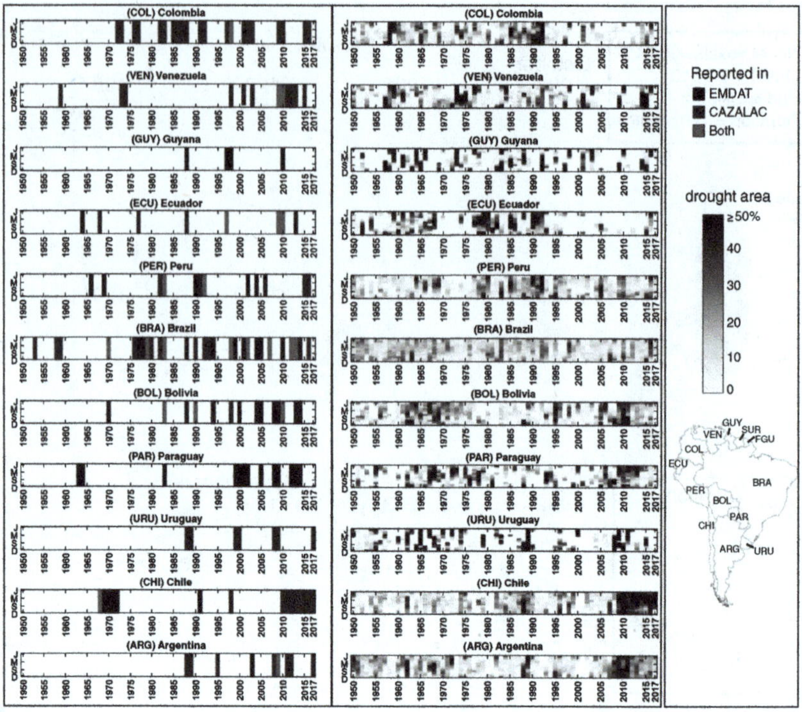

Figure 11.14 Reported droughts in the EMDAT and CAZALAC (left). Percentage of drought area calculated for each country (right).

The percentages of drought area calculated for each of the countries are also shown in Figure 11.14 (right). The results show that, in general, the occurrence of the calculated droughts coincides with the information reported. The results indicate some periods with important drought events in the 1960s, 1980s, 1990s, 2000s, and 2010s. In the most recent decades (the 2000s and 2010s), Brazil, Bolivia, Paraguay, Chile, and Argentina have experienced more droughts than in the rest of the period. The percentage of drought area shows that drought extent increases in the second semester of the year and concludes in the first semester of the following year, coinciding with the growing period of various crops of the region. These droughts can compromise the optimal soil moisture conditions necessary for the crops. According to EMDAT, Brazil is one of the countries that faced the most economic losses due to droughts in the period of analysis. The drought event presented in Figure 11.12 that lasted from 2006 to

2017 mainly encompassed Bolivia, Paraguay, Chile, Argentina, and part of Brazil, as shown by the percentages of drought area in Figure 11.14 (right). The megadrought in Chile and Argentina during the period of 2010–2017 (Garreaud et al., 2020) is also captured.

11.4. Summary and Conclusions

This research introduces a method of calculating the optimal cluster size filter for the 3-D characterization of drought. The combined effect of the cluster size filter and different drought indicator thresholds to calculate drought is also presented. The methodology was tested in South America with data from the Latin American Flood and Drought Monitor (LAFDM) for 1950–2017.

The following conclusions are drawn:

1. The drought indicator thresholds of 0, −1, − .5, and −2 were tested. In general, the −1.5 and −2 thresholds produce more drought events with shorter durations and smaller magnitudes (sizes) than the threshold of −1. The threshold of 0 produces a long-lasting drought event and some small events. Based on the results, this threshold is not recommended as a method of calculating 3-D drought clusters.
2. The optimal cluster size filter depends on the spatial resolution of the data and the threshold used.
3. Durations and magnitude (number of voxels) are mainly concentrated in less than 20 months and 10,000 voxels. Few events lasting around 24 months or longer are observed. A linear relationship between these two characteristics is found ($R^2 = 0.89$).

The main findings for South America are as follows:

1. Droughts of 2 to 6 months are observed practically throughout the entire study area, as well as 12-month droughts.
2. Some regions show the little occurrence of droughts, such as the Amazon basin, the upper Magdalena River basin in Colombia, the north of Uruguay, the northwest of Argentina, and the south of Chile. One of the areas with the least occurrence is the lower Amazon basin, near the discharge.
3. The 1957–1972, 2006–2017, and 1984–1994 droughts were the most extreme.

In general, the occurrence of the calculated droughts coincides with the information reported. The 1960s, 1980s, 1990s, 2000s, and 2010s were the periods with more droughts.

In the most recent decades (the 2000s and 2010s), Brazil, Bolivia, Paraguay, Chile, and Argentina have shown a greater occurrence of droughts than in the rest of the period.

Further research may include extension and testing of this methodology on other types of drought indicators. A more detailed study of the characteristics of droughts is also recommended. The relationship between drought and the South American Low-Level Jet phenomenon (Montini et al., 2019), the moisture transporter from the Amazon to the subtropics, is another interesting topic that could be explored in further studies. The results of this study are important for the calculation and characterization of drought and better monitoring and the construction of future drought forecasting systems in the region.

Acknowledgments

Vitali Diaz thanks the Mexican National Council for Science and Technology (CONACYT) and Alianza FiiDEM for the study grand 217776/382365. Henny A. J. Van Lanen is supported by the H2020 ANYWHERE project (Grant Agreement No. 700099). Dimitri P. Solomatine acknowledges the grant No. 17-77-30006 of the Russian Science Foundation, and the Hydroinformatics research fund of IHE Delft in whose framework some research ideas and components were developed. The study is also a contribution to the UNESCO IHP-VII program (Euro FRIEND-Water project) and the Panta Rhei Initiative on Drought in the Antropocene of the International Association of Hydrological Sciences (IAHS).

References

Andreadis, K. M., Clark, E. A., Wood, A. W., Hamlet, A. F., and Lettenmaier, D. P. (2005). Twentieth-Century Drought in the Conterminous United States. *Journal of Hydrometeorology*, *6*(6), 985–1001. https://doi.org/10.1175/JHM450.1

Corzo Perez, G. A., van Huijgevoort, M. H. J., Voß, F., and van Lanen, H. A. J. (2011). On the spatio-temporal analysis of hydrological droughts from global hydrological models. *Hydrology and Earth System Sciences*, *15*(9), 2963–2978. https://doi.org/10.5194/hess-15-2963-2011

Diaz, V., Corzo Perez, G. A., Van Lanen, H. A. J., and Solomatine, D. (2018). *Intelligent drought tracking for its use in Machine Learning: implementation and first results*. (G. La Loggia, G. Freni, V. Puleo, & M. De Marchis, Eds.), *HIC 2018. 13th International Conference on Hydroinformatics* (Vol. *3*). Palermo: EasyChair. https://doi.org/10.29007/klgg

Diaz, V., Corzo Perez, G. A., Van Lanen, H. A. J., Solomatine, D., and Varouchakis, E. A. (2020a). An approach to characterize spatio-temporal drought dynamics.

Advances in Water Resources, *137*, 103512. https://doi.org/10.1016/j.advwatres.2020.103512

Diaz, V., Corzo Perez, G. A., Van Lanen, H. A. J., Solomatine, D., and Varouchakis, E. A. (2020b). Characterisation of the dynamics of past droughts. *Science of The Total Environment*, *718*, 134588. https://doi.org/10.1016/j.scitotenv.2019.134588

Garreaud, R. D., Boisier, J. P., Rondanelli, R., Montecinos, A., Sepúlveda, H. H., and Veloso-Aguila, D. (2020). The Central Chile Mega Drought (2010–2018): A climate dynamics perspective. *International Journal of Climatology*, *40*(1), 421–439. https://doi.org/https://doi.org/10.1002/joc.6219

Guha-Sapir, D. (2019). EM-DAT: The Emergency Events Database - Université catholique de Louvain (UCL) - CRED. Retrieved from www.emdat.be

Haralick, R. M., and Shapiro, L. G. (1992). *Computer and Robot Vision, Volume I*. Addison-Wesley.

Herrera-Estrada, J. E., and Diffenbaugh, N. S. (2020). Landfalling Droughts: Global Tracking of Moisture Deficits From the Oceans Onto Land. *Water Resources Research*, *56*(9). https://doi.org/10.1029/2019WR026877

Herrera-Estrada, J. E., Satoh, Y., and Sheffield, J. (2017). Spatio-Temporal Dynamics of Global Drought. *Geophysical Research Letters*, 2254–2263. https://doi.org/10.1002/2016GL071768

Hosseini, A., Ghavidel, Y., and Farajzadeh, M. (2021). Characterization of drought dynamics in Iran by using S-TRACK method. *Theoretical and Applied Climatology*, *145*(1–2), 661–671. https://doi.org/10.1007/s00704-021-03656-3

Lloyd-Hughes, B. (2012). A spatio-temporal structure-based approach to drought characterisation. *International Journal of Climatology*, *32*(3), 406–418. https://doi.org/10.1002/joc.2280

Mckee, T. B., Doesken, N. J., and Kleist, J. (1993). The relationship of drought frequency and duration to time scales. *AMS 8th Conf. Appl. Climatol.*, *179*(January), 179–184. https://doi.org/citeulike-article-id:10490403

Montini, T. L., Jones, C., and Carvalho, L. M. V. (2019). The South American Low-Level Jet: A New Climatology, Variability, and Changes. *Journal of Geophysical Research: Atmospheres*, *124*(3), 1200–1218. https://doi.org/10.1029/2018JD029634

Sheffield, J., Andreadis, K. M., Wood, E. F., and Lettenmaier, D. P. (2009). Global and continental drought in the second half of the twentieth century: Severity-area-duration analysis and temporal variability of large-scale events. *J. Clim.*, *22*(8), 1962–1981. https://doi.org/10.1175/2008JCLI2722.1

Sheffield, J., Wood, E. F., Chaney, N., Guan, K., Sadri, S., Yuan, X., … Ogallo, L. (2014). A Drought Monitoring and Forecasting System for Sub-Sahara African Water Resources and Food Security. *Bulletin of the American Meteorological Society*, *95*(6), 861–882. https://doi.org/10.1175/BAMS-D-12-00124.1

UNESCO and CAZALAC (2018). *Atlas de sequías en Amercia Latina y El Caribe*. (J. Núñez Cobo & K. Verbist, Eds.). Retrieved from http://unesdoc.unesco.org/images/0026/002658/265894s.pdf

van Huijgevoort, M. H. J., Hazenberg, P., van Lanen, H. A. J., Teuling, A. J., Clark, D. B., Folwell, S., … Uijlenhoet, R. (2013). Global Multimodel Analysis of Drought in Runoff for the Second Half of the Twentieth Century. *Journal of Hydrometeorology*, *14*(5), 1535–1552. https://doi.org/10.1175/JHM-D-12-0186.1

Vernieuwe, H., De Baets, B., and Verhoest, N. E. C. (2020). A mathematical morphology approach for a qualitative exploration of drought events in space and time. *International Journal of Climatology*, *40*(1), 530–543. https://doi.org/10.1002/joc.6226

Wood, E. F., Fisher, C., Sheffield, J., and Chaney, N. (2016). *Overview of the Latin American and Caribbean Flood and Drought Monitor (LACFDM)*. Foz de Iguazú, Brazil. Retrieved from https://www.cazalac.org/mwar_lac/fileadmin/imagenes2/Remote_Sensing/Princeton_Day1_Overview_and_Station_Merging_light.pdf

World Meteorological Organization (WMO). (2006). *Drought monitoring and early warning: concepts, progress and future challenges. WMO-No. 1006*. Geneva, Switzerland. Retrieved from http://www.droughtmanagement.info/literature/WMO_drought_monitoring_early_warning_2006.pdf

World Meteorological Organization (WMO). (2012). *Standardized Precipitation Index user guide. WMO-No. 1090*. Geneva, Switzerland. Retrieved from https://library.wmo.int/doc_num.php?explnum_id=7768

Yevjevich, V. (1967). *An objective approach to definitions and investigations of continental hydrologic droughts. Hydrology Paper 23* (Vol. 23). Fort Collins, Colorado. Retrieved from https://dspace.library.colostate.edu/bitstream/handle/10217/61303/HydrologyPapers_n23.pdf

12

DEEP LEARNING OF EXTREME RAINFALL PATTERNS USING ENHANCED SPATIAL RANDOM SAMPLING WITH PATTERN RECOGNITION

Han Wang[1] and Yunqing Xuan[2]

[1] *China Institute of Water Resources and Hydropower Research, Beijing, China*
[2] *Zienkiewicz Centre for Computational Engineering, Swansea University, Swansea, United Kingdom*

An effective pattern recognition technique based on incorporating an unsupervised clustering algorithm into a spatial random sampling toolbox (SRS-GDA; Wang & Xuan, 2020) is used to identify and classify extreme rainfall patterns in England, Wales, and Scotland in different sizes over the last century. The spatial features of the rainfall patterns, such as their geographic location, size, shape, and orientation, can be automatically extracted from the clustering analysis. The top three of the dominating patterns are identified and the temporal variation of their attributes are also investigated. The enhanced toolbox presents great potential in auto-labeling clusters to support deep learning of complex environmental spatial-temporal features over large data sets, demonstrated by an example of convolution neural network, which is able to pick up the labeled rainfall patterns with high accuracy.

Advanced Hydroinformatics: Machine Learning and Optimization for Water Resources, Special Publications 78, First Edition. Edited by Gerald A. Corzo Perez and Dimitri P. Solomantine.

DOI: 10.1002/9781119639268.ch12

12.1. Introduction

Rainfall is one of the most essential variables in climate and water resources research, which has complex and nonlinear relationship with various meteorological and climatological variables and topographic conditions (Kirono et al., 2010). Many approaches and models have been developed for simulating and predicting rainfall based on such relationship. For example, the generalized linear model (GLM) is a widely used tool for quantifying the relation between rainfall and other hydroclimatic predictors (Urdiales et al., 2018; Yan et al., 2002; Yunus et al., 2017), such as temperature, potential evaporation, wind speed, El Niño, La Niña, and Southern Oscillation indices, as well as spatial predictors (Beecham et al., 2014; Kenabatho et al., 2017; Wang & Xuan, 2021), such as altitude, geographic location, size, and shape. However, as the topography does not frequently change over time and the large-scale structure of atmospheric circulation remains relatively stable, the spatial and temporal rainfall distribution can be bounded within several specific patterns. For example, the rainfall patterns in the windward regions are normally different from those in the leeward region, especially in the mountainous area (Lin et al., 2017); desert areas usually receive limited rainfall because of high temperature heating the descending air, which hinders the formation of clouds to cause rainfall. Therefore, identification and classification of such rainfall patterns is of great value to help understand underlying mechanism and improve the forecasting capability, especially for those based on deep learning methods, which have seen a fast-growing number of applications recently. For example, in hydroclimatic science, some researchers applied machine learning methods such as artificial neural networks (ANN) to extract features of spatiotemporal climatic variables (Qiu et al., 2017); others parameterize those features (Gentine et al., 2018; O'Gorman & Dwyer, 2018) to realize weather predictions (Liu et al., 2016; Petersik & Dijkstra, 2020).

However, appropriate training is essential and often challenging in the application of deep learning techniques, in order to achieve an accurate and reliable prediction, and to expend the applications in climate science. Unlike the data traditionally used to develop neural network algorithms such as face images for facial recognition and classification, which have commonalities and specific features that can be easily classified, the hydroclimatic data from model projections or observations are often highlighted by strong and intermittent spatiotemporal variation, and revealed as being chaotic, nonstationary, multiscale, and correlated.

The other main challenge is due to the amount of training data required by deep learning techniques. For example, a convolution neural network

(CNN) usually demands a large data set for training, which should be all precatalogued into different labels/classes. This apparently requires huge time and labor if done manually. To address these issues, we incorporated an unsupervised clustering technique, the K-means method, into an existing spatial random sampling toolbox for grid-based data analysis (SRS-GDA toolbox; Wang & Xuan, 2020). The SRS-GDA toolbox can help to reduce drastically the overhead of sampling data for training; and the new addition of the machine learning component makes it possible to classify and extract spatial temporal features from large environment observations data sets. The enhanced toolbox is tested using the daily rainfall images extracted from the GEAR data set (Tanguy et al., 2016), which is a grid-based data set (1 km × 1 km) of daily rainfall covering Great Britain (700 × 1,250 grids) over the period of 1898–2010. The daily rainfall is the total rainfall amount over a 24 hr period from 9 a.m. of the reporting day until 9 a.m. on the following day. The chosen data set shares many features of other large environmental modeling results and observations, for example, climate reanalysis and projections, such as those from ERA40 (Uppala et al., 2005), CMIP5 (Taylor et al., 2012). Therefore, the example given in this chapter can be readily expanded to cover many other similar cases.

There are three main indices used for quantifying the patterns of the variable (daily rainfall in this study), (1) geographic location, (2) pattern size, and (3) orientation. The rainfall patterns over the last century in Great Britain (GB) was first extracted using the toolbox automatically before several significant patterns in England, Wales, and Scotland were recognized, respectively. The efficacy of the toolbox in deep learning is demonstrated by the training of a CNN model to automatically recognize the rainfall patterns from a test set of 1,166 daily rainfall in GB.

The remainder of this chapter is organized as follows: section 12.2 describes the toolbox and pattern identification and extraction; section 12.3 shows the rainfall patterns classified over ahundred years in GB and the pattern recognition by CNN. Finally, the conclusions and recommendations of further study are given in section 12.4.

12.2. Toolbox

12.2.1. Spatial Random Sampling Toolbox for Grid-Based Data Analysis (SRS-GDA)

The Spatial Random Sampling toolbox for Grid-Based Data Analysis (SRS-GDA) was originally designed by Wang and Xuan (2020), which is used to generate arbitrary regions of interests (ROIs) from grid-based hydroclimatic data sets, aiming to address the needs of quantifying hydroclimatic responses that vary with location and scale. The features of the sampled ROIs can be either randomly set or customized by a variety of constraints such as shape, size, location, dominant orientation, and resolution. The sampling process is conducted within a boundary where there are valid records, over the ROIs defined by the three main parameters, which are the centroid coordinates defining the geographic location, the ROI size, and the shape parameter *sp* defined as a ratio between the dimension with the north-south direction D_{NS} and the west-east direction D_{WE} $\left(sp = \frac{D_{NS}}{D_{WE}}\right)$.

In this study, we continue to use the first two parameters to define the ROIs but improve the *sp* to be a ratio of the major axis and the minor axis (instead of N-S versus W-E). A new parameter ω is also added to define the orientation of this ROI, that is, the angle between the major axis of the ROI and the north direction.

12.2.2. Pattern Recognition and Feature Extraction of Rainfall Data

The original SRS-GDA toolbox does not have the capacity to directly support the deep learning of grid-based environment data set. A new pattern recognition with feature labeling is implemented in this study. The structure of the enhanced SRS-GDA toolbox is shown in Figure 12.1. The improvement lies in that it can now extract important features/attributes from the input images or data sets by an unsupervised clustering algorithm.

As shown in Figure 12.1, there are two different types of inputs to the toolbox: images or data sets. If the inputs are rainfall images, there is a transformation carried out before detecting rainfall patterns. The color scale of the images is normally represented by RGB (red, green, and blue channels) in image processing, which has some drawbacks because it produces a nonlinear and discontinued space and thus is affected by illumination changes. In this study, a transformation is applied to convert the colors from the RGB space to the CIELab color space (Tkalcic & Tasic, 2003), where a color is specified by three values: *L* for the lightness from black (0) to white (100); *a* from green to red; and *b* from blue to yellow. The use of the CIELab color space has helped improve the performance of the toolbox in quantifying the visual differences of colors.

The inputs fed to the toolbox are organized in a format of 3-D arrays whose third dimension are set as the time reference. Therefore, a 2-D slice of these arrays represents the spatial distribution (a 2-D image) of the hydroclimatic variable, and in this case, daily rainfall. In the example shown in

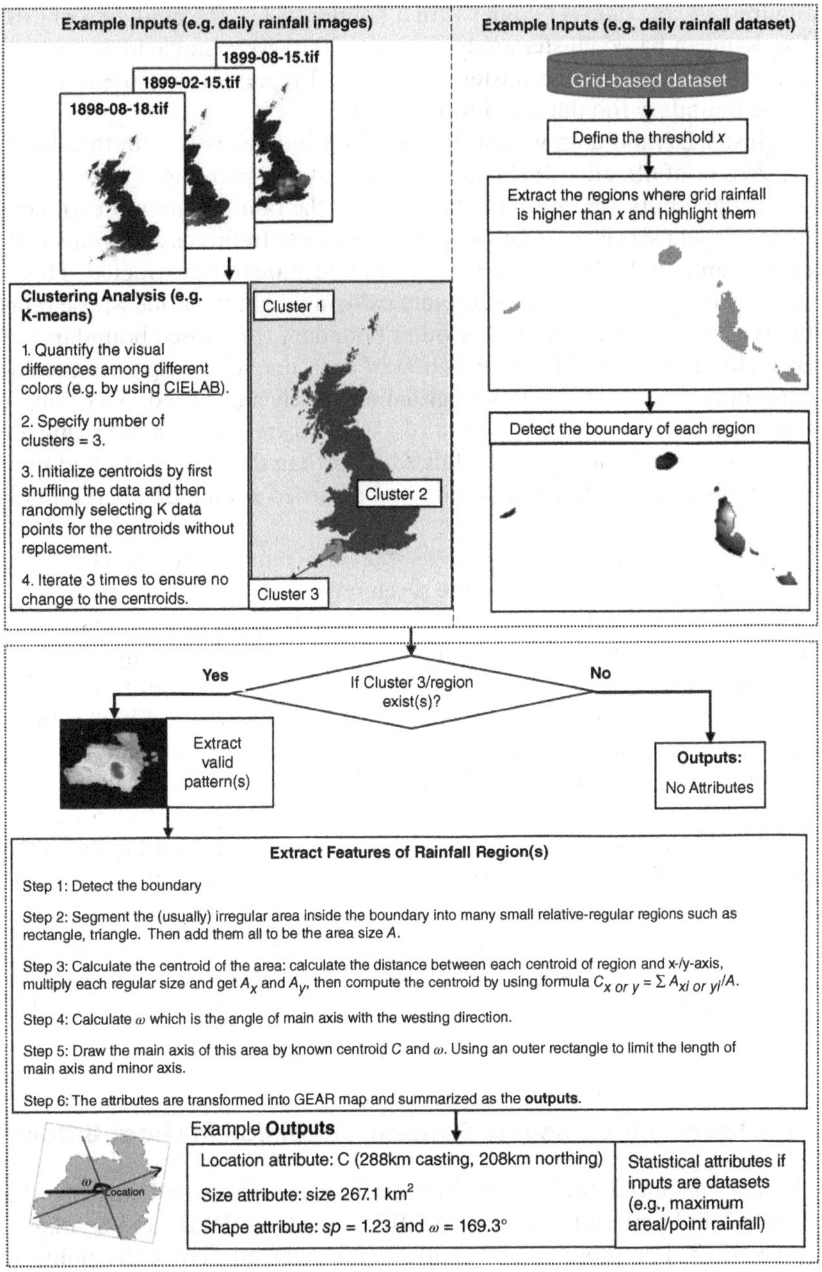

Figure 12.1 The new cluster analysis and pattern segmentation component designed for the new SRS-GDA toolbox.

Figure 12.1, the data set comes with a boundary, i.e., the land area of GB. The K-mean based cluster analysis is applied to every single image, which normally generates three clusters: (1) cluster 1 represents and area outside of the boundary (no data, or invalid data);
(2) cluster 2 (valid data within the boundary but below certain threshold, e.g. zero rainfall), and (3) cluster 3 of valid rainfall patterns.

If the inputs are rainfall data sets, only the rainfall amount extracted from the data sets is used for pattern recognition. In this case, regions with rainfall amount higher than a given threshold x are to be extracted. This is done by using the toolbox to automatically highlight the grids whose rainfall is over x before detecting the outer boundary (the closed boundary) of each cluster of grids. The information of each boundary (e.g., the coordinates of the boundary grid) is recorded separately and stored. An example is given in the right panel of Figure 12.1 where there are seven different patterns recognized whose grid rainfall is higher than the threshold x and each of them is within a closed boundary and regarded as an individual rainfall event.

Each rainfall pattern, such as the one represented by cluster 3, extracted from images or from the seven regions detected from grid-based data sets, is then polygonised using the original function provided by the toolbox. The resultant polygons (regions) of rainfall are usually large and irregular. They are then segmented into small regular shapes to calculate the centroid and the moments of the compound shape. Finally, three attributes are generated for the segmented shapes (regions): the location attribute, the size attribute, and the shape attribute, which includes both the sp and the major angle ω which indicates the angle between easting and the major axis (northward; see the arrow in Fig. 12.2) and in the range of 0° and 180°. The three attributes are then used to label and classify the rainfall pattern. Besides, when the inputs are data sets, in addition to the three spatial attributes, other statistical attributes of interest can be extracted from the detected regions for quantitative analysis (e.g., extracting the maximum areal or point rainfall within the boundary of the region).

12.3. Identify the Specific Patterns of Rainfall in Great Britain

The enhanced toolbox is then used to recognize and identify the daily rainfall pattern from 1898 to 2010 in England, Wales and Scotland, respectively. For those rainfall patterns that spread across the political boundaries, their detected geometric centroids are used to determine which country they belong to. For example, a pattern is deemed to be of England

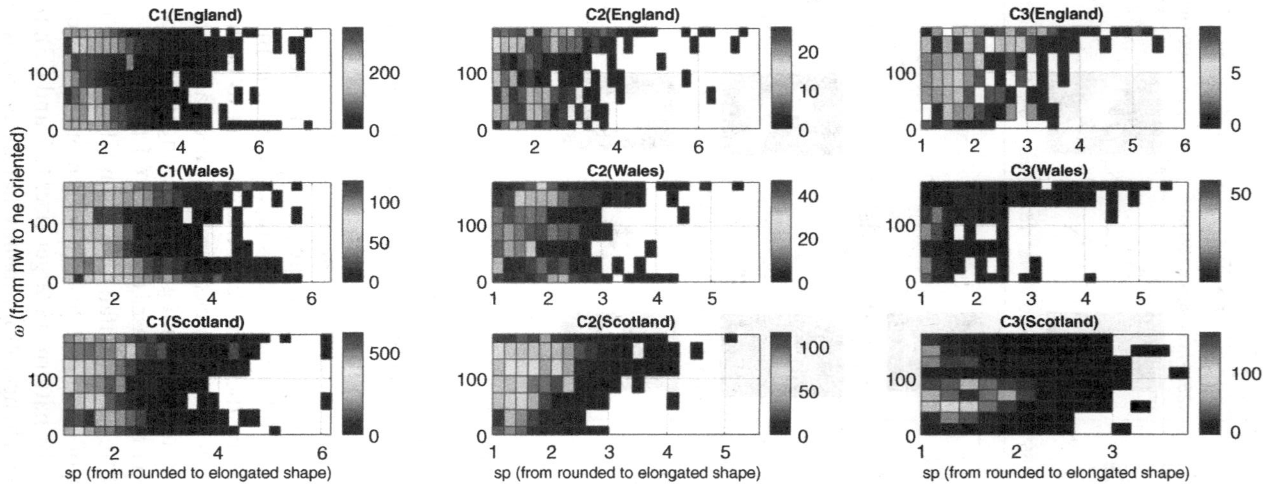

Figure 12.2 The histograms of the number of daily rainfall pattern in England, Wales, and Scotland displayed by different colors with respect to *sp* and *ω* at three sizes (C1, C2, and C3). The largest occurrence number of each category indicated as the darkest color at the top of the color bar is 45 ~ 75,135 ~ 165 and = 1 ~ 1.2 for C1(England); 0 ~ 15 and = 1 ~ 1.4 for C1(Wales); 15 ~ 45 and = 1 ~ 1.6 for C1(Scotland); 15 ~ 45, 45 ~ 75 and = 1 ~ 1.6, 1.4 ~ 1.8 for C2(England); 135 ~ 165 and = 1 ~ 1.4 for C2(Wales); 15 ~ 45 and = 1 ~ 1.4 for C2(Scotland); 135 ~ 165 and = 1.6 ~ 2.6 for C3(England); 105 ~ 135 and = 1 ~ 1.2 for C3(Wales); 60 ~ 80 and = 1.4 ~ 1.6 for C3(Scotland).

Table 12.1 Specific patterns of rainfall in three countries cataloged at three sizes.

Location	C1 size ≤ 250 km²	C2 250 km² < size ≤ 500 km²	C3 size > 500 km²
England	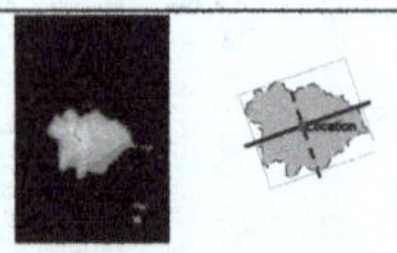 *sp*: 1.0~1.2; *ω*: 135°~165°	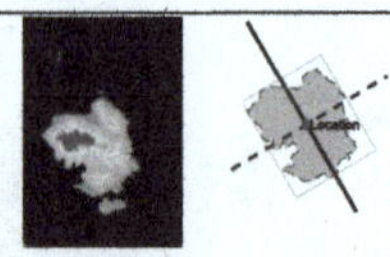 *sp*: 1.0~1.6; *ω*: 45°~75°	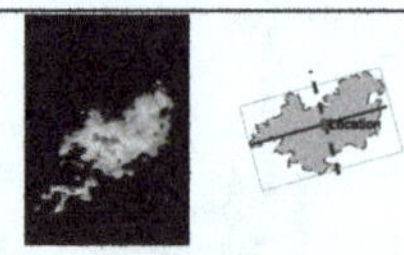 *sp*: 1.6~2.6; *ω*: 135°~165°
	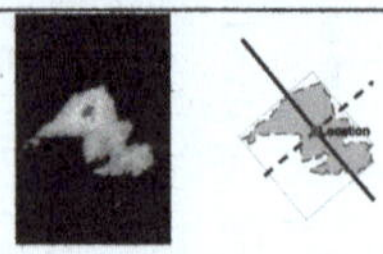 *sp*: 1.0~1.2; *ω*: 45°~75°	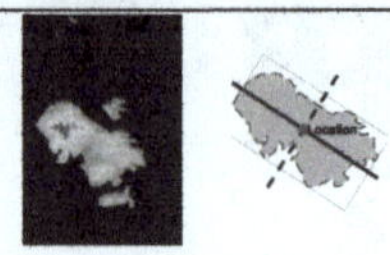 *sp*: 1.4~1.8; *ω*: 15°~45°	-
Wales	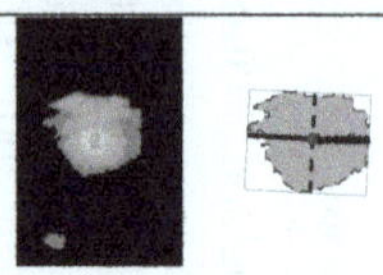 *sp*: 1.0~1.4; *ω*: 0~15°	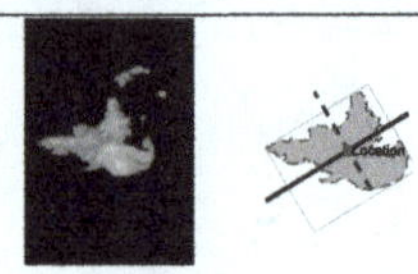 *sp*: 1.0~1.4; *ω*: 135°~165°	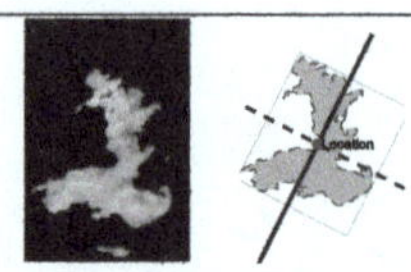 *sp*: 1.0~1.2; *ω*: 105°~135°
Scotland	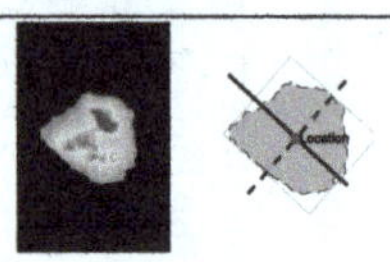 *sp*: 1.0~1.6; *ω*: 15°~45°	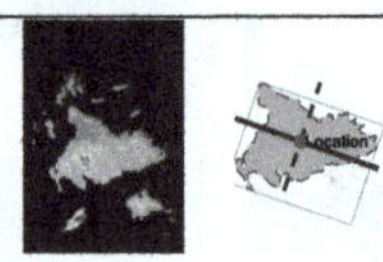 *sp*: 1.0~1.4; *ω*: 15~45°	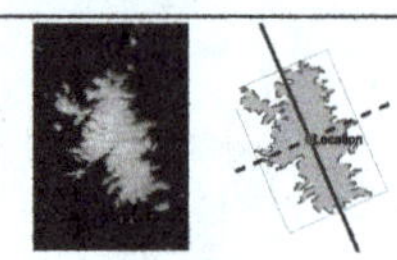 *sp*: 1.4~1.6; *ω*: 60°~80°

if its geometric centroid is within England. The rainfall patterns have been divided into three classes according to their sizes: (1) C1 (size $\leq 250\,\text{km}^2$), (2) C2 ($250\,\text{km}^2 < \text{size} \leq 500\,\text{km}^2$) and (3) C3 ($\text{size} > 500\,\text{km}^2$). The patterns with the most frequent occurrence are recorded and summarized in Figure 12.2 and several specific patterns are presented in Table 12.1.

Figure 12.2 shows the number of patterns that occurred in the period of over 100 yr, represented by the color scale with respect to both *sp* and *ω*. For all three countries in general, a clear trend of decrease in patterns associated with large *sp* values (indicating narrow, elongated shapes) can be seen as the size of the pattern increases (i.e., from C1 to C3).

In England, the most frequent rainfall patterns in C1 are those mainly oriented in the direction of northeast to east with more rounded shape ($1.0 < sp < 1.2$). However, for patterns in larger sizes (C2 and C3), the orientation of the elongation changes from mainly northwest (ω is around 45°) to northeast-by-east ($\omega \geq 135°$). This finding may be explained by the dynamics of the rainfall systems in various sizes. Small size of rainfall distribution is likely to be driven by localized convective storms, which tend to have a more regular shape. Large rainfall system may come as a result of frontal systems, hence having a narrow band shape. Another very interesting feature is that intensive rainfall usually happened in a small region. The other two typical patterns of middle size in England are usually located in Lake District and nearby, while the one with a size larger than 500 km^2 is usually across the whole south of England.

In Wales, the typical pattern of the less than 250 km^2 category is more rounded with a slight east-west orientation ($\omega \leq 15°$), while the pattern of middle size becomes northeast-by-east-orientated and less rounded, usually with a concave in the north-west direction because of topographic effect and this pattern has a relative fixed location at northwest of Cardiff near the boundary of Wales and England (location index is easting 350 km, northing 250 km). The pattern larger than 500 km^2 is also with a very common *L* shape, which is almost across the north and south of Wales. It should be noted that the underlying data set is in fact masked by the land boundary; therefore, patterns of large size (C3) are artificially limited by the shape of the land.

In Scotland, the pattern of less than 125 km^2 is also more rounded and usually located north of the Lake District near the boundary of England and Scotland. The middle-sized pattern, whose orientation is the same as the smaller rainfall pattern, has two common locations: one is the north of Lake District and the other is the north of Highland. The pattern with the size over 500 km^2 is northwest-orientated and normally covers the whole north and west of Highland.

As for presenting the temporal change of rainfall patterns in the last century, the entire duration is divided into five groups. Table 12.2 lists the dominating daily rainfall pattern(s) of each group at three different sizes in England, Wales, and Scotland, where the percentage in bracket presents the proportional number of patterns.

In England, the percentage of the patterns in C2 stays stable while that for C1 decreases and C3 increases after 1947. For the patterns in C1, their shapes change slightly from a more rounded shape to an ellipse and the orientation shifts from northeast-by-east to northwest-by-north after 1922. For the patterns of the C2 category, both their shapes and orientations are stable and in the range of 1 to 1.8 and northwest, respectively. For the patterns in C3, their shapes are more elongated and the orientation changes from northeast-by-east to northwest after 1997.

In Wales, a very different temporal variation has been detected. First, the fractions of the patterns in different size categories remain almost unchanged. Second, for the individual size group, patterns in

Table 12.2 Temporal change of specific rainfall patterns in three countries in three size categories.

		1898–1922	1923–1947	1948–1972	1973–1997	1998–2010
England	C1	(89.2%) sp: 1 ~ 1.2 ω: 135° ~ 165°	(90.6%) sp: 1 ~ 1.2 ω: 45° ~ 75°	(89%) sp: 1.2 ~ 1.4 ω: 45° ~ 75°	(89%) sp: 1.2 ~ 1.4 ω: 45° ~ 75°	(88.9%) sp: 1.2 ~ 1.4 ω: 45° ~ 75°
	C2	(8%) sp: 1.6 ~ 1.8 ω: 15° ~ 45° sp: 1 ~ 1.2 ω: 45° ~ 75°	(7.8%) sp: 1.4 ~ 1.6 ω: 45° ~ 75°	(7.9%) sp: 1 ~ 1.4 ω: 15° ~ 75°	(8%) sp: 1.8 ~ 2.0 ω: 15° ~ 45° sp: 1.0 ~ 1.2 ω: 45° ~ 75°	(7.9%) sp: 1.6 ~ 1.8 ω: 45° ~ 75°
	C3	(2.8%) sp: 2.4 ~ 2.6 ω: 135° ~ 165°	(1.6%) sp: 1 ~ 2.8 ω: 75° ~ 165°	(3.1%) sp: 1.6 ~ 1.8 ω: 135° ~ 165°	(3%) sp: 1 ~ 1.2 ω: 105° ~ 135°	(3.2%) sp: 1.4 ~ 1.6 ω: 45° ~ 75°
Wales	C1	(77.2%) sp: 1.2 ~ 1.4 ω: 0 ~ 15°	(80%) sp: 1.2 ~ 1.4 ω: 0 ~ 15°	(79.8%) sp: 1.4 ~ 2.0 ω: 0 ~ 45°	(80.3%) sp: 1.8 ~ 2.0 ω: 0 ~ 15° sp: 1.4 ~ 1.6 ω: 45° ~ 75°	(80.6%) sp: 2 ~ 2.2 ω: 15° ~ 45°
	C2	(17.6%) sp: 1 ~ 1.2 ω: 135° ~ 165°	(16.2%) sp: 1.2 ~ 1.4 ω: 135° ~ 165°	(13.7%) sp: 1.2 ~ 1.4 ω: 105° ~ 135° sp: 2.2 ~ 2.4 ω: 0 ~ 15°	(13.8%) sp: 1.6 ~ 1.8 ω: 135° ~ 165°	(14%) sp: 1 ~ 1.2 ω: 105° ~ 135°
	C3	(5.2%) sp: 1 ~ 1.2 ω: 105° ~ 135°	(5.8%) sp: 1 ~ 1.2 ω: 105° ~ 135°	(6.5%) sp: 1 ~ 1.2 ω: 105° ~ 135°	(5.9%) sp: 1 ~ 1.2 ω: 105° ~ 135°	(5.4%) sp: 1 ~ 1.2 ω: 15° ~ 45° sp: 1 ~ 1.2 ω: 75° ~ 135°
Scotland	C1	(74.2%) sp: 1 ~ 1.2 ω: 165° ~ 180°	(76.8%) sp: 1.2 ~ 1.4 ω: 15° ~ 45°	(83.8%) sp: 1.4 ~ 1.6 ω: 15° ~ 45°	(83.8%) sp: 1.4 ~ 1.6 ω: 15° ~ 45°	(83.8%) sp: 1.2 ~ 1.4 ω: 135° ~ 165°
	C2	(12.6%) sp: 1 ~ 1.2 ω: 15° ~ 45°	(11.9%) sp: 1 ~ 1.2 ω: 15° ~ 45°	(8.9%) sp: 1 ~ 1.2 ω: 45° ~ 75°	(8.8%) sp: 1.4 ~ 1.6 ω: 45° ~ 75°	(8.9%) sp: 1.4 ~ 1.6 ω: 45° ~ 75°
	C3	(13.2%) sp: 1.2 ~ 1.4 ω: 30° ~ 60°	(11.3%) sp: 1.5 ~ 1.8 ω: 60° ~ 90°	(7.3%) sp: 1.5 ~ 1.8 ω: 60° ~ 90°	(7.4%) sp: 1.2 ~ 1.5 ω: 60° ~ 90°	(7.3%) sp: 1.2 ~ 1.8 ω: 60° ~ 90°

C1 become more and more elongated while their main orientation remains northwest-by-west; patterns in C2 do not change much in their elongated shape, but their orientation shifts from northeast-by-east to northeast-by-north. For patterns in the C3 category, all have a similar shape and orientation (as shown in Table 12.1) and do not change over time. Again, the boundary impact mentioned previously may be the cause of the behavior.

In Scotland, the situation of percentage is just the opposite to what has been shown in England. The percentage of patterns in C1 increases while those of C2 and C3 both decrease, especially after 1947. The shape of the patterns in C1 varies from a more rounded one to rectangular, which is similar to that of England, but the orientation changes from northwest-by-west to northeast-by-east. For the patterns in C2, their shapes appear to be more and more elongated and the orientation shifts from northwest-by-west to northwest-by-north after 1947. As in Wales, the shape and orientation of the patterns in C3 are almost unchanged during this time.

12.4. Generating the Training Sets for Deep Learning Studies

The new integrated classification capacity of the SRS-GDA toolbox makes it possible to automatically produce labeled training sets from tens of thousands of large-scale environmental images for deep-learning applications in feature recognition and forecasting. In this process, the attributes such as those spatial features (location, size, and orientation) can be quickly identified to precatalogue the data or images into different labels/classes. To present this, an Alexnet model of convolution neural network (CNN) is built to demonstrate the process of auto-identifying the daily rainfall patterns in GB, using the training sets auto-labeled by the enhanced SRS-GDA toolbox. There are four labels used in this test: (1) no rainfall (L0), (2) concentric pattern (L1), (3) elongated pattern (L2), and (4) compound pattern (L3), as shown in Figure 12.3.

As the overhead demand for computation (e.g., more than 40,000 daily rainfall images), Alexnet, which is a type of CNN, was constructed for training because it reduces the overfitting and allows for multi-GPU training, which reduces the training time (Krizhevsky et al., 2012). There are eight layers in the Alexnet model, with the first five being convolutional layers with the kernel size of 11×11, 5×5, and 3×3 and the last three fully connected layers. The structure is designed as the first convolutional layer (11×11) following by an overlapping max pooling layer, connected to the second convolutional layer (5×5), which is also followed by one overlapping max pooling layer. The remaining three convolutional layers

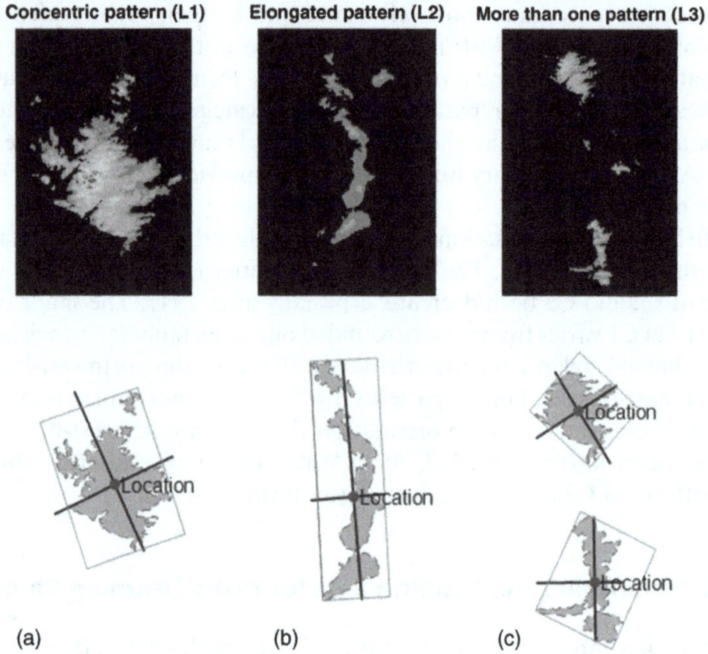

Figure 12.3 Attribute for labels (a) L1, (b) L2, and (c) L3 used in CNN.

(3 × 3) are connected directly and the end links to one overlapping the max pooling layer. The final link is to the last three fully connected layers. ReLU, a nonlinear activation function, is applied after all the convolution and fully connected layers to get activation values corresponding to neurons. To apply the Alexnet model, we first converted all images to the acceptable size, which is 227-by-227-by-3 where the last dimension (size 3) indicates the three color channels (e.g., R, G, B). Then we modified the last three layers to specify four classes for classification and set a faster learning rate for these newly modified layers and a slower learning rate for the transferred layers in order to obtain an effective and better transfer learning. Finally, 1,166 daily rainfall images are randomly selected from all images while the rest are used for training the Alexnet model.

During the training phase, the Alexnet of CNN model learns the training sets and selects 30% for self-validation randomly; the model validates the network every three iterations. The testing sets (1,166 images) are applied for predicting their classifications. The self-validation accuracy is 91.3% when reaching the final iteration, and the accuracy of prediction of the testing sets has been evaluated as 93.4% (1,089 out of 1,166 new figures have been classified correctly). However, the inherent complexities of the environmental data cause complicated patterns as well. Further work is recommended to make the toolbox more robust to process climate projection data at various, often low resolutions. A downscaling or down-sampling method may need to be integrated. Certainly, optimizing the CNN algorithm should be carried out for recognizing more complex patterns other than the simple example shown in this paper.

12.5. Conclusion

We present in this chapter the development of pattern recognition component in an enhanced random sampling toolbox SRS-GDA for automatic identification and classification of extreme rainfall patterns by its location, size, and shape (orientation), using a daily rainfall data set of Great Britain over more than one century. The specific rainfall patterns in England, Wales, and Scotland of different sizes are then derived and analyzed. We also demonstrate how the new function can support deep learning of spatial features form complex environmental data sets, using an example of the CNN model, which is able to pick up typical daily rainfall patterns from this century-long data set. Further work is recommended to investigate closely the underlying data sets with respect to potential inconsistency in the resolution of the data. In addition, a further study with long-term, finer spatiotemporal resolution observations, as well as event orientated study such as stratiform events in the winter and convective events in the summer is likely to make conclusions more robust.

References

Beecham , S., Rashid, M., & Chowdhury, R. K. (2014). Statistical downscaling of multi-site daily rainfall in a South Australian catchment using a Generalized Linear Model. *International Journal of Climatology, 34*(14), 3654–3670.

Gentine, P., Pritchard, M., Rasp, S., Reinaudi, G., & Yacalis, G. (2018). Could machine learning break the convection parameterization deadlock? *Geophysical Research Letters, 45*(11), 5742–5751.

Kenabatho, P., Parida, B., & Moalafhi, D. (2017). Evaluation of satellite and simulated rainfall products for hydrological applications in the Notwane Catchment, Botswana. *Physics and Chemistry of the Earth Parts A/B/C, 100,* 19–30.

Kirono, D. G., Chiew, F. H., & Kent, D. M. (2010). Identification of best predictors for forecasting seasonal rainfall and runoff in Australia. *Hydrological Processes: An International Journal, 24*(10), 1237–1247.

Krizhevsky, A., Sutskever, I., & Hinton, G. E. (2012). Imagenet classification with deep convolutional neural networks. *Adv. Neural Inf. Process Syst., 25*, 1097–1105.

Lin, F.-R., Wu, N.-J., & Tsay, T.-K. (2017). Applications of cluster analysis and pattern recognition for typhoon hourly rainfall forecast. *Advances in Meteorology*.

Liu, Y., Racah, E., Correa, J., Khosrowshahi, A., Lavers, D., Kunkel, K., et al. (2016). *Application of deep convolutional neural networks for detecting extreme weather in climate data sets. arXiv*:1605.01156.

O'Gorman, P. A., & Dwyer, J. G. (2018). Using machine learning to parameterize moist convection: Potential for modeling of climate, climate change, and extreme events. *Journal of Advances in Modeling Earth Systems, 10*(10), 2548-2563.

Petersik, P. J., & Dijkstra, H. A. (2020). Probabilistic Forecasting of El Niño Using Neural Network Models. *Geophysical Research Letters, 47*(6), e2019GL086423.

Qiu, M., Zhao, P., Zhang, K., Huang, J., Shi, X., Wang, X., & Chu, W. (2017). *A short-term rainfall prediction model using multi-task convolutional neural networks*. Paper presented at the 2017 IEEE International Conference on Data Mining (ICDM).

Tanguy, M., Dixon, H., Prosdocimi, I., Morris, D. G., & Keller, V. D. J. (2016). *Gridded estimates of daily and monthly areal rainfall for the United Kingdom (1890–2015)* [CEH-GEAR]. NERC Environmental Information Data Centre.

Taylor , K. E., Stouffer, R. J., & Meehl, G. A. (2012). An overview of CMIP5 and the experiment design. *Bulletin of the American Meteorological Society, 93(4)*, 485–498.

Tkalcic, M., & Tasic, J. F. (2003). *Colour spaces: perceptual, historical and applicational background (Vol. 1)*. The IEEE Region 8 EUROCON 2003. Computer as a Tool., Ljubljana, Slovenia, 2003 (pp. 304–308). https://doi.org/10.1109/EURCON.2003.1248032

Uppala, S. M., Kållberg, P. W., Simmons, A. J., Andrae, U., Bechtold, V. D. C., Fiorino, M., . . . Kelly, G. A. (2005). The ERA-40 re-analysis. *Quarterly Journal of the Royal Meteorological Society*, *131*(612), 2961–3012.

Urdiales, D., Meza, F., Gironás, J., & Gilabert, H. (2018). Improving stochastic modeling of daily rainfall using the ENSO index: Model development and application in Chile. *Water, 10*(2), 145.

Wang, H., & Xuan, Y. (2020). SRS-GDA: A spatial random sampling toolbox for grid-based hydro-climatic data analysis in environmental change studies. *Environmental Modeling & Software, 124*, 104598.

Wang, H., & Xuan, Y. (2021). Spatial variation of extreme rainfall observed from two century-long data sets. *Geophysical Research Letters, 48*(8). https://doi.org/10.1029/2020GL091933

Yan, Z., Bate, S., Chandler, R. E., Isham, V., & Wheater, H. (2002). An analysis of daily maximum wind speed in northwestern Europe using generalized linear models. *Journal of Climate, 15(15)*, 2073–2088.

Yunus, R. M., Hasan, M. M., Razak, N. A., Zubairi, Y. Z., & Dunn, P. K. (2017). Modeling daily rainfall with climatological predictors: Poisson-gamma generalized linear modeling approach. *International Journal of Climatology, 37(3)*, 1391–1399.

13

TELECONNECTION PATTERNS OF RIVER WATER QUALITY DYNAMICS BASED ON COMPLEX NETWORK ANALYSIS

Jiping Jiang[1,2], Sijie Tang[1], Bellie Sivakumar[3], Tianrui Pang[2], Na Wu[4], and Yi Zheng[1]

[1] *School of Environmental Science and Engineering, Southern University of Science and Technology, Shenzhen, China*
[2] *State Key Laboratory of Urban Water Resource and Environment, School of Environment, Harbin Institute of Technology, Harbin, China*
[3] *Department of Civil Engineering, Indian Institute of Technology Bombay, Mumbai, India*
[4] *College of Environmental Science and Engineering, Tongji University, Shanghai, China*

Water quality in rivers is influenced by natural factors and human activities that interact in complex and nonlinear ways, which make water quality modeling a challenging task. The concept of complex networks (CN), a recent development in network theory, seems to provide new avenues to unravel the connections and dynamics of water quality phenomenon, including clandestine teleconnections. This study explores the spatial patterns of water quality using CN concepts at both catchment scale and larger national scale. Three major water quality parameters–dissolved oxygen, permanganate index (CODMn), and ammonia

Advanced Hydroinformatics: Machine Learning and Optimization for Water Resources, Special Publications 78, First Edition. Edited by Gerald A. Corzo Perez and Dimitri P. Solomantine.

DOI: 10.1002/9781119639268.ch13

nitrogen (NH_3-N)-measured weekly for 12 years at 91 monitoring stations across China, are analyzed. The results show that the degree centrality and clustering coefficient values for water quality indicators is DO > NH_3-N > CODMn at both basin scale and national scale. The findings improve understanding of water quality dynamics and suggest new methods for environment system analysis and watershed management.

13.1. Introduction

Complexity is an important non-linear characteristic of a river water system (Kirchner & Neal, 2013). The hydrological and water quality processes of rivers exhibit complexities in temporal (Jiang et al., 2020) and spatial connection (Fang et al., 2017). Because of the terrain, land use, surface-ground water interaction, regional meteorology, and human activities, the spatial patterns in water quality time series in a watershed is bound to have complex correlations. Such correlations and connections can be due to events that happen far away geographically, which is called teleconnection (Ashok et al., 2007). Analysing the regularity and possible causality of these correlations will help deepen our understanding of the energy and solute transport in river basins from the perspective of system theory, and help significantly advance our ability to model, predict, and manage river water quality dynamics. In this regard, complex network (CN) theory, which has been developing rapidly in recent years, is an effective modelling tool to help us gain knowledge on the teleconnection in water quality dynamics.

The concepts of CNs (Watts & Strogatz, 1998) emerged from earlier developments in network theory or graph theory, topology, trees, and random graph theory ((Erdős & Rényi, 1960: König, 1936). Complex networks have been applied in many different fields and associated problems, including DNA transcription in molecular biology (Zheng & Flanagan, 2017), biodiversity CN for a better understanding of species interactions (Bascompte, 2009), predictive power of the behaviour of techno-social systems (Vespignani, 2009), sustainability analysis of social-ecological systems (Ostrom, 2009), analysis of vaccine efficacy in the vaccination behaviour (Huang et al. 2020), human information processing (Lynn et al., 2020), mathematical epidemic analysis (Wang et al., 2019), and the graph learning (Lynn & Bassett, 2020).

In the field of hydrology, CN is a powerful tool to extract information from large high-dimensional datasets (Kurths et al., 2019). It, thus, can be used to unravel the connections in a variety of systems, as has been demonstrated for the global influence of the El Niño–Southern Oscillation

(ENSO) on regional rainfall (Agarwal, 2019; Ferster et al., 2018), influence of the Atlantic Meridional Overturning Circulation (AMOC) on air surface temperature (Agarwal et al., 2019), dominant climate modes (Agarwal et al., 2018; Halverson & Fleming, 2015), catchment classification indicating hydrologic similarity (Fang et al., 2017), short/long range spatial connections of rainfall (Agarwal et al., 2018; Boers et al., 2014, Jha et al., 2015), and spatial and temporal hydrologic patterns (Halverson & Fleming, 2015; Konapala & Mishra, 2017; Sivakumar & Woldemeskel, 2014).

The variability in river water quality parameters is the result of a complex and nonlinear influences of environmental factors (external factors) and regional characteristics (internal factors) and, therefore, connections are the key to understand the water quality dynamics. In this regard, CN provides strong theoretical foundations and new ideas and facilitate water environment planning and management. With data-intensive investigations becoming a new scientific discovery paradigm, the concepts of CN can go a long way in our ability to model, predict, and manage water quality and river systems.

This work collected intensive water quality data in plenty water quality stations and utilize network indexes to explore the connections and teleconnections of typical water quality parameters. The Huaihe River basin was selected as a large watershed scale study area and large rivers covering east China were selected as a large region scale study area. Based on the linear correlation between stations, the degree centrality and clustering coefficient were calculated to characterize the statistical characteristics of water quality CNs, and to reveal the potential teleconnection of water quality processes.

13.2. Study Areas and Methods

13.2.1. Study Areas and Monitoring Campaign

In this study, two scales of water quality system in China are studied: watershed scale and national scale. For the watershed scale, the Huaihe River is considered, and the major rivers in China are considered for the national (or large region) scale. They are both significantly impacted by human activities.

Huaihe River Basin

The Huaihe River basin (30°55′~36°36′ N, and 111°55′~121°25′ E) is located in eastern China (Fig. 13.1), between the Yangtze River basin and the Yellow River basin. It flows through five provinces of China, namely

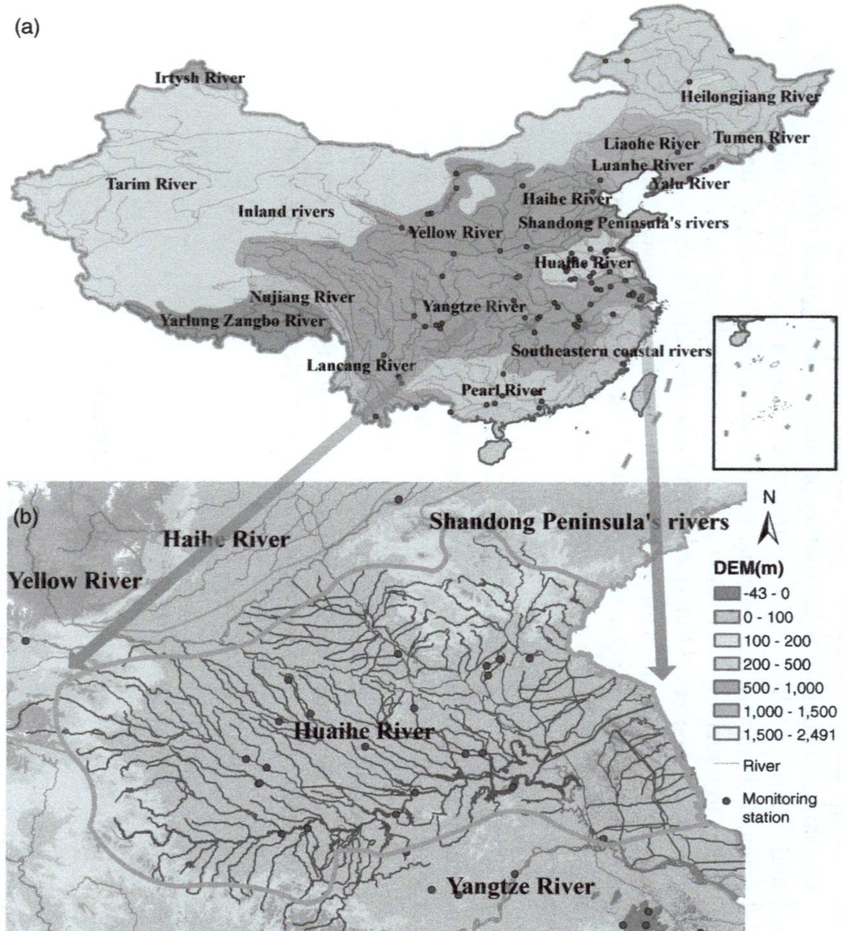

Figure 13.1 Study area and monitoring campaign in China and Huaihe River basin: (a) The major basins in China and the location of Huaihe River basin; (b) river network in Huaihe River basin.

Hubei, Henan, Anhui, Shandong, and Jiangsu. It is the seventh largest river basin in China, with a drainage area of 270,000 km^2 approximately. The population living in the basin is 165 million. The basin average annual precipitation is about 894 mm, of which more than 70% occurs during the flood season from June to October. The spatial distribution of precipitation is quite uneven, decreasing from 1,400 mm in the southern mountain region to less than 700 mm in the northern region near the Yellow River. Although the basin-average annual runoff is about 230 mm, the temporal and spatial distributions are highly uneven due to climatic variability.

China's Major Rivers

The large region scale (or national scale) study area includes a total of 63 major river systems in China as shown in Figure 13.1. Some of these are the Yellow River, Yangtze River, Pearl River, Haihe River, Huaihe River, Heilongjiang River, Liaohe River, Taihu Lake, Chaohu Lake, and Dianchi Lake.

13.2.2. Monitoring Campaign and Data Processing

The water quality data for this study are obtained from the China National Environmental Monitoring Centre that operates 148 automatic water quality monitoring stations all over China, 91 of which are available and effective displayed as dark dots in Figure 13.1. The weekly monitoring values are used for analysis. The water quality parameters monitored consist of pH, dissolved oxygen (DO), total organic carbon (TOC), permanganate index (COD_{Mn}), and ammonia nitrogen (NH_3-N). Among these, we mainly focus on DO, COD_{Mn}, and NH_3-N in this study. Most of the stations considered in this study are located on the east side of Heihe-Tengchong line, in an economically developed area.

Data from 91 active stations are used for national scale research, and the time span is from 1 Jan 2004 to 2 Jun 2016. Twenty-five stations are used for the study of the Huaihe River basin, though 27 actually exist. The time span of data considered is from the 29th week of 2008 to the 22d week of 2016. For any missing data, the value is estimated by interpolating the continuous monitoring results in the previous and following 3 weeks. If data for such weeks are not available, then the average value of the same weeks in the previous and following years are taken.

13.3. Theory of Complex Network

13.3.1. Concept of Network

Basically, a network is a set of points and lines, in which various types of nodes or vertices may be connected together by all sorts of links or edges. A series of indexes are introduced to characterize the pattern of CN, including degree centrality, clustering coefficient, average shortest path length, degree distribution, and betweenness. These simple statistics can be used to quantify the properties of networks and to define the types (e.g., regular graph, random graph, small-world network, and scale-free network).

The network shown in Figure 13.2, for example, can be represented as $G = \{P, E\}$, where P is a set of 6 nodes and E is a set of 7 links, namely $P = \{1,2,3,4,5,6\}$ and $E = \{\{1,2\},\{1,6\},\{2,6\},\{3,6\},\{4,6\},\{5,6\},\{3,4\}\}$. Figure 13.2a shows the simplest form of network, which contains identical nodes and links; however, the network may have different types of nodes and/or links (Fig. 13.2b). The nodes and links may have a variety of properties, such as weight (Fig. 13.2c); and the connection may be directed, pointing in one direction (Fig. 13.2d). Besides, more particular features, such as multilink, self-link, and hyperlink, also exist in more CNs of real systems.

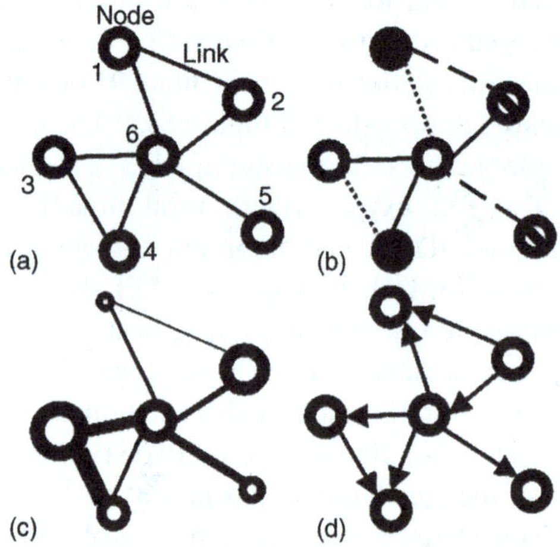

Figure 13.2 Different networks: (a) An undirected network with identical nodes and links; (b) a network with more than one type of nodes and links; (c) a network with weighted nodes and links; and (d) a directed network (adapted from Sivakumar, 2015).

13.3.2. Indexes of Complex Network

The number of links k_i connected to a node i is called a *node degree*. Then the degree centrality is defined as follows:

$$dc_i = \frac{k_i}{N-1},$$

which indicates the influence of node i in the network. The degree distribution, $P(k)$, represents the probability that a randomly selected node has exactly k neighbors (i.e., node degree of k). The common types of degree distribution are exponential distribution, power-law distribution, and Poisson distribution (Krapivsky et al., 2001).

Another important property of networks is the tendency to cluster, which is quantified by the clustering coefficient (Wasserman & Faust, 1994). Considering a selected node i with the node degree of k_i, there is supposed to be $k_i(k_i\text{-}1)/2$ links at most between the k_i neighbors. Then the clustering coefficient of node i can be given as follows:

$$C_i = \frac{2E_i}{k_i(k_i - 1)},$$

where E_i is the actually existing links in the cluster. Further, the clustering coefficient of a network is the average of the clustering coefficient of all the nodes.

13.3.3. CNs-Based Teleconnection Analysis

Taking the monitoring stations as the network nodes and the correlations (of their temporal water quality dynamics) as the potential links, a virtual CN is thereby constructed. The hidden teleconnections and network patterns can be revealed by indexes of CNs, including degree centrality and clustering coefficient. The links, meanwhile, are depicted by the Pearson correlation coefficient (Benesty et al., 2009), which is calculated between each pair of stations. The identification of links between nodes is done by assuming different correlation threshold (CT) values, which are arbitrarily chosen. In general, the threshold values can significantly influence the properties of CNs.

In this study, the differences of network features between several water quality indicators and basins/regions are analyzed. However, such teleconnections may be indirect, and result from similar drivers, such as pollution release, climate pattern, and hydrologic conditions in a basin. Investigations on the details of such mechanisms/drivers are beyond the scope of this study and, therefore, are not attempted here.

13.4. Results and Discussion

13.4.1. Teleconnection Pattern on Huaihe River Basin

Linear Correlation Analysis

Figure 13.3 presents the results from the linear correlation analysis. As expected, the correlation analysis between pairwise water quality monitoring stations shows the number of links of networks (i.e., the neighbors of nodes) decreases with an increase of CT value. Besides, for different stations, the distribution of the number of neighbors is uneven; some stations have many links to other stations and some others do not. Specifically, the correlation for COD_{MN} is significantly weaker than that for NH_3-H and DO. Even when the CT is as low as 0.3, the teleconnection is still not obvious in the COD_{MN} network, let alone the higher CT values.

Degree Centrality Analysis

Figure 13.4 shows the degree centrality results for the three water quality networks. It is clear that for all the three water quality parameters, the

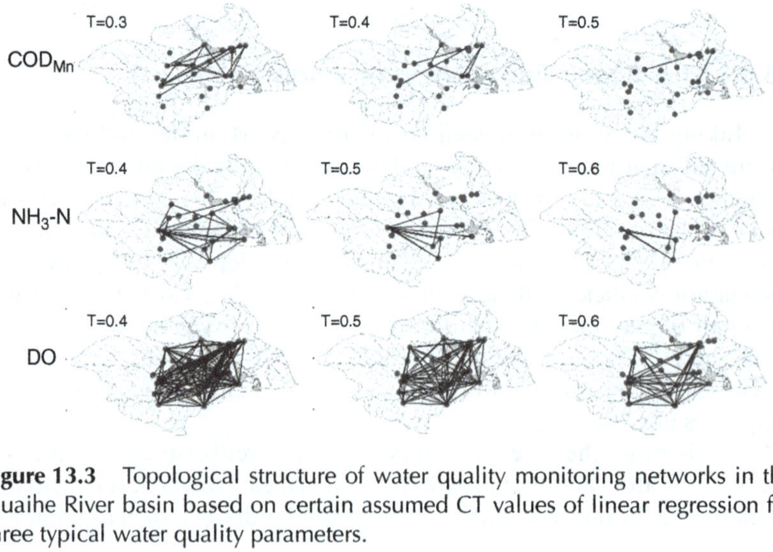

Figure 13.3 Topological structure of water quality monitoring networks in the Huaihe River basin based on certain assumed CT values of linear regression for three typical water quality parameters.

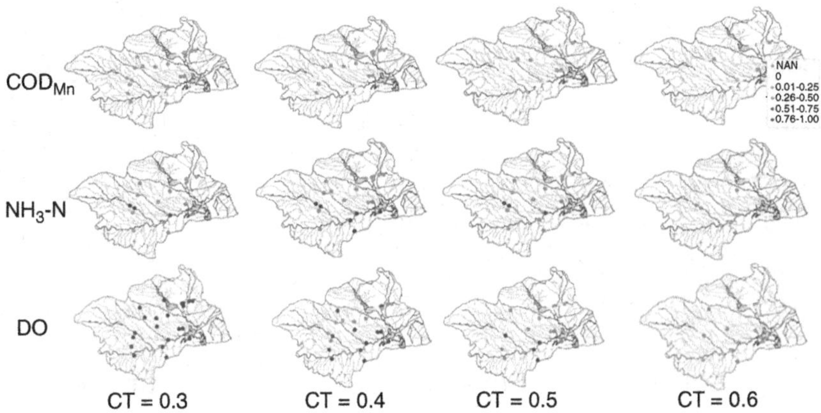

Figure 13.4 Degree centrality pattern of water quality monitoring networks in the Huaihe River basin based on certain CT values of linear correlation for three typical water quality parameters.

degree centrality of nodes is not necessarily related to the spatial proximity; even the stations with close geographical location can have completely different degree centrality and vice versa. Though some stations with high degree centrality are near to the secondary tributaries, there is no evidence of a convincing distribution characteristic. Furthermore, the degree centrality of the DO networks is relatively higher than COD_{MN} and NH_3-H in line with the linear correlation results.

Clustering Coefficient Analysis

Similar with the results of degree centrality analysis, clustering coefficients of stations help to understand the teleconnection of water quality dynamics as well. The spatial locations of stations are not considered to be relative with the clustering coefficient, as shown in Figure 13.5. For the three water quality parameters, the average clustering coefficient decreases with an increase in the CT values, except for a few specific stations whose clustering coefficients slightly increase. Moreover, the highest clustering coefficient is still for the DO network, and the next is for NH_3-H.

13.4.2. Teleconnection Analysis on China's Major Rivers

Linear Correlation Analysis

The same procedure, which was carried out for the Huaihe River data, is implemented to look into the teleconnection within the national scale

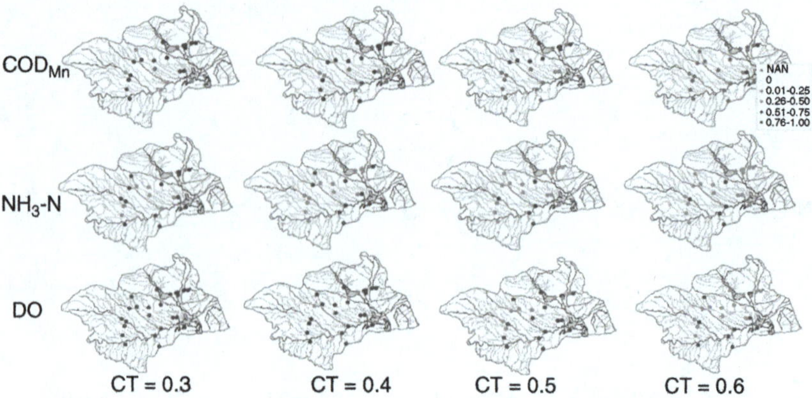

Figure 13.5 Clustering coefficient pattern of water quality monitoring networks in the Huaihe River basin based on certain CT values of linear correlation for three typical water quality parameters.

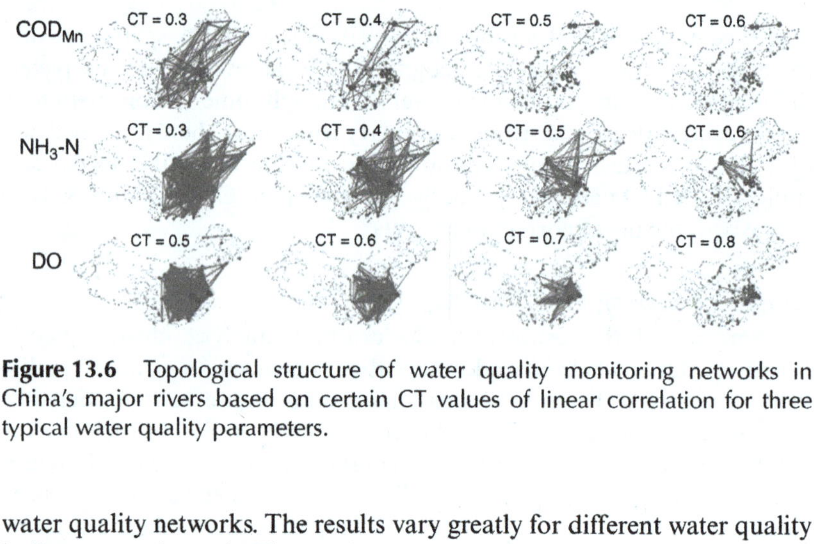

Figure 13.6 Topological structure of water quality monitoring networks in China's major rivers based on certain CT values of linear correlation for three typical water quality parameters.

water quality networks. The results vary greatly for different water quality indicators as shown in Figure 13.6.

For COD_{MN}, there are only 30 pairs of stations (0.73% of total pairs) whose correlation is deemed to have a link with a CT = 0.4. Most of them are from the same or homologous basin, where the climatic, hydrologic, and COD_{MN} pollution status are close. For NH_3-H, the correlation is more apparent with 264 pairs of stations (6.45%) having values greater than 0.4 and 23 pairs (0.56%) having values greater than 0.6. It should be noted, however, that 10 out of the 23 pairs are transbasin pairs. The high correlation values, therefore, can be attributed to their similar emission and degradation conditions. The correlation coefficients based on the DO data are most significant, averagely with 1,184 pairs of stations (28.91%) having greater than values of 0.5 and 36 pairs (0.88%) having values greater than 0.8. Among the 36 pairs, 20 are located in the Taihu Lake basin and Poyang Lake basin, meaning that large lakes may play an important role in the DO dynamics of surface water system. Nevertheless, overall, the teleconnection in the national scale water quality monitoring network is not obvious, and closer stations usually have stronger correlations.

Degree Centrality Analysis

Since simple networks with too few links, which usually occur for high threshold values, will undermine the validity of the CN analysis, CT value is 0.2, 0.3, 0.4, and 0.5 to perform degree centrality analysis. Again, the average degree centrality for the DO monitoring data is greater than that for COD_{MN} and NH_3-H, as shown in Figure 13.7. When CT = 0.2, most of the nodes in the COD_{MN} network are less than 0.5, while in the DO network, it is greater than 0.75. Even when CT is up to 0.5, the latter can be around 0.25~0.75 as degree centrality of the COD_{MN} network declines to almost 0. The NH_3-H network falls in between.

Clustering Coefficient Analysis

As can be seen from Figure 13.8, the clustering coefficient pattern of the national scale water quality monitoring network is similar to that of the

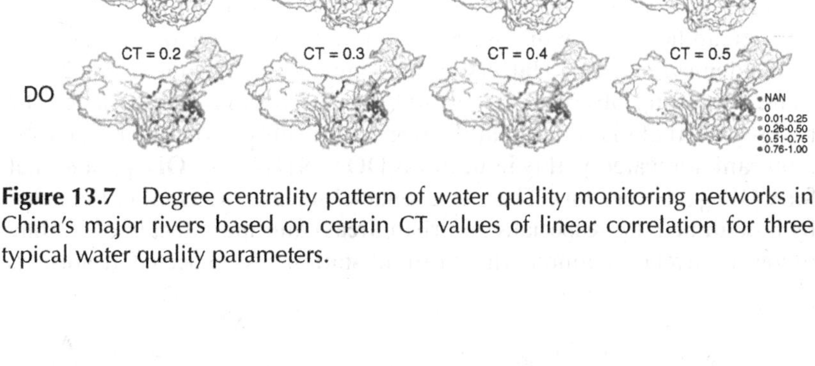

Figure 13.7 Degree centrality pattern of water quality monitoring networks in China's major rivers based on certain CT values of linear correlation for three typical water quality parameters.

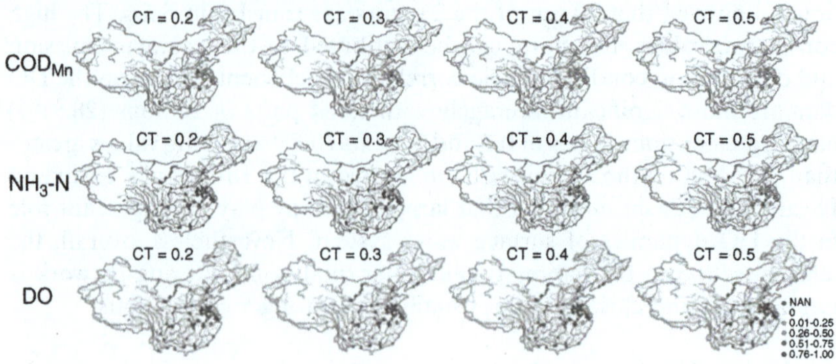

Figure 13.8 Clustering coefficient pattern of water quality monitoring networks in China's major rivers based on certain CT values of linear correlation for three typical water quality parameters.

Huaihe River basin. The teleconnection of COD_{MN} is not apparent, and only three stations in the Songhua River basin have clustering coefficient values when CT = 0.5, because of the regional differences in upstream pollution emission. Meanwhile, 18 stations' clustering coefficients are greater than 0.5 for NH_3-H network when CT = 0.5. There are two stations each in the Yellow River basin and the Yangtze River basin, and one station each in the Songhua River basin and Huaihe River basin, which have a clustering coefficient greater than 0.75 and can be seen as key stations of NH_3-H control. For DO, approximately 90% of nodes have a clustering coefficient greater than 0.75 even when CT = 0.5.

13.5. Comparison of Teleconnection Features Between Scales and Parameters

Although the preceding results indicate that the average degree centrality and clustering coefficient values are much smaller than those for a streamflow system of a watershed or across large regions (Sivakumar & Woldemeskel, 2014; Halverson & Fleming, 2015; Fang et al., 2017), some interesting observations are still worth reporting at this stage. The basin-scale analysis shows that the degree centrality and clustering coefficient rank for water quality indicator is DO > NH_3-N > COD_{MN}, and that for COD_{MN} is very low. This is reasonable, since COD is more impacted by human activity (especially in urban areas) and more localized leading to lower correlation among those distant stations. As expected, it presents similar degree and clustering characteristics for some distant stations and quite a few differences for some nearby stations. However, NH_3-N is more relative to stormwater-induced nonpoint pollution, leading to stronger upstream-downstream relationship. The high teleconnection in the DO network may be attributed to the overall connectivity of surface water systems and strong diffusion of DO. A similar distribution pattern of degree centrality and clustering coefficient is also found at the national scale for all parameters. With a detailed inspection on the range of the Huaihe River in the national scale results (Fig. 13.3 vs. Fig. 13.6), NH_3-N shows that the important nodes with higher degree centrality and clustering coefficient in Huaihe River basin still tend to play a vital role in the national-scale network. But the pattern for DO presents a difference between the two scales even in the same river. It reflects the different drivers of teleconnection among different parameters.

13.6. Study Limitations and Future Works

We conducted the CN methods to analyze the teleconnections in water quality monitoring networks. However, the number of stations considered in this study may be insufficient to provide strong interpretations and conclusions, from the perspective of CN. For instance, the small number of stations, yielding small number of nodes and links, cannot produce accurate CN indexes, especially in the context of teleconnection patterns. It should be noted, however, that the water environment is getting more attention from government and the public. Concepts such as Smart City requires a large on-line monitoring network for real-time management. The rapid development of water quality monitoring technology and instruments makes the big data of water quality possible, leading to more widespread applications of the CN method.

Besides, there should be a time lag between the water quality dynamics at upstream and downstream stations. In this study, however, we ignored the drivers from upstream water quality process, which may hide the local correlation, and exaggerate the teleconnection crossing the basin/region.

The teleconnection analysis can certainly help us understand the intuitive rules of water quality dynamics. However, more fundamental and intrinsic laws need to be brought in to enlighten the dependence on attribution analysis, for further improving the reasonability of this work. The physical causes of the CN results, especially for their spatial difference, should be discussed further. Usually, water quality is mainly controlled by the geographical conditions in a basin, including the climatic conditions, soil characteristics, geological conditions, land-use/land-cover, and many others. How they directly influence the spatial pattern of CN and if these factors will be identified as important as well at national scale should be a focus. Furthermore, human activity has achieved a dominant position in environmental changes. Research on to what extent artificial factors (e.g., population density and point source pollution) affect the features of water quality network and teleconnection by reshaping the geographical environment can be promising. Besides, some natural factors such as climate type, longitude, and latitude may also influence the teleconnection of water quality dynamics in certain important ways on a larger scale. The insight into water quality process can enhance our ability to more properly manage the river systems.

13.7. Conclusion

In this study, the concept of CN was used to look for potential spatial patterns in water quality dynamics at the basin and national scales. Weekly monitoring data of three major water quality indicators (i.e., DO, NH_3-N, and COD_{Mn}), with a time span longer than 10 yr, were used for the calculation of CN indexes, including degree centrality and clustering coefficient. The following key findings have emerged, which will help understand the teleconnection of water quality dynamics and guide relevant works in the future.

Though the degree centrality and average clustering coefficient values at the national scale are relatively smaller than that at the basin scale, and much smaller than that for the runoff system, the existence of teleconnection of water quality dynamics is worthy of consideration.

The correlations in water quality dynamics are not related to the spatial location. Even close stations can have completely different water quality features, and distant stations may have strong correlations (i.e., teleconnection, of water quality).

In general, the degree centrality and clustering coefficient rank for water quality indicators is DO > NH_3-N > COD_{MN}, which is reasonable because COD_{MN} is more impacted by human activity in urban areas, NH_3-N is relative to nonpoint pollution, and DO has stronger diffusion trend.

The explosive growth of water quality monitoring data is conducive to explore the applications of CN methods further, among which attribution analysis may deepen our understanding of the teleconnection of water quality dynamics.

Acknowledgments

This work was supported by the National Natural Science Foundation of China (Grant No. 51979136), the Open Project of State Key Laboratory of Urban Water Resource and Environment, Harbin Institute of Technology (Grant No. ESK201901), and Science, Technology and Innovation Commission of Shenzhen Municipality (Grant No. JSGG20180508151852303). We thank Baoyu Wang and Meng Yuan from Harbin Institute of Technology for the raw data collection and treatment, and anonymous reviewers for constructive suggestions.

References

Agarwal, A. (2019). *Unraveling spatio-temporal climatic patterns via multi-scale complex networks*. Universität Potsdam Potsdam.

Agarwal, A., Caesar, L., Marwan, N., Maheswaran, R., Merz, B., & Kurths, J. (2019). Network-based identification and characterization of teleconnections on different scales. *Scientific Reports, 9*(1), 1–12.

Agarwal, A., Marwan, N., Maheswaran, R., Merz, B., & Kurths, J. (2018). Quantifying the roles of single stations within homogeneous regions using complex network analysis. *Journal of Hydrology, 563*, 802–810.

Ashok, K., Behera, S. K., Rao, S. A., Weng, H., & Yamagata, T. (2007). El Niño Modoki and its possible teleconnection. *Journal of Geophysical Research: Oceans, 112*(C11).

Bascompte, J. (2009). Disentangling the web of life. *Science, 325*(5939), 416–419.

Benesty, J., Chen, J., Huang, Y., & Cohen, I. (2009). *Noise reduction in speech processing*. Springer.

Boers, N., Rheinwalt, A., Bookhagen, B., Barbosa, H. M., Marwan, N., Marengo, J., et al. (2014). The South American rainfall dipole: A complex network analysis of extreme events. *Geophysical Research Letters, 41*(20), 7397–7405.

Erdös, P., & Rényi, A. (1960). On the evolution of random graphs. *Institute of Mathematics, Hungarian Academy of Science, 5*(1), 17–60.

Fang, K., Sivakumar, B., & Woldemeskel, F. M. (2017). Complex networks, community structure, and catchment classification in a large-scale river basin. *Journal of Hydrology, 545*, 478–493.

Ferster, B. S., Subrahmanyam, B., & Macdonald, A. M. (2018). Confirmation of ENSO-Southern Ocean teleconnections using satellite-derived SST. *Remote Sensing, 10*(2), 331.

Halverson, M., & Fleming, S. (2015). Complex network theory, streamflow, and hydrometric monitoring system design. *Hydrology and Earth System Sciences, 19*(7).

Huang, J., Wang, J., & Xia, C. (2020). Role of vaccine efficacy in the vaccination behavior under myopic update rule on complex networks. *Chaos, Solitons & Fractals, 130*, 109425.

Jha, S. K., Zhao, H., Woldemeskel, F. M., & Sivakumar, B. (2015). Network theory and spatial rainfall connections: An interpretation. *Journal of Hydrology, 527*, 13–19.

Jiang, J., Zheng, Y., Pang, T., Wang, B., Chachan, R., & Tian, Y. (2020). A comprehensive study on spectral analysis and anomaly detection of river water quality dynamics with high time resolution measurements. *Journal of Hydrology, 125175*.

Kirchner, J. W., & Neal, C. (2013) Universal fractal scaling in stream chemistry and its implications for solute transport and water quality trend detection. *Proceedings of the National Academy of Sciences, 110*(30), 12213–12218.

Konapala, G., & Mishra, A. (2017). Review of complex networks application in hydroclimatic extremes with an implementation to characterize spatio-temporal drought propagation in continental USA. *Journal of Hydrology, 555*, 600–620.

König, D. (1936). *Theorie der endlichen und unendlichen Graphen: Kombinatorische Topologie der Streckenkomplexe*, Akademische Verlagsgesellschaft, mbh.

Krapivsky, P. L., Rodgers, G. J., & Redner, S. (2001). Degree distributions of growing networks. *Physical Review Letters, 86*(23), 5401–5404.

Kurths, J., Agarwal, A., Shukla, R., Marwan, N., Rathinasamy, M., Caesar, L., et al. (2019). Unraveling the spatial diversity of Indian precipitation teleconnections via a non-linear multi-scale approach. *Nonlinear Processes in Geophysics, 26*, 251–266.

Lynn, C. W., & Bassett, D. S. (2020). How humans learn and represent networks. *Proceedings of the National Academy of Sciences, 117*(47), 29407–29415.

Lynn, C. W., Papadopoulos, L., Kahn, A. E., & Bassett, D. S. (2020). Human information processing in complex networks. *Nature Physics, 16*(9), 965–973.

Ostrom, E. (2009). A general framework for analyzing sustainability of social-ecological systems. *Science, 325*(5939), 419–422.

Sivakumar, B. (2015). Networks: a generic theory for hydrology? *Stochastic Environmental Research and Risk Assessment, 29*(3), 761–771.

Sivakumar, B., & Woldemeskel, F. M. (2014). Complex networks for streamflow dynamics. *Hydrology and Earth System Sciences, 18*(11), 4565–4578.

Vespignani, A. (2009). Predicting the behavior of techno-social systems. *Science, 325*(5939), 425–428.

Wang, X., Wang, Z., & Shen, H. (2019). Dynamical analysis of a discrete-time SIS epidemic model on complex networks. *Applied Mathematics Letters, 94*, 292–299.

Wasserman, S., & Faust, K. (1994). *Social network analysis: Methods and applications,* Cambridge University Press.

Watts, D. J., & Strogatz, S. H. (1998). Collective dynamics of "small-world" networks. *Nature, 393*(6684), 440–442.

Zheng, Y., & Flanagan, S. V. (2017) The case for universal screening of private well water quality in the U.S. and testing requirements to achieve it: Evidence from arsenic. *Environmental Health Perspectives, 125*(8), 085002.

14

PROBABILISTIC ANALYSIS OF FLOOD STORAGE AREAS MANAGEMENT IN THE HUAI RIVER BASIN, CHINA, WITH ROBUST OPTIMIZATION AND SIMILARITY-BASED SELECTION FOR REAL-TIME OPERATION

Xingyu Zhou[1], Andreja Jonoski[2], Ioana Popescu[2], and Dimitri P. Solomatine[2,3,4]

[1] *Shanghai Investigate, Design & Research Institute Co., Ltd., Shanghai, China*
[2] *Department of Hydroinformatics and Socio-Technical Innovation, IHE Delft Institute for Water Education, Delft, The Netherlands*
[3] *Water Resources Section, Delft University of Technology, Delft, The Netherlands*
[4] *Water Problems Institute of the Russian Academy of Sciences, Moscow, Russia*

Large-scale river basins in the southern part of China suffer frequently from flooding problems, and storage areas are often used for minimizing downstream flooding risk, which in turn causes damage to economic activities within these areas. Such damages need to be minimized, even though this is often conflicting with the main objective of reducing downstream flooding risk. This problem is especially highlighted in the middle part of the Huai River basin. Previous studies focused on solving this multi-objective optimization problem using generic algorithms coupled with hydrodynamic models. This study further develops probability analysis using a robust optimization and similarity-based

Advanced Hydroinformatics: Machine Learning and Optimization for Water Resources, Special Publications 78, First Edition. Edited by Gerald A. Corzo Perez and Dimitri P. Solomantine.

DOI: 10.1002/9781119639268.ch14

selection method for investigating the possibility to do real-time operations. The results show that, in most cases, both robust strategies and a similarity selection method can effectively reduce flood risk downstream, and, at the same time, reduce storage area damage. This suggests that the strategies from similarity selection are more effective in reducing flooding risk but cause more storage area damage and vice versa.

14.1. Introduction

Flooding problems are increasingly drawing attention worldwide because of their large impact on life and society. Over the last few decades, in many areas, flood events have been occurring with increasing frequency, showing the tendency of causing greater damages (Saghafian et al., 2010; Pachauri et al., 2014). Traditionally, structural measures, such as use of flood storage areas upstream, were widely taken in order to reduce flood risk of important downstream areas (e.g., cities). However, this strategy often neglects possible damages that can be caused in the upstream areas used for flood storage, especially because floods are rare events; and in between these events, socioeconomic development takes place in these areas. Therefore, research on addressing such conflicting problems within a multi-objective optimization framework were introduced in the water-related fields (Cheng & Chau, 2002; Dittmann et al., 2009; Maier et al., 2014; Luo et al., 2015; Jonoski et al., 2019). However, the multi-objective optimization and decision-making analysis often does not take the uncertainty into consideration, in which cases the developed solutions may not easily adapt to changed environmental conditions. Therefore, accounting for uncertainties using probability analysis is needed when analyzing flooding problems with multi-objective optimization, in order to offer better management of flood risk.

This study focuses on the middle section of Huai River in China, covering about 110 km of the river section upstream of Bengbu City, where spatially distributed precipitation causes frequent flooding inundation and tremendous damage and loss. Four storage areas (Shouxi, Dongfeng, Tangyu, and Jingshan lake) were built along the Huai River from Lutaizi station to Bengbu City (Fig. 14.1). When potential danger of flooding in Bengbu is detected, these storage areas would be operated for diversion of flood waters, resulting in reduction of the flood peak at Bengbu. However, the same areas are used by the local population for housing and agriculture activities. Any use of storage areas would therefore inevitably produce

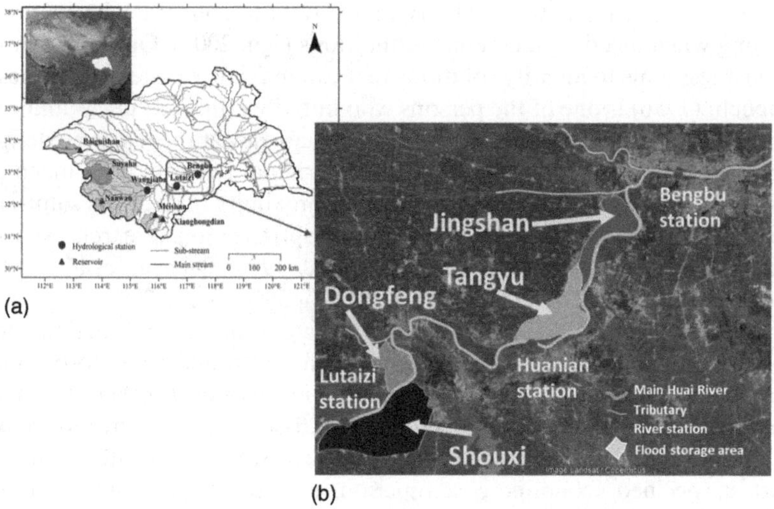

Figure 14.1 (a) Huai River basin (adapted from Wu et al., 2015) and (b) scheme of the case study area (Jonoski et al., 2019).

damage. During operations, this damage is also expected to be minimized, even while being in conflict with the main objective of reducing flood risk of Bengbu City.

Previous studies have investigated optimal strategies for storage area gates operation in this study area. Jonoski et al. (2018) identified optimal solutions using a coupled multi-objective optimization algorithm and a hydrodynamic model, while Jonoski et al. (2019) updated the objective function formulations and the geometric setting of hydrodynamic model (to include 2-D analysis of the flooding) in order to make the coupled simulation-optimization model more realistic. These previous studies were both based on a particular flood event of 2007, from 24 June to 7 September. The research presented here builds on these existing studies, and further develops storage area operation strategies by taking uncertainty of inflow into consideration.

When it comes to dealing with flooding uncertainty, two types of methods are often considered. One effective way is to develop robust solutions that are able to cope with a large spectrum of uncertain conditions, still resulting in relatively small or manageable impact. Another approach is to make real-time decisions regarding strategy of actions (e.g., gate operations of flood storage areas) when certain flood-generating inflow is forecasted, which could guarantee that the selected strategy would be effective or even optimal for the given situation.

Robustness refers to the ability of a system to continue with its functioning when meeting uncertainty conditions (Jen, 2003). Optimization of robustness aims to identify solutions that can maximize system robustness. Taguchi (1986) is one of the persons who initially estimated uncertainty to optimize robustness by minimizing mean square deviation of a given design objective. Such a method that regularizes original function (e.g., mean or variance of objective functions value in the proximity of a certain solution) has widely been used later in robustness optimization research even in water-related fields (Andricevic & Kitanidis, 1990; Babayan et al., 2005; Hollinshead & Lund, 2006; Malekmohammadi et al., 2009). Another popular method is to optimize two different objective functions for solving multi-objective problems, which often introduces Pareto-optimal method when these objectives are conflicting with each other (Beyer & Sendhoff, 2007). For example, Jung et al. (2013) proposed three formulas of quantifying tunnel pressure robustness, which were separately coupled with a specified economic cost function, in order to find out the most suitable way of describing system robustness. The authors compared three Pareto fronts and identified the function representing the sum of nodal pressure robustness index as the best one to generate least vulnerable solutions. The mentioned approaches of robust optimization solve given standard optimization problem, but they are unable to present the impact of uncertainties (Marquez-Calvo & Solomatine, 2019). To address this, Solomatine (2012) proposed an approach explicitly presenting the effect of uncertainty in multi-objective optimization problems by generating multiple Pareto fronts, which is the so-called Robust Optimization and Probabilistic Analysis of Robustness (ROPAR). Marquez-Calvo and Solomatine (2019) further developed this method so that decisions can be made depending on the uncertainty analysis. This is quite recent methodology, which needs to be applied in more practical case studies for further confirmation of its value.

Another area of increasing importance is real-time operation. Over the last decades, many researchers combined optimization and real-time operation in order to seek real-time optimal solutions, which is also introduced in the field of water management. There are two prominent categories of approaches to use optimization in the real-time system. The first one is to integrate optimization algorithm with real-time simulation models in which case a forecasting model is often introduced (Shamir & Salomons, 2008; Fallah-Mehdipour et al., 2012; Che & Mays, 2015; Odan et al., 2015). With this method, decisions are determined based on present observation and future possible situations. However, such real-time operation involving advanced forecasts and modeling of upstream conditions often brings additional uncertainties. The second approach is based on usage of an existing set of optimal solutions, derived from analysis of historical events (commonly available in some databases), from which the real-time decision is chosen using some criteria. For example, Malekmohammadi et al. (2009) built a hydrodynamic model for Bakhtiari and Dez River-Reservoir systems in Iran, used to derive optimal solutions from historical floods using genetic algorithm. When current observation comes, similar events as well as corresponding optimal solutions would be found by the K nearest neighbor (K-NN) algorithm. However, such research that considers real-time operation is mostly related to reservoir operation or water distribution systems operations, while studies that are considering flood storage area management are absent. To our knowledge, this study is the first one to study both real-time operation and robust optimization of flood storage areas for developing operation strategies.

This chapter proposes robust optimization strategies and near-real-time operation method for operating multiple gates of the flood storage areas considered. Before the start of strategy development, uncertainty of inflow was analyzed, which was realized by sampling synthetic hydrographs, generating 100 possible inflow hydrographs with the characteristic of different severity levels. In order to produce robust strategies, the method of ROPAR was used and the probability of inundation of the considered storage areas was derived. For developing potential real-time operation method, these sampled hydrographs and corresponding optimal solutions were taken as pre-exiting set of solutions to be considered during actual strategy selection and extraction. In order to identify the most similar inflow conditions, four methods were proposed and tested. The testing was carried out by three amplified hydrographs (considered actual for real-time operations), which were derived by a standard approach taken in China to represent hydrographs with different return periods. Finally, the robust strategies and real-time decisions were compared so that the suitability of the different methods could be assessed. The analysis of storage area operation could hopefully provide suggestions for local administration with the aim of reducing flood risk in Bengbu City while simultaneously lowering storage area damage.

14.2. Materials and Methods

14.2.1. Available Data and Models

As this study is a continuation of previous work of Jonoski et al. (2018) and Jonoski et al. (2019), some of their methodologies and results need to be briefly introduced here. These past studies, as well as the one presented here, use a hydrodynamic model and a coupled simulation-optimization model, described below.

14.2.2. HEC-RAS Hydrodynamic Model

The hydrodynamic model was built using HEC-RAS software, covering the whole study area, from upstream Lutaizi station to downstream Bengbu City, including four main storage areas. This model simulates the propagation of hydrographs from Lutaizi to Bengbu providing results in terms of water levels in both Bengbu City and four storage areas. These results depend on the upstream hydrograph and certain operations of inlet and outlet gates that control the usage of the flood storage areas. In previous studies, the inflow hydrograph was from observed discharge from 24 June to 7 September 2007 (Fig. 14.2).

14.2.3. Coupled Simulation-Optimization Model

The coupled simulation-optimization was set up by embedding the HEC-RAS model in the optimization process using the Nondominated Sorting Genetic Algorithm II (NSGA II).

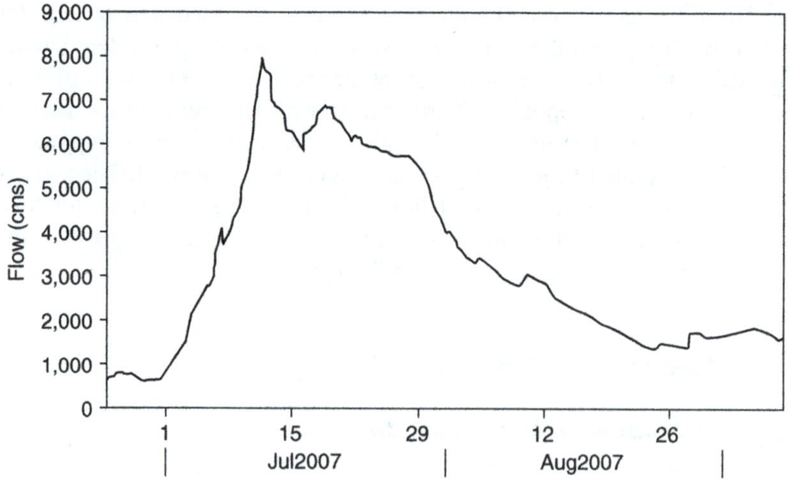

Figure 14.2 Inflow hydrograph at Lutaizi station from 24 June to 7 September 2007.

The decision variables $\boldsymbol{X}$ in this optimization problem are the values of stage differences between the river and the storage areas, which in HEC-RAS triggers the opening/closing of the control gates. These values are in fact determining the operational strategies for the gates. In order to decrease complexity in the optimization, only the operation of four storage area inlet gates was taken into consideration. (Stage differences for the outlet gates were set at 0.1 m for opening and 0 m for closing, as their use is only after the passage of the flood peak.) For each inlet gate, the stage difference for gate opening is an absolute value while the stage difference for closing uses a variable that is a fraction (from 0 to 1), which multiplies the value of stage difference for gate opening (to ensure that stage differences for closing are always smaller than those for opening). Therefore, there are eight decision variables ($\boldsymbol{X} = (X_1, X_2, X_3, X_4, X_5, X_6, X_7, X_8)$) in total (four inlet gates for four storage areas, each with one opening and one closing stage difference value).

The conflicting objectives were described by two objective functions, downstream flood risk (Obj_1) at the location of Bengbu City, and storage area damage index (Obj_2). The first objective consists of three components: (subobjective 1) the duration when stage in Bengbu is higher than predetermined warning level (H_W, 20.3 m), which needs to be as short as possible; (subobjective 2) the difference between maximum stage and warning level, which should be as small as possible; (3) because stage at Bengbu should be strictly below a predetermined safeguard level (H_S, 22.3 m), it becomes a constraint in the optimization problem by introducing a penalty function. The function of the first objective is

$$Min\, F_1(\boldsymbol{X}) = w_1 \cdot O_1(h(\boldsymbol{X}), H_W) + w_2 \cdot O_2(h(\boldsymbol{X}), H_W) + P(h(\boldsymbol{X}), H_S), \tag{14.1}$$

$$O_1(h(\boldsymbol{X}), H_W) = T(h(\boldsymbol{X}) > H_W), \tag{14.2}$$

$$O_2(h(\boldsymbol{X}), H_W) = \max(h(\boldsymbol{X})) - H_W, \tag{14.3}$$

$$P(h(\boldsymbol{X}), H_S) = 10^{17(\max(h(\boldsymbol{X}))-H_S)}, \tag{14.4}$$

where, $F_1(\boldsymbol{X})$ is downstream flood risk; $\boldsymbol{X}$ is the vector of decision variables; $h(\boldsymbol{X})$ is stage hydrograph at Bengbu City; H_W is warning level (20.3 m); H_S is safeguard level (22.3 m); O_1 is subobjective 1 (in unit of day); O_2 is subobjective 2 (in unit of decimeter); P is a penalty function; w_1 and w_2 are weights for subobjective 1 and 2 respectively (both equal to 0.5). T is duration of stage $h(\boldsymbol{X})$ above H_W.

The calculation of subobjective 1 and 2 is indicated in Figure 14.3 with the example of a hydrograph in Bengbu City in 2007.

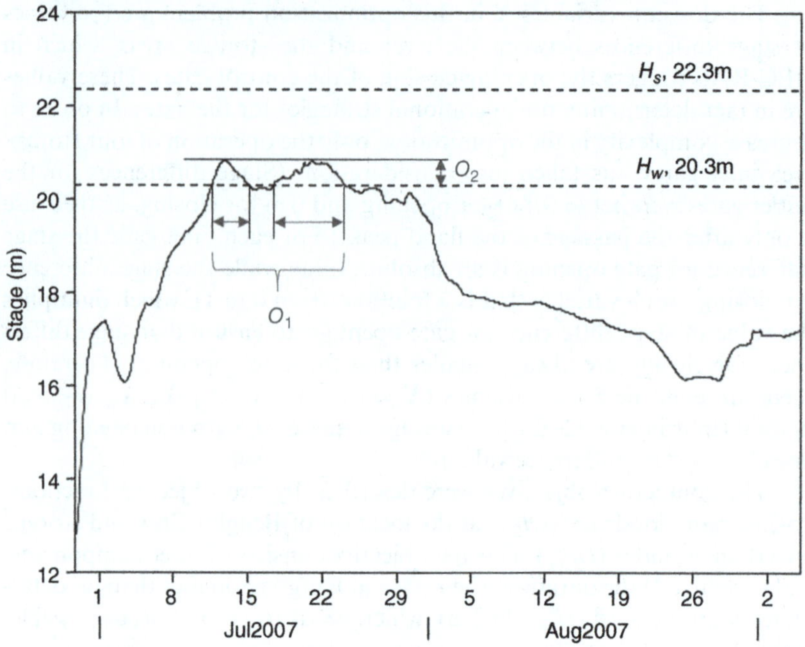

Figure 14.3 Schematic diagram of calculation for downstream risk.

The second objective can be derived from the damage curves developed by Jonoski et al. (2019), presented in Figure 14.4, which are obtained using different calculation methods for different land-use types present in the four storage areas, described by a detailed DEM and land-use maps. At each time step, the curve is used to obtain damage, given certain flooding level, and these damages are summed as total damage for each storage area at the end of the simulation period. The final estimation of the overall damage index is the sum of weighted storage area damage. The function is

$$Min\,DI = \sum_{i=1}^{4} w_i \cdot \left(\frac{1}{T} \cdot \sum_{t=1}^{T} D_i^t \right), \tag{14.5}$$

where, DI is storage area damage index; w_i is weight for each storage area ($w_1 = 0.37$, $w_2 = 0.25$, $w_3 = 0.19$, $w_4 = 0.19$) – values proportional to population living in each storage area; D_i^t is damage in a specific storage area at time t; T is total number of time steps.

The optimization process generates sets of decision variables, which are delivered to the hydrodynamic model. With storage area operation

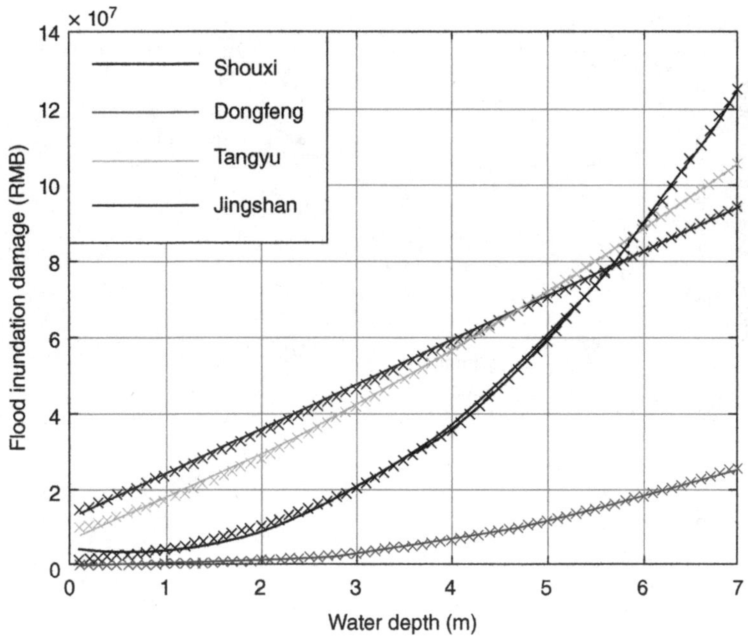

Figure 14.4 Damage curves for four storage areas (Jonoski et al., 2019).

parameters (stage differences), the hydrodynamic model provides water stage information so that the objective functions are calculated. The NSGA II algorithm continues this process toward obtaining improved values of the objective functions, until the end when it identifies the final Pareto set of optimal solutions (Fig. 14.5). Figure 14.6 shows the actual operation with one optimal strategy as well as the results of applying that strategy. The top-left subfigure of Figure 14.6 shows the maximum inundation in the flood storage areas, and the top-right shows the downstream water level at Bengbu with and without using that strategy. The last four subfigures show the gate operations for the four storage areas.

14.2.4. Methodology

Based on the previous work and obtained results, the methodology of this study consists of three components. The first one is the determination of uncertain inflow. Synthetic hydrographs were proposed with similar characteristics as the real one used previously (from year 2007), on which

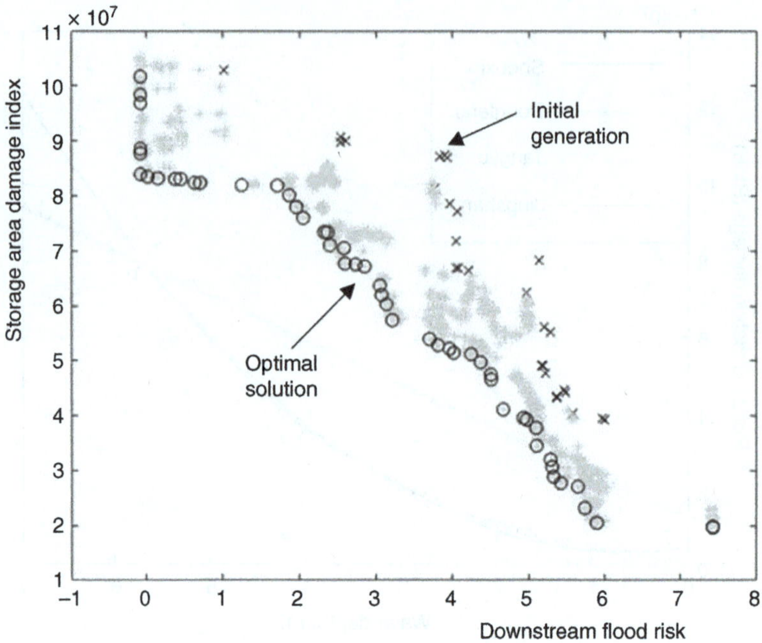

Figure 14.5 Example of Pareto front of the coupled simulated-optimization model.

Latin Hypercube Sampling (LHS) was performed. The obtained results were used for developing robust strategies as well as a similarity-based selection method for real-time operation. The method of Homogeneous Multiple Enlargement (HME) was used to develop hydrographs for verifying the methodologies. The second component is to develop robust strategies using the method of Robust Optimization and Probabilistic Analysis of Robustness (ROPAR). The last component proposed four methods to do similarity selection as ways for selecting real-time optimal strategies, which finally would be compared with the results from ROPAR under the same flood conditions.

14.2.5. Uncertainty Analysis of Inflow

Due to the limitation of availability of enough inflow data in the study area, it was decided to sample synthetic hydrographs using parametric functions instead of analysing historical records. The function is

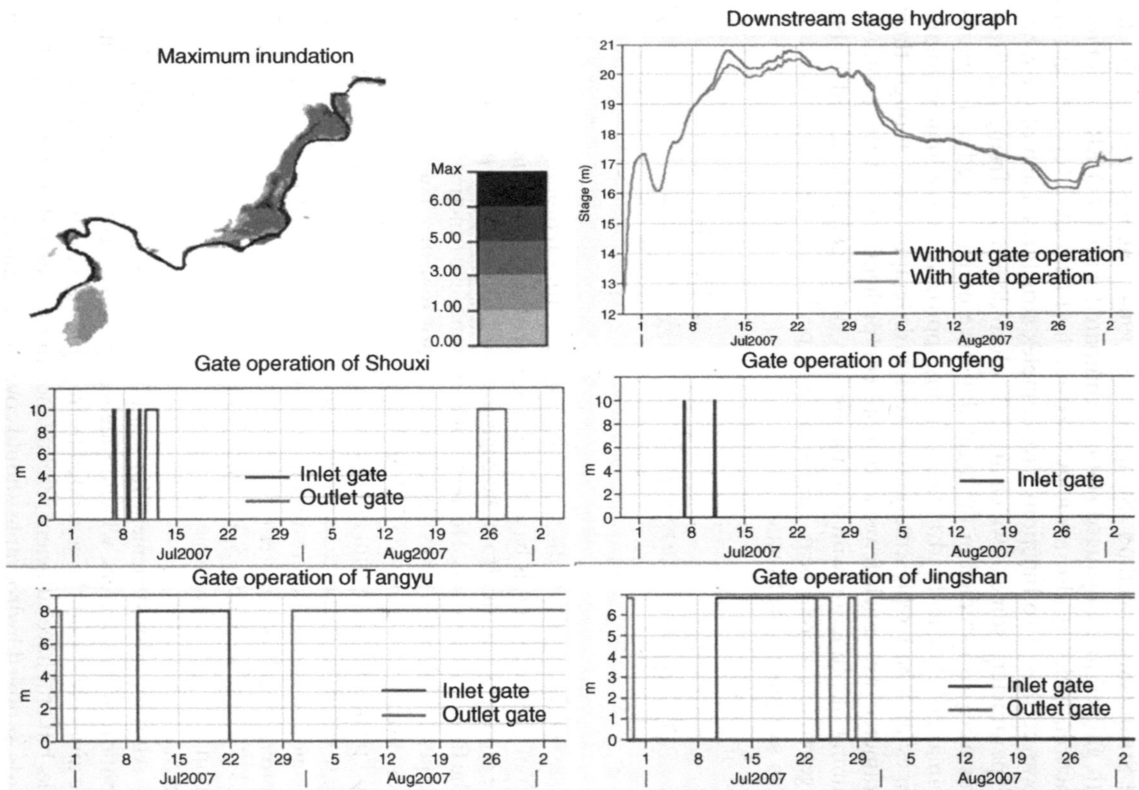

Figure 14.6 Example result of optimal strategy operation (Jonoski et al., 2019).

expected to present different flood peaks and shapes and thus different severities. A large amount of research has studied parametric functions of hydrographs. For example, the study of Gądek et al. (2017) showed that the Strupczewski method (Strupczewski, 1964) has better suitability to both gauged and ungauged stations compared with UPO ERR Gamma function, which is also a widely used method (Baziak & Gądek, 2019). However, these two functions have complex structures. Each of them has five parameters, some of which are difficult to determine. Triangular functions were also studied in the areas that have undetermined hydrographs (Gama, 2012). The disadvantages of this approach are the limited graphs shapes and bad representation of real hydrographs, even though their parameters are much fewer and more comprehensible. On the contrary, Gamma-like function with four parameters shows good shape and is able to generate different peak values, time to peak, skewness, and base flow when sampling. The function is

$$Q_t = Q_b + (Q_p - Q_b) \times \left(\frac{t}{T_p}\right)^{\frac{1}{S_k - 1}} \times exp\left(\frac{1 - \left(\frac{t}{T_p}\right)}{S_k - 1}\right), \tag{14.6}$$

where, Q_t is calculated discharge at time t; t is time (counting from 0 with unit of time step); Q_b is base flow; Q_p is peak discharge; T_p is time to peak (the interval counting from the beginning of hydrograph to peak time); S_k is the factor of skewness. However, one Gamma-like function can have only one peak, which for many flood events is not realistic. The available data of the hydrograph from 2007 (Fig. 14.1) indicates two possible subsequent flood peaks. Therefore, an alternative method was proposed, to assign different values to parameters for different periods, selected in such a way as to generate two peaks formed within one hydrograph (Fig. 14.7). In that approach, each parameter (base flow discharge Q_b, peak value Q_p, time to peak T_p, and skewness S_k) is sampled twice, and therefore there are eight parameters in total for one hydrograph: Q_{b1}, Q_{b2}, Q_{p1}, Q_{p2}, T_{p1}, T_{p2}, S_{k1}, and S_{k2}.

Another critical element for sampling is the selected probability distribution of parameters, which influences the diversity of sampled results. The aim here is that the generated hydrographs are diverse as much as possible, but still have sufficient resemblance to the real hydrographs. In other words, sampled hydrographs should cover the conditions that may happen in future, which in fact represents the uncertainty. Following these requirements, the base flow and time to peak were designed to be concentrated on a specific value (using normal distribution), while the peak discharge

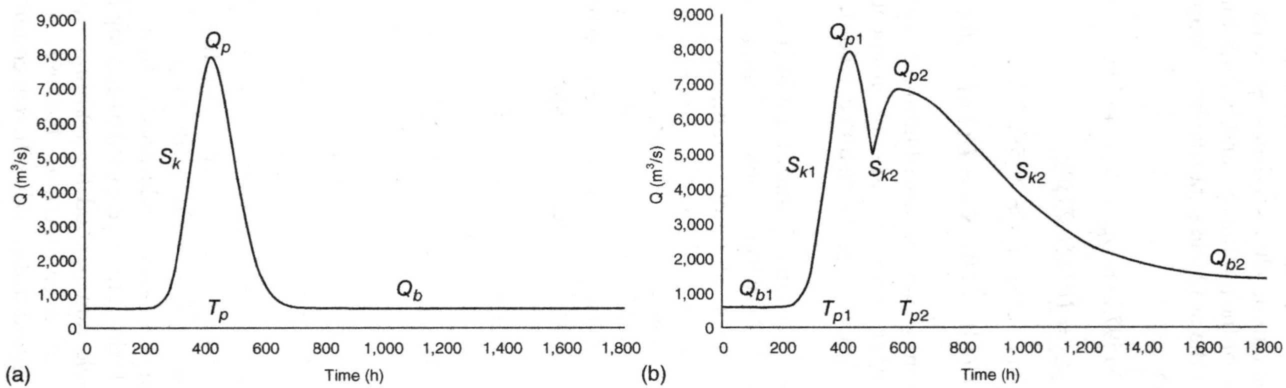

Figure 14.7 Hydrographs generated by Gamma-like function of (a) one and (b) two peaks.

and skewness were selected to be uniformly distributed (as they can be any value, still related to the natural characteristics of the river, precipitation, etc.). The determination of the parameters and bounds for the chosen distributions was based on the available data (the hydrograph of 2007), with final selection as follows:

$$Q_{b1} = N(597{,}25);\ Q_{b2} = N(1340{,}25); \tag{14.7}$$

$$Q_{p1} = Q_{p2} = U(6850{,}7950); \tag{14.8}$$

$$T_{p1} = N(422{,}25);\ T_{p2} = N(589{,}25);\ . \tag{14.9}$$

$$S_{k1} = U(1.07, 1.20);\ S_{k2} = U(1.01, 1.20); \tag{14.10}$$

where, N is the notation of normal distribution; U is the notation of uniform distribution; other terms are same as in equation (14.6). The period of hydrographs is the same as the historical record, from 24 June to 8 September.

The sampling method used in this study is Latin Hypercube Sampling (LHS) (Swiler & Wyss, 2004), which is able to sample parameters more diversely with the same quantity of sampling numbers compared with some frequently used methods such as Monte Carlo Sampling (Kapelan et al., 2005; Kang et al., 2009; Jung et al., 2013). Because each experiment of multi-objective optimization with one hydrograph would take nearly 1 day, considering time saving, the number of samples needs to be relatively small but still large enough to cover the uncertainty spectrum. In this study, it was set at 100 samples.

After the LHS, an additional operation was performed in which the flood volume of each hydrograph was modified to a value that is approximately the same as the available real hydrographs (from 2007). This was done by changing the value of the skewness parameter. This operation was introduced mainly because in the analysis the same flood volume needs to be distributed over the four storage areas and the main river. Too varying patterns (including variation of flood volume) result in some unrealistic hydrograph shapes and introduce difficulties in analyzing the effect of the synthetic hydrograph parameters to results (Zhou, 2019). That means the uncertainty studied in this research is restricted to the condition that the flood volume of all sampled hydrographs during the considered period is the same (a restriction that could be addressed in future research).

These 100 sampled hydrographs were used to develop robust strategies and in the similarity selection method for real-time operation, as will be elaborated in the subsequent sections. However, the performance assessment of the so produced strategies in different conditions needs to be carried out with hydrographs that are different from sampled ones and with different characteristics of severity. For this purpose, the Homogeneous

Multiple Enlargement method was used, which is frequently used for generating design hydrographs (Yue et al., 2002; MWR, 2006; Xiao et al., 2009; Tan et al., 2017; Yin et al., 2018; Rizwan et al., 2019). In this method, a typical hydrograph needs to be selected first, whose discharge would be amplified with a multiplier, which is calculated as the ratio of flood volume or peak discharge for a particular return period to that of the typical hydrograph. Usually, the volume ratio is used for storage areas. Considering that the use of the four storage areas in this study may last for a month, the volume is chosen as the sum of successive 30 days' largest discharge, so that the formula of the multiplier is

$$K = \frac{V_{30,R}}{V_{30,T}}, \tag{14.11}$$

where, K is the multiplier; $V_{30,R}$ and $V_{30,T}$ are the 30 days' volume for a particular return period and a typical year, respectively.

For the sake of data availability, the typical hydrograph is chosen as the 2007s. The generated hydrographs have the same shape as 2007 but with different flood volume, depending on the return period. Two most severe records were found in Chinese history, the years 1954 and 2003, with 30 and 60 return years, respectively (see Table 14.1). Their flood volumes are used for calculating the multipliers. It should be noted that with this approach the optimal strategies will be tested with very different hydrographs compared with the ones produced with the sampling. These multiplied hydrographs are presented in section 14.3.1.

14.2.6. Robust Optimization with ROPAR

Robust Optimization and Probabilistic Analysis of Robustness (ROPAR) is a method to develop optimal solutions, as well as to identify

Table 14.1 Historical most severe flood records at Lutaizi station and calculated multipliers.

Year	30 d's flood volume ($10^9 m^3$)	Return period (year)	Multiplier K
1954	22.02	60	1.54
1956	14.62	20	1.03
1968	11.95	7	0.84
1991	12.99	9	0.91
2003	15.10	30	1.06
2007	14.26	12	1.00

Note: Severe flood records at Lutaizi station from Lu et al. (2016).

robust ones among optimized options (Marquez-Calvo & Solomatine, 2019). In this study, the method was used to develop robust optimal strategies of storage area operation under a specific downstream risk level. It was carried out following these steps: under the inflow uncertainty quantified as described in section 14.2.5, optimal strategies for each possible hydrograph are first identified by providing the hydrographs to the simulation-optimization model and performing the optimization. This step results in 100 Pareto fronts, with two dimensions of minimum downstream risk and minimum storage area damage, corresponding to 100 inflows. When assembling all these fronts in one coordination system, a scatter diagram would be formed, in which each point contains information of optimal strategy (gate operations) and the corresponding hydrograph; then a cluster of points would be picked by a certain criterion, which in this study is that they have same downstream risk. The risk could be a spectrum (e.g., with a span of 1), and the points included in mean potential robust strategies; a probability density function (PDF) can be built by collecting the storage area damage of those points, meaning the probability of storage area inundation loss under the chosen risk level; in order to find out the most robust one among these optimal strategies, each of these potential optimal strategies is simulated with 100 sampled hydrographs. Two criteria are calculated for electing optimal ones. They are (1) worst case and (2) expected value. The functions are

$$E\left(X_i\right) = \frac{1}{n}\sum_{j}^{n} Obj_2(X_i, HG_j), \tag{14.12}$$

$$W(X_i) = \max\left[Obj_2(X_i, HG_j)\right], \tag{14.13}$$

where, E is expected value; W is worst case; X_i is the ith solution; HG_j is the jth hydrograph; Obj_2 is the value of storage area damage index (the objective 2); n is sampling number ($n = 100$).

It can be understood that the worst case selection criterion refers to maximum damage from the 100 simulated results, and the expected value selection criterion is average damage from the same results. Therefore, each optimal strategy has two criteria to estimate its robustness. The selection of most robust strategy is done by choosing the one with either minimum worst case or minimum expected value. In many cases, the same robust strategy is chosen by two different criteria, which means the strategy can lead to minimum damage under different flood conditions no matter if it is calculated by average or maximum value.

14.2.7. Similarity Selection Method for Real-Time Operation

In the case of developing real-time operation, lead time is an important consideration. According to Pan (2002), the forecasting scheme, which is already in use by Huai River commission, is able to predict flow condition (hydrograph) at Lutaizi station using the Muskingum method from an upstream station called Hengpaitou station. The propagation time from Hengpaitou to Lutaizi is 36 hr (Pan, 2002), which can be considered as lead time for real-time operation in this study. As described in Jonoski et al. (2018), the optimization of the storage area gate operation is dependent on the simulated water level results, which are obtained from the simulation model with input of nearly 3 months discharge hydrograph (Fig. 14.2). Therefore, implementation of the optimization process with incoming real-time data is not adopted, and the use of optimal strategies obtained from preanalysis described previously (using historical data) is used in this study.

Using the database of optimized strategies obtained from the sampled hydrographs (already developed with ROPAR), the decision to be implemented is made according to the most similar hydrograph in that database to the expected real hydrograph, which is called *similarity selection.* As described previously, having forecasts about the whole hydrograph is practically impossible, given its very long duration, so the similarity assessment is restricted to the area of the hydrograph around the peaks. A threshold value of Q_0, 5,500 m^3/s, is adopted, and only discharges above this value are considered in the similarity assessment. The proposed criteria for estimating this similarity are as follows:

1. The first criterion compares the volume of flood peaks where discharge is larger than 5,500 m^3/s ($Q > Q_0$). This method was proposed because the volume of the upper part of a hydrograph is one of the most critical elements influencing flood severity downstream. If part of this volume is allocated to the storage areas, the flood peak(s) downstream (at Bengbu City) will be reduced. The criterion of selecting the most similar hydrograph is

$$\min FV = \left| V^R_{Q>Q_0} - V^S_{Q>Q_0} \right|, \tag{14.14}$$

where, FV is the notion of the flood volume method; $V_{Q>Q_0}$ is the flood volume of discharge larger than Q_0; Q_0 is a threshold value (5,500 m^3/s); the superscripts R and S mean real hydrograph and sampled hydrograph, respectively.

2. The second criterion uses Nash-Sutcliffe efficiency coefficient (NSE), which gives an indicator about closeness between two hydrographs. Larger NSE means better match. Focusing again only on the upper part of the hydrograph, the equation used for similarity assessment is

$$\max NSE = 1 - \frac{\sum_{t=t_1}^{t_2} \left(Q_t^S - Q^R_{t,Q>Q_0} \right)^2}{\sum_{t=t_1}^{t_2} \left(Q^R_{t,Q>Q_0} - \overline{Q^R_{Q>Q_0}} \right)^2}, \tag{14.15}$$

where, NSE is Nash-Sutcliffe efficiency coefficient; Q_t is the discharge at time t; $\overline{Q^R_{Q>Q_0}}$ is the average value of discharge larger than Q_0 of a real hydrograph; t_1 and t_2 are the start and ending time for discharge larger than Q_0, respectively; other terms are the same as in equation (14.13).

3. The third criterion also uses one indicator for similarity, which is the weighted root mean square error (wRMSE). A weight is added in each time step, decided by the proportion of discharge at time t to the sum of discharges across all considered time steps, which is a way to introduce higher importance (weight) to higher flow discharges. The equation used, which also focuses on upper part of hydrograph, is

$$\min wRMSE = \sqrt{\frac{1}{t_2 - t_1}\sum_{t=t_1}^{t_2} w_t\left(Q_t^S - Q^R_{t,Q>Q_0}\right)^2}, \tag{14.16}$$

$$w_t = \frac{Q^R_{t,Q>Q_0}}{\sum_{t=t_1}^{t_2} Q^R_{t,Q>Q_0}}, \tag{14.17}$$

where, $wRMSE$ is weighted root mean square error; w_t is weight at time t; other terms are same as equation (14.14).

4. The last criterion considers the factors affecting downstream risk and storage area damage. According to the two objective functions (see section 14.2.1.2), the downstream risk is affected by the flood duration (the time when the stage is larger than warning level) and peak value (difference between maximum stage and warning level) while the storage area damage is calculated with the water levels inside, which are largely influenced by peaks of hydrographs. Similarly, the last criterion is to compare the duration and flow peaks of two hydrographs. The threshold discharge is same as previously defined. Each term is normalized, using following equation:

$$\min PD = nor(PD_1) + nor(PD_2) + nor(PD_3) \tag{14.18}$$

$$nor(PD_i) = \frac{PD_i^j - \min\left(PD_i^j\right)}{\max\left(PD_i^j\right) - \min\left(PD_i^j\right)}, \tag{14.19}$$

$$PD_1^j(Q^R, Q^{S,j}) = \left|Q^R_{p1} - Q^{S,j}_{p1}\right|, \tag{14.20}$$

$$PD_2^j(Q^R, Q^{S,j}) = \left|Q^R_{p2} - Q^{S,j}_{p2}\right|, \tag{14.21}$$

$$PD_3^j(Q^R, Q^{S,j}) = \left|D\left(Q^R_{Q>Q_0}\right) - D\left(Q^{S,j}_{Q>Q_0}\right)\right|, \tag{14.22}$$

where, PD is the notion of the peak and duration method; nor is the function of normalization; D is duration when discharge is larger than Q_0; Q_{p1} and Q_{p2} are first and second peak, respectively; j means the jth sampled hydrograph; i means the ith term in PD ($i = 1, 2, 3$); other symbols are same as equation (14.14).

With the four criteria presented above some similar hydrographs would be selected, for which the corresponding optimized strategies need to be identified for real-time operation. As for each inflow hydrograph, there are a number of different strategies (represented in the Pareto front), a certain level of the downstream risk needs to be first chosen, and the corresponding strategy is selected as final for implementation.

It needs to be noted that, even though by similarity selection the time-consuming process of optimization is eliminated (as mentioned above), the duration of discharge larger than 5,500 m^3/s is still very long (around 20 days), much more than the 36 hr available lead time. Therefore, to generate knowledge about the real hydrograph top (to be used in the similarity assessment), additional work would be required, for example, using information about precipitation in the upstream part of the catchment, and associated rainfall-runoff modeling. The methodology presented here is based on the assumption that there is enough time to perform such analysis.

14.3. Results and Discussion

14.3.1. Possible Upstream Hydrographs

With two methods of developing possible upstream hydrographs, two corresponding results were generated.

The sampled hydrographs are presented in Figure 14.8, with characteristics of varying peaks and skewness but the same flood volume, which will be used later in the ROPAR method and the strategy selection for real-time operation. It should be noted that although two peaks are created by setting different parameters for the Gamma-like function, avoiding the shape difference with the 2007's hydrograph, the single peak or multiple peaks may appear in reality. Because of the unavailability of a real record for reference, these phenomena are not discussed in this study, which could be addressed in future research.

The enlarged hydrographs to be used for testing are shown in Figure 14.9. As the multiplier of the year 1954 is much larger than the others, the hydrograph of 1954 has much higher peaks and will be named

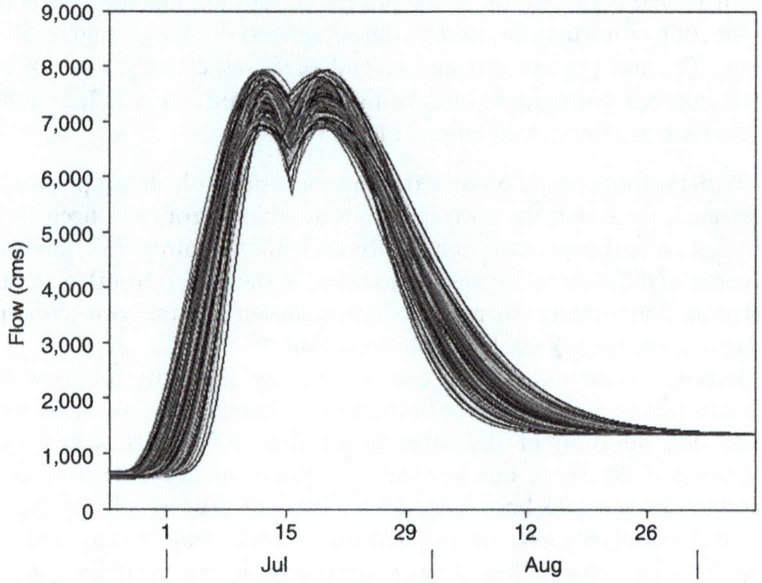

Figure 14.8 One hundred sampled hydrographs using LHS.

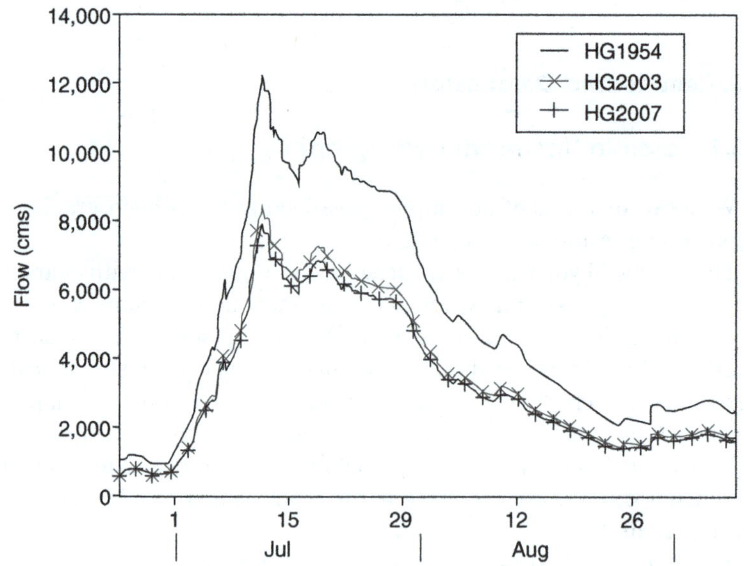

Figure 14.9 Enlarged hydrographs using Homogeneous Multiple Enlargement.

extreme condition in later analysis. It is worth mentioning that the names of the hydrographs are only for indicating the year used for calculating the multipliers, rather than the actual hydrographs of that particular year.

14.3.2. Robust Strategies Resulting From ROPAR

For three chosen downstream risk values 10, 11, and 12, the corresponding probability distribution functions (PDFs) of storage area damage were found (Fig. 14.10). It can be seen that when downstream risk is higher, there is larger likelihood of small storage area damage. That means that when making decisions to flood storage areas, it should be considered how large is the expected risk for the downstream city of Bengbu.

As described earlier when introducing the ROPAR method, in order to identify robust optimal strategies, the strategies corresponding to a certain value of downstream risk were simulated with each possible upstream inflow, and the criteria of selection were calculated. One example result is depicted in Figure 14.11, for the case of downstream risk 10. The solution with minimum value of average damage (expected value) is marked with × while the solution with minimum maximum damage (worst case) is marked with +. It is observed that a strategy leading to small expected value can often be the same as the one with small worst case, forming an asterisk mark, in which case the robust optimal strategy is often identified as being the same regardless of which criterion is used.

The final decision of robust strategies with different risk levels is listed in Table 14.2. When downstream risk is 10 or 12, the two criteria lead to the same strategy selection, while two different robust strategies were identified for downstream risk of 11.

The four robust strategies from Table 14.2 were tested with inflow of the three enlarged hydrographs (Fig. 14.9) by calculating the two objective functions (see section 14.2.1.2). It is worth mentioning that to prevent

Table 14.2 Robust strategies for different downstream risk levels.

Downstream risk	Type of criterion	Strategy ID
10	Expected value	S70
	Worst case	S70
11	Expected value	S93
	Worst case	S96
12	Expected value	S40
	Worst case	S40

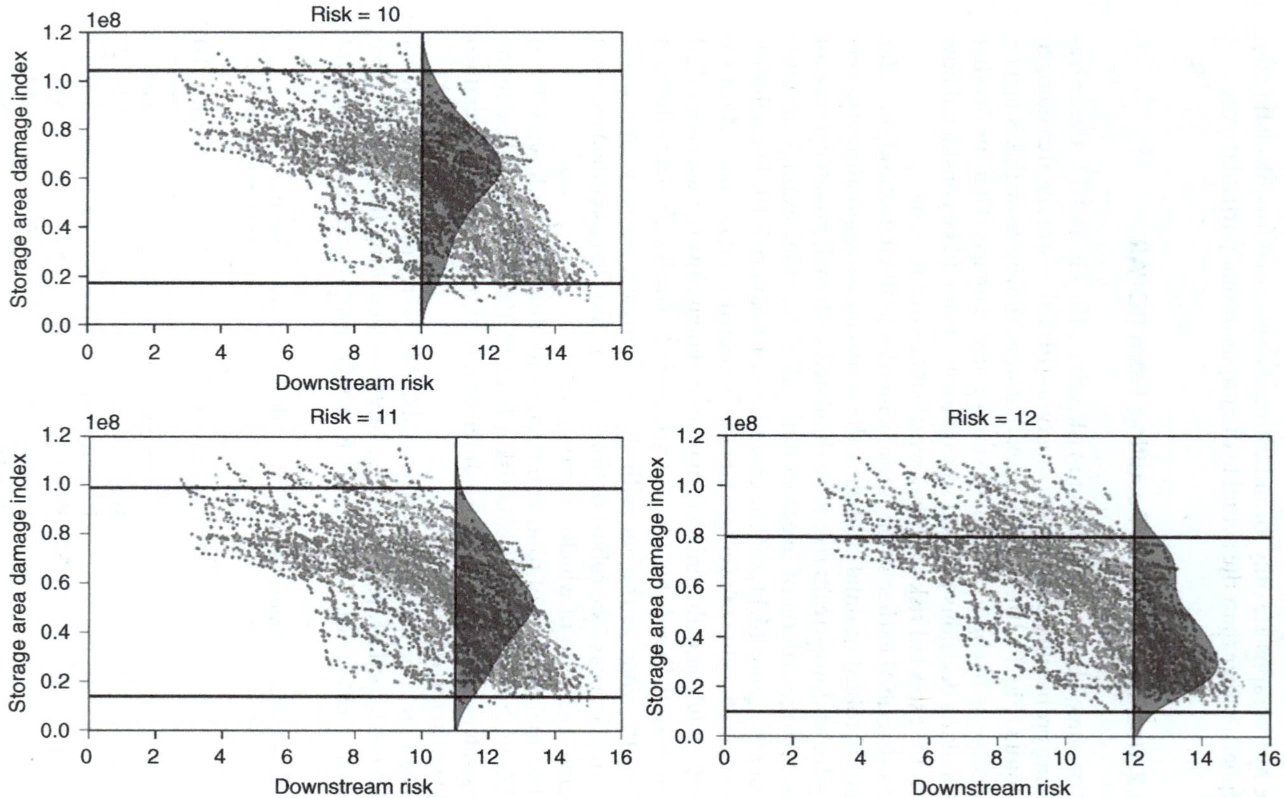

Figure 14.10 PDFs of storage area damage under certain downstream risk.

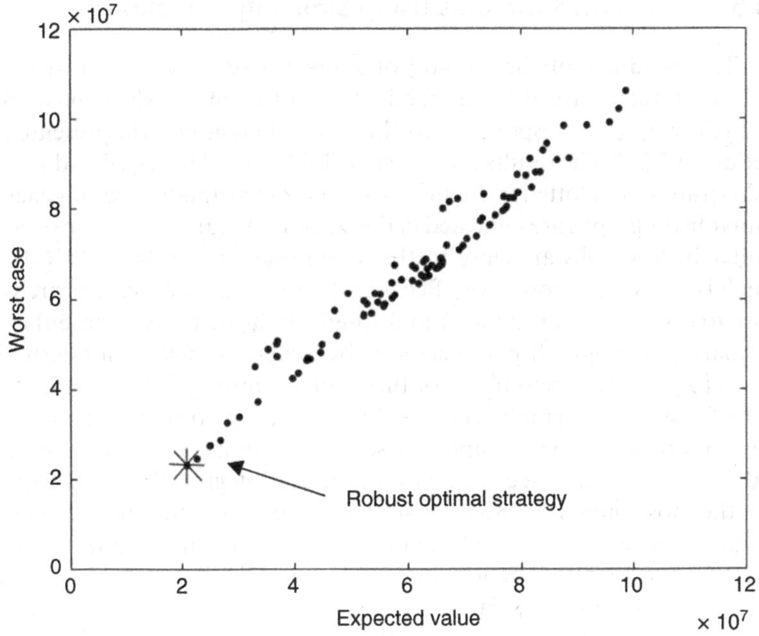

Figure 14.11 Estimation of worst case and expected value of potential robust strategies.

the work of the penalty function in the objective function, which would cause a very large value, the penalty function was not used here. As shown in Figure 14.12, downstream risks are all larger than 20 corresponds to HG1954, in which case no strategy can prevent the flooding of Bengbu City. The identified robust strategies are still unable to deal with such extreme floods. As to the comparison of the performance of strategies, S93 is the best one among them to reduce downstream risk for the two high-risk hydrographs, HG1954 and HG2003, and still efficient to protect Bengbu with HG2007. With respect to storage area damage, S40 is best to alleviate damages in storage areas when meeting HG2003 and HG2007. Therefore, for the sake of insurance of downstream safety, S93 is recommended to be implemented; if we already know that coming inflow would not severely threaten downstream safety, S40 can be used in order to reduce storage area damage. However, it should be noted that it is difficult to determine inflow severity in advance due to the short lead time in this study area.

14.3.3. Resulting Strategies Using Similarity Selection

The goodness of the method of similarity selection was also tested with the three enlarged hydrographs. Each of them has their most similar hydrographs corresponding to the four different criteria presented in section 14.2.2.3. The results are shown in Table 14.3. The tested and similar hydrographs are plotted in Figure 14.13 and the methods used to generate similar hydrographs are indicated in the legends. As can be seen, the selected similar hydrographs are close to the tested ones except for the HG1954, which is an extreme condition. Because these similar hydrographs are very close to each other, it is difficult to distinguish the best fitted one, either by comparing hydrograph parameters or by ocular estimate. An alternative way is to plot the Pareto fronts of these similar hydrographs together with that of tested hydrographs (Fig. 14.14), which can to some extent reflect that a more severe hydrograph is associated with larger downstream risk and storage area damage even with optimal strategies. This way helps us find the most similar hydrograph so that the best way to estimate similarity can be extracted. It should be noted that when doing optimization with HG1954, the penalty function for objective 1 (see section 14.2.1.2) always works and leads to a very large result, which makes it difficult to compare the Pareto fronts. Considering that the optimization result is only for estimating the similarity, the penalty function is not used in the optimization process for HG1954.

The Pareto front of HG1954 is located at the upper right side of the plotting area (as shown in Fig. 14.14). Therefore, the HG8, using the method of flood volume, is closer to this tested case. As for HG2003, the HG93, derived from the method of NSE and wRMSE, is obviously most similar, compared with other similar hydrographs. Likewise, HG45 is most similar to HG2007.

When analyzing the results from the application of the four criteria, it is found that the first method (flood volume) gives different results in different cases. That is because this method only focuses on the volume but neglects the effect of the hydrograph shape. The NSE and wRMSE criteria tend

Table 14.3 Similar hydrographs chosen by four criteria.

Selection criterion	HG1954	HG2003	HG2007
Flood volume	HG8	HG19	HG45
NSE	HG66	HG93	HG45
wRMSE	HG66	HG93	HG45
Peak & duration	HG66	HG86	HG42

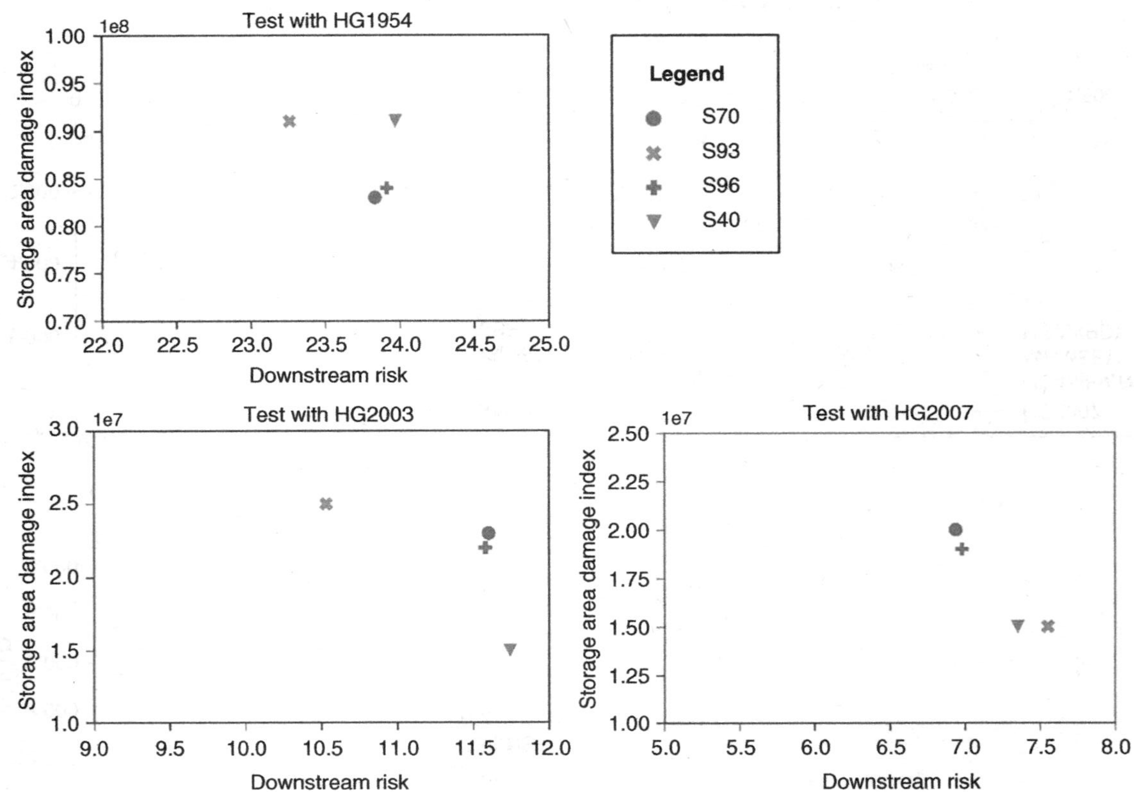

Figure 14.12 Test of robust strategies with enlarged hydrographs.

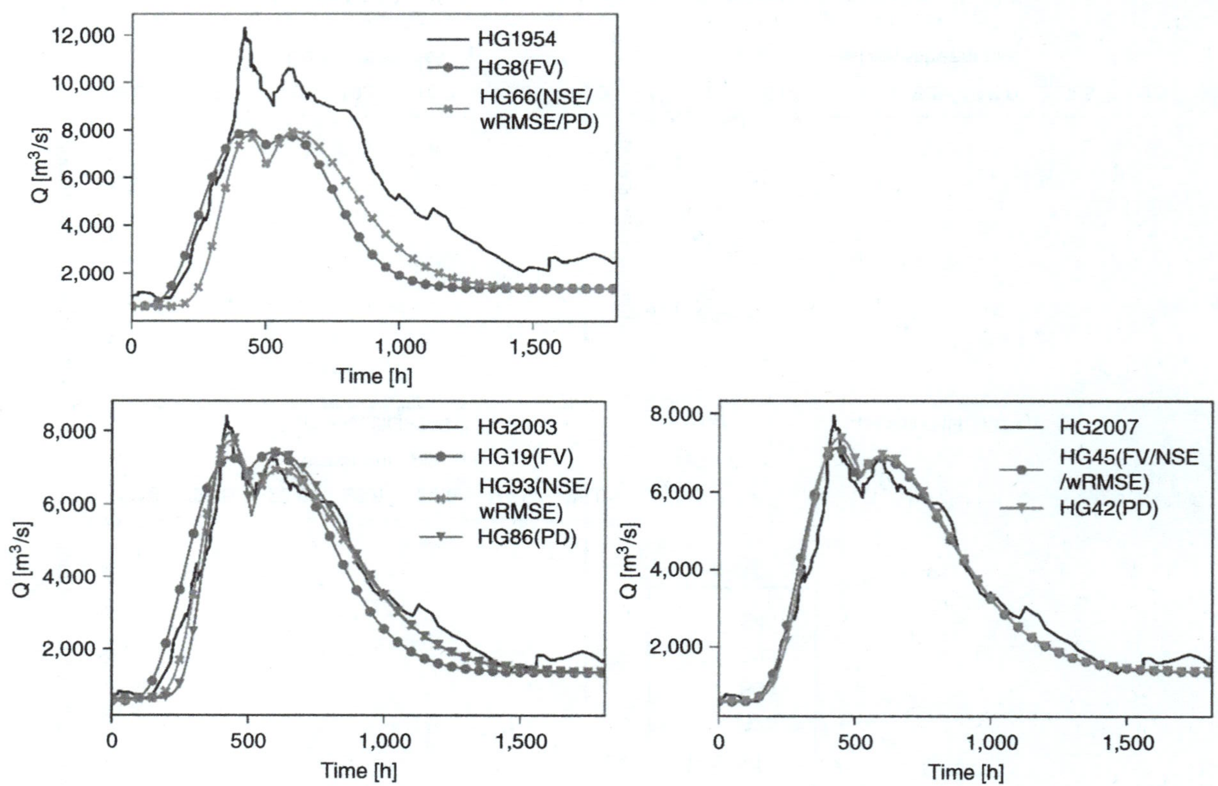

Figure 14.13 Comparison of tested and similar hydrographs.

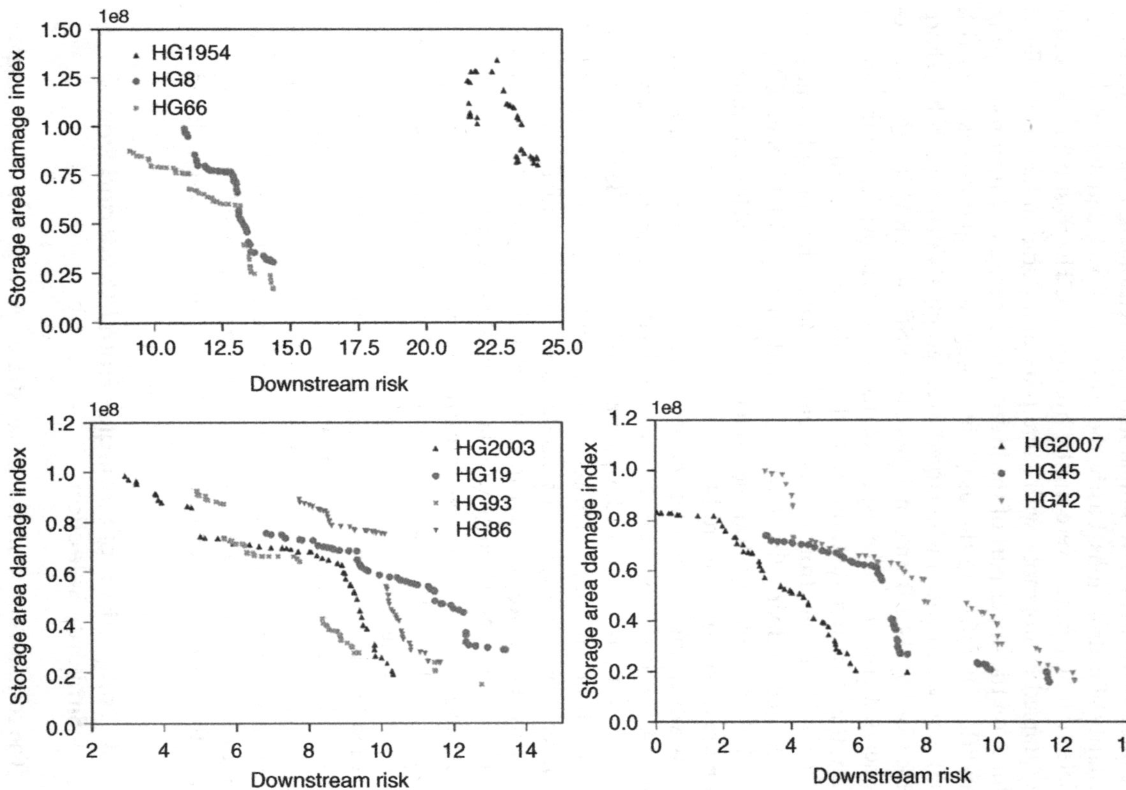

Figure 14.14 Comparison of Pareto fronts of tested and similar hydrographs.

to identify the same hydrograph as the most similar one in different cases. The definition of NSE includes sensitivity to large values of discharge, and wRMSE also gives high weight to large discharges, which leads to the fact that these two criteria give similar and quite reasonable results in terms of similarity of hydrographs. Lastly, the criterion of peak and duration results in selection of a more severe similar hydrograph. The reason for this is that the sampled hydrographs with same peak and duration have larger flood volume on the upper part of the hydrographs because of their round peak shape (largely affecting the usage of storage areas and downstream risk level), compared with the enlarged hydrographs, which have rather sharp shape (see Fig. 14.13). In conclusion, the NSE and wRMSE are more recommended criteria in the aspect of accuracy. Meanwhile, for the sake of safety, the peak and duration method can also be considered for usage.

From the Pareto fronts of the identified similar hydrographs, one point was selected by fixing the downstream risk at value of 11. Similarly, the results of these selected strategies that can be considered for real-time operation are depicted in Figure 14.15.

By observing the testing results (Fig. 14.15) and the Pareto fronts (Fig. 14.14), it can be seen that with a fixed downstream risk level, the strategies leading to small downstream risk (safer for Bengbu City) are selected from the upper cluster of potential strategies (in Fig. 14.14). However, these safe strategies would cause larger storage area damage, obviously. Therefore, in order to ensure Bengbu City not to be flooded, it is better to choose a strategy with larger possibility to flood storage areas. Furthermore, if encountering an extreme condition, such as the year of 1954, another effective way is to fix a smaller downstream risk (e.g., 9 or 10) when making real-time decisions.

Finally, it is worth mentioning again that the proposed real-time operation is all based on the assumption that enough lead time (approximately 20 days) is available before implementation. The possibility of practical use of the proposed methodology still needs to be investigated. Further study is necessary, for example, by either increasing the lead time or making decision with reliable knowledge of a smaller part of the hydrograph top.

14.3.4. Comparison of Strategies From ROPAR and From Similarity Selection

The comparison of the two sets of strategies (Fig. 14.16) shows that the strategies from the similarity selection method are more effective to reduce downstream risk in most cases. Especially for HG1954, one of the real-time strategies results in smaller flood risk as well as less storage area damage.

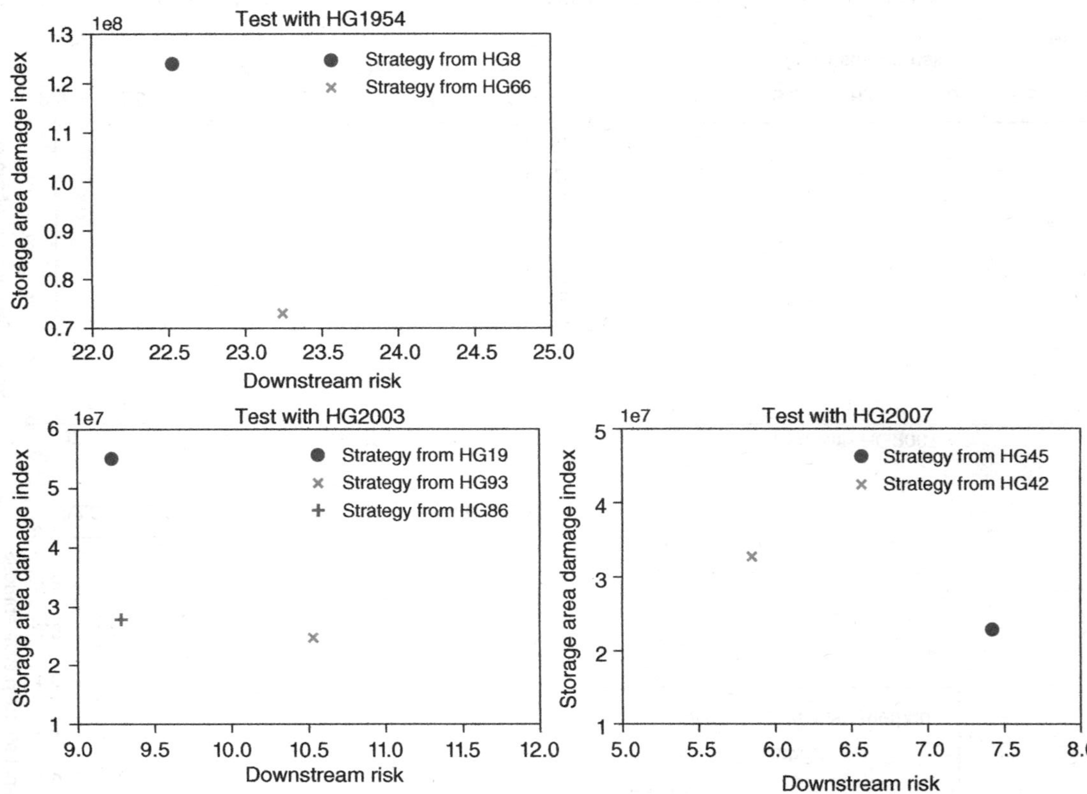

Figure 14.15 Test of real-time strategies with three enlarged hydrographs.

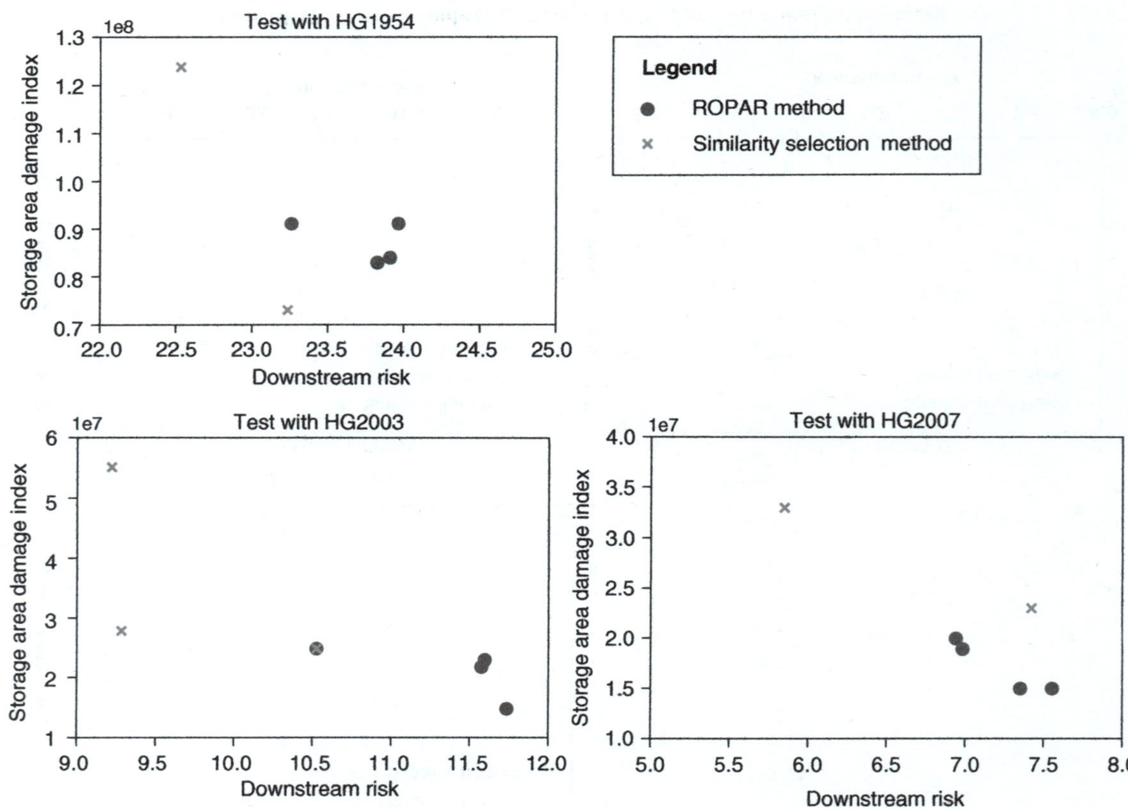

Figure 14.16 Comparison of robust strategies and similarity selection method.

However, although the similarity selection performs better, it would still produce larger storage area damage in most cases while the robust strategies can also prevent Bengbu City from flooding (the peak stage is much less than safeguard level) without compromising the land and properties of storage areas. In that sense, the robust strategies are good enough in most flooding situations; when the inflow is highly likely to cause flooding in Bengbu City (such as some extreme inflow), similarity selection is better to use. It should be again noted that neither the robust nor the similarity selection method is able to protect Bengbu City in case of HG1954. This means that under extreme conditions, these two methods are unsuitable for use. One reason for this is that the sampled hydrographs are less diverse because the parameters setting for Gamma-like function limits the possibility to produce extreme flood events like HG1954. Therefore, further study may enlarge the spectrum of sampled hydrographs, for example, by modifying the parameters of synthetic hydrographs, or by choosing another method that does not need to take the restriction that the sampled hydrographs have same flood volume. However, in any case, extreme conditions may still lead to flooding because of limits of the existing infrastructure, no matter what optimal strategies would be identified.

14.4. Conclusions

A conflicting problem exists in the middle of the Huai River between protecting Bengbu City from flooding and reducing loss in flood protection storage areas. Previous studies developed a simulation-optimization method for developing strategies of operating storage area gates but without considering uncertainty. Based on the model of previous studies, this study analyzed the uncertainty of inflow hydrographs and proposed robust optimal strategies and a similarity-based selection method for potential real-time operation. For robust optimization, the newly developed ROPAR method was used. The method for consideration in real-time operation is to choose similar inflow conditions from the database of possible hydrographs as well as their optimal strategies. The conclusions are as follows:

1. The distribution of optimal strategies results when considering the uncertainty of inflow shows that when a smaller downstream risk is expected, the possibility of large storage area damage would be higher even with optimal strategies. The robust strategies developed by ROPAR perform differently under three enlarged hydrographs of different return periods. Strategy S40 is good to minimize storage

area damage in most cases but for safety of Bengbu City, S93 is recommended to be used.

2. The criteria NSE and wRMSE can estimate similarity between hydrographs most accurately, so they are recommended to be used when making real-time decision. From another aspect, the peak and duration criteria can also be used, as they often provides safer strategies for downstream, even though they may cause larger storage area damage in some cases. After determining similar hydrographs, a strategy with larger storage area damage can be chosen for implementation, given that it is more efficient to reduce downstream risk.
3. With the same upstream inflow hydrograph, the strategies from similarity selection often lead to less downstream flood risk, but larger storage area damage compared with robust strategies, which are better for the safety of Bengbu especially when encountering extreme floods. However, in most cases, the robust strategies are also effective to protect Bengbu City.

In this study, the uncertainty analysis is somewhat restricted as the flood volume of sampled hydrographs is kept almost the same. However, different severities can still be identified among them, in which case the strategies developed using either robust strategies or similarity selection strategies are effective to protect Bengbu City in most cases, except under extreme conditions (e.g., the return period of 60 yr). In order to improve the analysis and development of optimal strategies, further studies can experiment with more diverse sampled hydrographs by setting different parameter values or without restriction to flood volume variations. In any case, the shape of the sampled hydrographs should be realistic. Another issue that should be noted is that the proposed real-time operation is based on the assumption that sufficient lead time is available before implementation. One of the recommendations here is to integrate the current hydraulic model with a hydrological forecasting scheme in upstream Huai River, which is under development by Cheng (2019), so that the lead time can be largely prolonged. A second suggestion is that a data-driven model may be built in the future, which can produce downstream flood risk and storage area damage quickly, given input of strategies and parameters of the inflow hydrograph. This model development is attainable, given the huge amount of data that is already generated in the process of robust strategies development. With a data-driven model, the elapsed time of optimization can be largely shortened (the hydrodynamic model will not be necessary any more), in which case the optimal strategies could be identified in real-time.

References

Andricevic, R., & Kitanidis, P. K. (1990). Optimization of the pumping schedule in aquifer remediation under uncertainty. *Water Resources Research, 26*(5), 875–885.

Babayan, A., Kapelan Z., Savic D., & Walters G. (2005). Least-cost design of water distribution networks under demand uncertainty. *Journal of Water Resources Planning and Management, 131*(5), 375–382.

Baziak, B., & Gądek, W. (2019). The Pearson Type IV distribution function employed to describe the parametric flow hydrograph. *Acta Geophysica, 67*(5), 1419–1433.

Beyer, H.-G., & Sendhoff, B. (2007). Robust optimization–a comprehensive survey. *Computer Methods in Applied Mechanics Engineering, 196*(33–34), 3190–3218.

Che, D., & Mays, L. W. (2015). Development of an optimization/simulation model for real-time flood-control operation of river-reservoirs systems. *Water Resources Management, 29*(11), 3987–4005.

Cheng, C. (2019). *Can machine learning reduce uncertainties of distributed hydrological models? Case study of Dapoling- Wangjiaba catchment*. Master thesis, IHE Delft.

Cheng, C., & Chau, K.-W. (2002). Three-person multi-objective conflict decision in reservoir flood control. *European Journal of Operational Research, 142*(3), 625–631.

Dittmann, R., Froehlich, F., Pohl, R., & Ostrowski (2009). Optimum multi-objective reservoir operation with emphasis on flood control and ecology. *Natural Hazards & Earth System Sciences, 9*(6), 1973.

Fallah-Mehdipour, E., Bozorg Haddad, O., & Mariño, M. A. (2012). Real-time operation of reservoir system by genetic programming. *Water Resources Management, 26*(14), 4091–4103.

Gądek, W., Baziak, B., & Tokarczyk, T. (2017). Strupczewski method for parametric design hydrographs in ungauged cross-sections. *Archives of Hydro-Engineering Environmental Mechanics, 64*(1), 49–67.

Gama, M. C. (2012). *Development of an operation of reservoirs and flooding inference model: a case of study for the Yellow river*. Master thesis, Unesco-IHE.

Hollinshead, S. P., & Lund, J. R. (2006). Optimization of environmental water purchases with uncertainty. *Water Resources Research, 42*(8).

Jen, E. (2003). Essays & commentaries: Stable or robust? What's the difference? *Complex, 8*(3), 12–18.

Jonoski, A., Popescu, I., & Zhe, S. (2018). Optimal operation of flood storage areas in Huai River using coupled HEC-RAS River model and NSGAII global optimization algorithm. In G. L. Loggia et al. (Eds), *EasyChair* (pp. 1004–1012). HIC 2018, 13th International Conference on Hydroinformatics.

Jonoski, A., Popescu, I., Zhe, S., Mu, Y., & He, Y. (2019). Analysis of flood storage area operations in Huai River using 1D and 2D river simulation models coupled with global optimization algorithms. *Geosciences, 9*(12), 509.

Jung, D., Kang, D., Chung, G., & Kim, J. H. (2013). Comparison of the robustness-based optimal designs of water distribution systems in three different formulations. *Journal of Hydroinformatics, 15*(4), 1425–1436.

Kang, D. S., Pasha, M. F. K., & Lansey, K. (2009). Approximate methods for uncertainty analysis of water distribution systems. *Urban Water Journal*, *6*(3), 233–249.

Kapelan, Z. S., Savic, D. A., & Walters, G. A. (2005). Multiobjective design of water distribution systems under uncertainty. *Water Resources Research, 41*(11), 97–116.

Lu, C., Liu, B., Sun, Q., Li, C., Li, H., & Lu, C. (2016). Analysis of inner and outer flood encounter in north bank area of the Huai River midstream. *Journal of Nature Disasters, 25*(1), 19–25.

Luo, J., Chen, C., & Xie, J. (2015). Multi-objective immune algorithm with preference-based selection for reservoir flood control operation. *Water Resources Management, 29*(5), 1447–1466.

Maier, H., Kapelan, Z., Kasprzyk, J., Kollat, J., Matott, L., Cunha, M., et al. (2014). Evolutionary algorithms and other metaheuristics in water resources: Current status, research challenges and future directions. *Environmental Modelling and Software, 62*, 271–299.

Malekmohammadi, B., Zahraie, B., & Kerachian, R. (2009). A real-time operation optimization model for flood management in river-reservoir systems. *Natural Hazards, 53*(3), 459–482.

Marquez-Calvo, O. O., & Solomatine, D. P. (2019). Approach to robust multi-objective optimization and probabilistic analysis: The ROPAR algorithm. *Journal of Hydroinformatics, 21*(3), 427–440.

MWR (2006). Regulation for calculating design flood of water resources and hydropower projects. In *Electrical Ha*. Chinese Water Resources and Hydropower Press, Beijing.

Odan, F. K., Ribeiro Reis, L. F., & Kapelan, Z. (2015). Real-time multiobjective optimization of operation of water supply systems. *Journal of Water Resources Planning Management, 141*(9), 04015011.

Pachauri, R. K., Allen, M. R., Barros, V. R., Broome, J., Cramer, W., Christ, R., et al. (2014). *Climate change 2014: Synthesis report*. Contribution of Working Groups I, II and III to the fifth assessment report of the Intergovernmental Panel on Climate Change. IPCC.

Pan, G. (2002). *Practical hydrological forecast scheme for huaihe river system in Huaihe River basin*. Yellow River Press.

Rizwan, M., Guo, S., Yin, J., & Xiong, F. (2019). Deriving design flood hydrographs based on copula function: A case study in Pakistan. *Water Resources Management, 11*(8), 1531.

Saghafian, B., Ghermezcheshmeh, B., & Kheirkhah, M. M. (2010). Iso-flood severity mapping: A new tool for distributed flood source identification. *Natural Hazards, 55*(2), 557–570.

Shamir, U., & Salomons, E. (2008). Optimal real-time operation of urban water distribution systems using reduced models. *Journal of Water Resources Planning and Management, 134*(2), 181–185.

Solomatine, D. (2012). An approach to multi-objective robust optimization allowing for explicit analysis of robustness. https://www.un-ihe.org/sites/default/files/solomatine-ropar.pdf

Strupczewski, W. (1964). Equation of flood crest. *Wiadomosci Słuzby Hydrologicznej i Meteorologicznej, 2*(57), 35–58.

Swiler, L. P., & Wyss, G. D. (2004). *A user's guide to Sandia's latin hypercube sampling software: LHS UNIX library/standalone version*. Sandia National Laboratories.

Taguchi, G. (1986). *Introduction to quality engineering: Designing quality into products and processes*. Tokyo: Asian Productivity Organization.

Tan, Q. F., Wang, X., Liu, P., Lei, X.-H., Cai, S.-Y., Wang, H., et al. (2017). The dynamic control bound of flood limited water level considering capacity compensation regulation and flood spatial pattern uncertainty. *Water Resources Management, 31*(1), 143–158.

Wu, Y., Zhong, P.-A., Zhang, Y., Xu, B., Ma, B., & Yan, K. (2015). Integrated flood risk assessment and zonation method: A case study in Huaihe River basin, China. *Natural Hazards, 78*(1), 635–651.

Xiao, Y., Guo, S., Liu, P., Yan, B., & Chen, L. (2009). Design flood hydrograph based on multicharacteristic synthesis index method. *Journal of Hydrologic Engineering, 14*(12), 1359–1364.

Yin, J., Guo, S., Wu, X., Yang, G., Xiong, F., & Zhou, Y. (2018). A meta-heuristic approach for multivariate design flood quantile estimation incorporating historical information. *Hydrology Research, 50*(2), 526–544.

Yue, S., Ouarda, T. B., Bobée, B., Legendre, P., & Bruneau, P. (2002). Approach for describing statistical properties of flood hydrograph. *Journal of Hydrologic Engineering, 7*(2), 147–153.

Zhou, X. (2019). *Probabilistic analysis of optimal management of storage areas for flood risk reduction in Huai River, China*. Master thesis, IHE Delft.

15

MULTI-OBJECTIVE OPTIMIZATION OF RESERVOIR OPERATION POLICIES USING MACHINE LEARNING MODELS: A CASE STUDY OF THE HATILLO RESERVOIR IN THE DOMINICAN REPUBLIC

Carlos Tami[1], Gerald A. Corzo Perez[2], Fidel Perez[3], and Germán Santos[1]

[1] *Colombian School of Engineering Julio Garavito, Bogotá, Colombia*
[2] *IHE Delft Institute for Water Education, Delft, The Netherlands*
[3] *Mother and Teacher Pontifical Catholic University, Santo Domingo, Dominican Republic*

Finding a balance between conflicting interests in multi-purpose reservoirs is an important challenge for decision makers. This study assesses the use of different computational tools to obtain optimal reservoir operations at the Hatillo dam in the Dominican Republic. A multi-objective optimization approach is applied to models that simulate reservoir operations and three different machine learning (ML) models are employed to learn the real operation of the system. A general model is proposed to simulate daily reservoir operations (2009–2019), integrating water balances, physical constraints of the dam components, and the ML models, the latter defining daily controlled discharges. In the optimization process, the ML parameters are the decision variables, while the objectives evaluated

Advanced Hydroinformatics: Machine Learning and Optimization for Water Resources, Special Publications 78, First Edition. Edited by Gerald A. Corzo Perez and Dimitri P. Solomantine.

DOI: 10.1002/9781119639268.ch15

are irrigation, hydropower generation, and flood control. The results are compared with the actual operation of the reservoir. The flood control objective was found to have a wide room for improvement over the real operation of the reservoir, and several of the solutions were found to improve the real operation for the three proposed objectives. The multilayer perceptron models tended to generate the best results for this case study and the nondominated sorting generic algorithm (NSGA II) optimizer generated the best optimization results.

15.1. Introduction

Reservoir optimization is one of the most relevant study areas within the field of water resource systems (Momtahen & Dariane, 2007). With the effect of climate change, population growth, and basin dynamics, optimal operation of reservoirs has more relevance in water resources management.

Multipurpose reservoirs are often used to attend multiple demands for domestic, industrial, irrigation, environment, hydropower production and flood control, among others (Sun et al., 2018). Therefore, considering the different interests that may exist in decision making for the operation of reservoirs, as well as the mutual conflicts that these interests may present, finding operating policies to which all the involved parties will agree is not an easy task. For the flood control objective, for instance, it is better if there is less water in the reservoir, allowing to accumulate a large additional volume of water during the flood season (Zhou et al., 2018), while more water in the reservoir is better for irrigation purposes (Scola et al., 2014). Regarding hydropower production, the stored energy in the reservoir gradually increases with the rise in water level (Jansen et al., 2013), and outflows might support irrigation, however, there are also well-known cases where hydropower production reduces water availability for irrigation (Ruijie et al., 2017). Thus, finding optimal operating policies that use multi-objective approaches to refocus decision making toward a balance of interests is highly relevant to improve the integrated use of reservoirs. It is important to address trade-offs between multiple objectives to achieve the management goals (Myo Lin & Rutten, 2016).

The literature cites different approaches for tackling multi-objective optimization problems. Many of these are oriented toward the search for sets of nondominated solutions reflected in decision spaces or Pareto fronts. Traditional approaches aim to simplify multi-objective problems into several single-objective problems that can be addressed via standard optimization algorithms (Miettinen, 1999). For instance, the weights or weighting method converts objectives into scalar values by assigning subjective weights (Labadie, 2004), or the ε-constrained method (Haimes et al., 1971) transforms all objectives into constraints except for one, to which a standard optimization method is applied, thus simplifying the problem. Both methods are widely applied in reservoir optimization (see Heydari et al., 2016; Ko et al., 1992; Khattab & Al-Mohseen, 2020). Another method, the goal attainment (Gembicki & Haimes, 1975), consists of defining a set of specified targets associated with a set of objectives, which can be underachieved or overachieved and are controlled by a vector of weighting coefficients (Loucks & van Beek, 2017). In reservoir optimization, this method has been employed by Khattab and Al-Mohseen (2020), and Basu (2004). All these approaches, however, are limited when the complexity of the problem increases (e.g., greater number of objective functions, decision variables, or constraints).

Thanks to advancements in computational resources, better technologies, and more robust equipment, metaheuristic techniques have been gaining popularity in the field of water resources. Such is the case for evolutionary algorithms (EA) that use the mechanisms of Darwin's evolution of species (see Cámara, 2015; Loucks & van Beek, 2017). When applied to multi-objective optimization, EAs are efficient because of their ability to solve complex problems involving characteristics such as discontinuities, multimodality, disjoint feasible spaces, and noise when evaluating functions (Ma et al., 2015). These algorithms seek to generate a Pareto-approximate set of solutions to serve as a reference for decision making in multicriteria problems (see Kollat et al., 2011; Giuliani et al., 2014). Among these EAs, the Nondominated Sorting Genetic Algorithm II (NSGA II) (Deb et al., 2002) and the multi-objective evolutionary algorithm based on decomposition (MOEA / D) (Zhang & Li, 2007) stand out. Some authors have used these algorithms to optimize multipurpose reservoirs. Castelletti et al. (2012) used the NSGA II to optimize the Hoa Binh reservoir on the Da River in Vietnam for hydroelectric generation, water supply, and flood control. Myo Lin et al. (2020) also used the NSGA II within the scope of multi-objective model predictive control for real-time operation of a multireservoir system in the Sittaung River basin in Myanmar. Similarly, Ma et al. (2015) applied the MOEA / D to optimize the Ankang reservoir in China for flood control.

One approach that has gained popularity in recent years is *direct policy search* (DPS), a term employed by Rosenstein and Barto (2001). This concept is also known in water resource literature as parameterization-simulation-optimization, as referenced by Koutsoyiannis and Economou (2003). Per this method, optimizers are applied directly to approximation functions or models that simulate the reservoir's operation (Ritter, 2016; Ritter et al., 2020). With advances in artificial intelligence techniques, some recent studies have applied approximation functions or machine learning (ML) models that use artificial neural networks (ANNs) to represent reservoir operations (e.g., Ehsani et al., 2016; Zhang et al., 2018). Giuliani et al. (2014) used radial basis networks (RBNs) and a multi-objective evolutionary algorithm (MOEA) to obtain optimal operating policies in a multipurpose reservoir. The combination of DPS, approximation functions, and MOEA results in an approach named evolutionary multi-objective direct policy search (EMODPS), developed by Giuliani et al. (2016), who used it to improve operating policies of a multipurpose reservoir.

Based on the EMODPS approach, this study aims to develop a methodology to obtain optimal reservoir operation policies using ML models. That is, design ML models that can define the volume of water that must be released in a time interval, and which through multi-objective optimization can learn optimal operations (optimal policies) to satisfy certain objectives. For this purpose, three different machine learning (ML) models, namely, the multilayer perceptron (MLP), the radial basis network (RBN), and the linear function (LF) are compiled. As well as two EAs: NSGA II and MOEA / D. The Hatillo reservoir in the Dominican Republic is used as a case study that has three purposes in mutual conflict: hydroelectric generation, irrigation, and flood control. The main objective for this case study is to find optimal policies to improve the real operation of the reservoir (analyzed for the period 2009–2019) for the three proposed objectives. The following sections describe the case study and the proposed methodology, and later demonstrate the results and conclusions of the study.

15.2. Case Study

15.2.1. The Hatillo Reservoir

The Hatillo reservoir is in the Yuna River basin (Fig. 15.1), Dominican Republic, about 113 km northeast of Santo Domingo. This reservoir has been used for irrigation, hydropower generation, and mitigating downstream flooding. The Yuna River basin drains an area of 5,498 km^2, flowing downstream for approximately 210 km. It is the second most important river in the Dominican Republic; and with an average flow of 91 m^3/s at its mouth (35.4 m^3/s at the entrance of the reservoir), it is the river with the highest flow. Mean annual precipitation in the basin (in terms of spatial

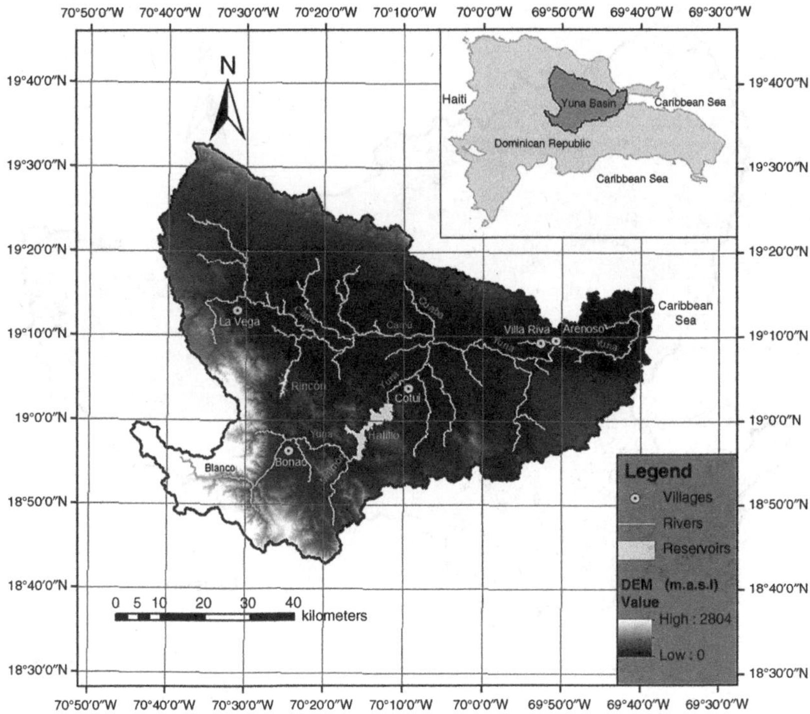

Figure 15.1 Map of the Yuna River basin.

variation) ranges between 1,080 and 3,120 mm, with the area of greatest precipitation in the mountain region upstream of the reservoir. Most of the annual rainfall in the Dominican Republic occurs between the end of April and the end of October, with a bimodal trend that has peaks in May and around the end of August to October (CEPAL, 2020; Jury et al., 2007). Due to its geographical location, the country is exposed to extreme weather events such as hurricanes, cyclones, and tropical storms. The hurricane season begins in June and ends in November, with the months of August to the end of October the periods of greatest occurrence (Jesús Encarnación, 2019).

During periods of prolonged rain, floods occur frequently in the lower Yuna basin. The rains generate an increase in river flow that consequently overflows and affects low-lying populations downstream of the reservoir. The lower Yuna plain is subject to flooding in general, characterized by a terrain of low slopes, extensive plains, and a meandering river course.

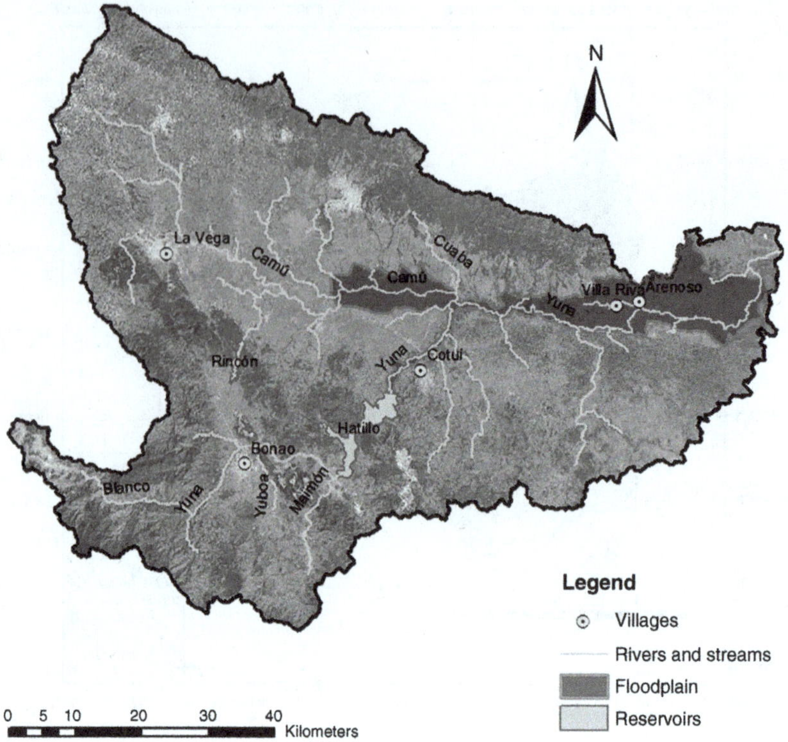

Figure 15.2 Floodplain in the lower Yuna basin.

Several flood events have been reported, such as flooding in October of 2007 that occurred because of Tropical Storm Noel. The event affected the entire Dominican territory and resulted in losses exceeding 439 million U.S. dollars (CEPAL, 2008). Another more recent event was the flood season between November 2016 and April 2017, which generated losses of 862.37 million US dollars, affecting sectors of public works, water and sewerage, agriculture, and housing among others (MEPyD, 2017). On the other extreme, periods of long droughts also affect the area, reducing hydroelectric generation and putting at risk the water meant to supply the irrigation districts in the lower Yuna.

At present, only 26% of the basin area is regulated by reservoirs. Hatillo is the most representative, regulating 23% of the total area of the basin (1,192 km^2). This reservoir is the largest body of surface water in the Dominican Republic, covering an average area of 27.4 km^2. The system has a dam built on a 50 m high embankment of earth and rock.

The reservoir has a total storage capacity of 441 million m^3 with a maximum length of 15 km. The minimum operating level is 70 m above sea mean level (MAMSL). The elevation of the crest of the spillway is 86.5 MAMSL, which corresponds to the maximum operating level, while the extraordinary maximum level is 100.75 MAMSL. The main components of the dam are presented in Figure 15.3. The reservoir has an uncontrolled spillway (i.e., no gates, only free overflow) with a maximum capacity of 650 m^3/s; there are two hollow jet valves for the bottom outlet, with a maximum capacity of 82 m^3/s each. A powerhouse with a Francis turbine has an installed capacity of 8 MW and an annual energy generation of 40 GWh; the maximum flow capacity of the turbine is 30 m^3/s. Hatillo's installed capacity is low compared with other hydroelectric plants in the Dominican Republic. Examples of other plants include Tavera (96 MW), Valdesía (54 MW), and Aguacate (52 MW). Downstream of Hatillo, large areas of land are used for agriculture in the lower Yuna River basin, predominantly for rice cultivation, which is the main food crop in the Dominican Republic (Moquete, 2004). The reservoir supplies water for the irrigation of approximately 27,000 ha of crops.

The dam is operated by Empresa de Generación Hidroeléctrica Dominicana (EGEHID) and the Dominican Institute of Water Resources (INDRHI). The reservoir operation under normal conditions consists of discharging 26 m^3/s through the turbine throughout the day; this water returns to the Yuna River and it is used for irrigation. However, in emergency scenarios, such as in heavy rainy seasons or hurricane seasons, the National Dam Committee (COPRE) defines the actions for the operation of the reservoir. Releases through hollow jet valves occur when more water needs to be released due to flood control strategies or irrigation demands. Unfortunately, the information about operating rules of the reservoir were not available, as well as target levels, power production targets, and irrigation discharges. Nevertheless, daily data of inflows, spills, turbine discharges, valve outlets, and total outflows for the period 2009–2019 were provided by EGEHID. From these data, it was observed that 54% of the reservoir outlets are through spill, 45% are through the turbine, and 1% are through the bottom valves. The fact that most of the reservoir releases have not been controlled (i.e., through spillway) allows us to hypothesize that the flood control objective can be improved.

It is important to mention that given the lack of reservoir operation rules, this study sought to emulate the real operation (2009–2019) using machine learning (ML) models, which is described in section 15.3. Additionally, due to the lack of target levels, power production targets, irrigation demands, and specific objectives of the reservoir, an operating model has

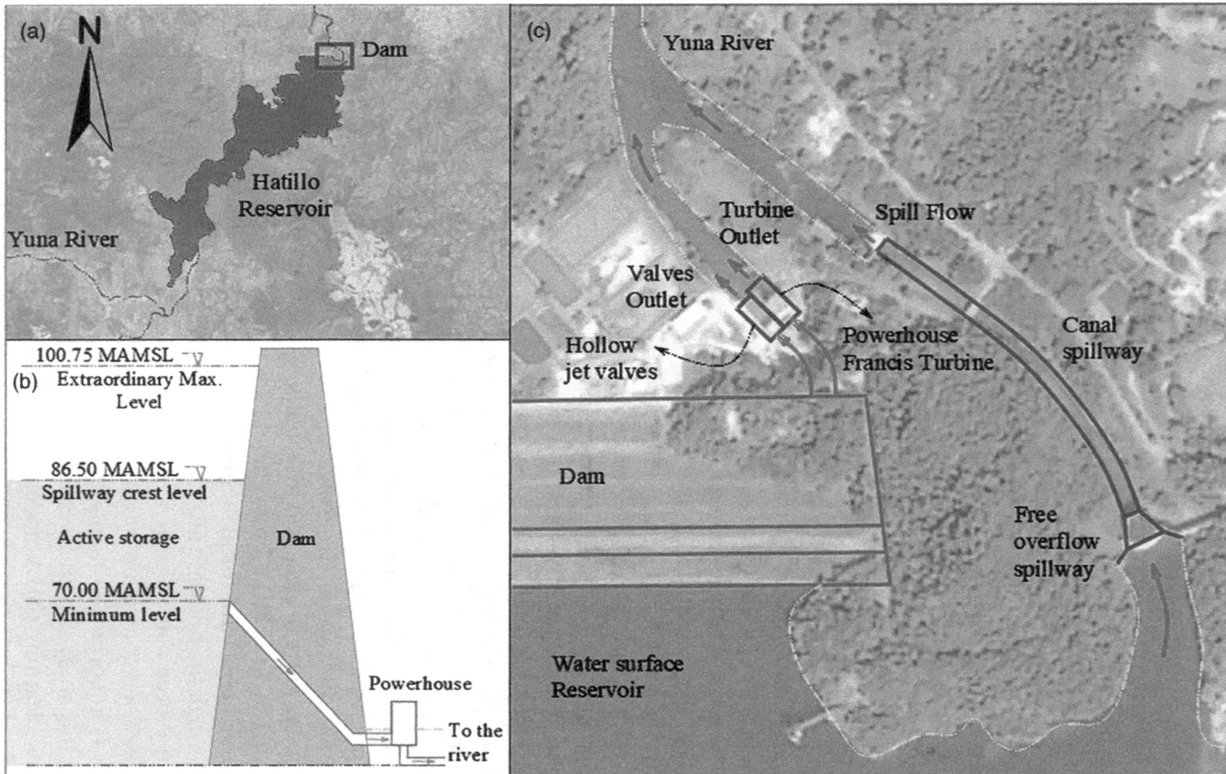

Figure 15.3 Dam components: (a) location of the dam, (b) operation levels, (c) schematic representation of the components of the dam.

been proposed (see section 15.2.3) to be optimized by applying three objectives with which it seeks to emulate the interests in the operation of the reservoir.

15.2.2. Data

The period of analysis for the case study was defined from the information available on the reservoir's operation; it corresponds to the period of 2009 to 2019, with daily time steps. These data consist of series of inflows, levels, and outflows of the reservoir (Fig. 15.4), in addition to storage elevation curve, area elevation curve, and evaporation losses. In machine learning models (which are developed in section 15.3), the normalization or standardization of data is often advisable. In this study, the data are standardized using the standard scoring method: $X_n = (X - \mu)/\sigma$, with X being the value of the variable without standardization, σ the standard deviation, and μ the average of the time series.

15.2.3. Reservoir Operation Model

A reservoir operation model was built from the data from the period of analysis. This model integrades water balances, physical characteristics of the dam components and a water release decision component that is integrated by machine learning models (which will be further discussed). The water balances determine the storage volume of the reservoir at each time step t, the level and the surface area are calculated by applying the storage-elevation curve and the area-elevation curve, respectively. The water balance equation is as follows:

$$S_{t+1} = S_t + IN_t - Q_{spill,t} - Q_{c,t} - EVP_t \tag{15.1}$$

where S_{t+1} (subject to $S_{t+1} \geq 0$) and S_t correspond, respectively, to the storage volume of the reservoir for the next time step $(t+1)$ and the current time step t; IN_t corresponds to the inflows of the reservoir for the current time step t, taking the inflow time series; EVP_t corresponds to the evaporation losses for the current time step t, which are calculated based on the surface area of the reservoir and the month of the year; $Q_{spill,t}$ corresponds to the spilled volume of the reservoir for the current time step t, which is calculated as a function of the height above the crest of the spillway; and $Q_{c,t}$ corresponds to the controlled outflows of the reservoir for the current time step t. At every time step (i.e., day), the model computes the water balance.

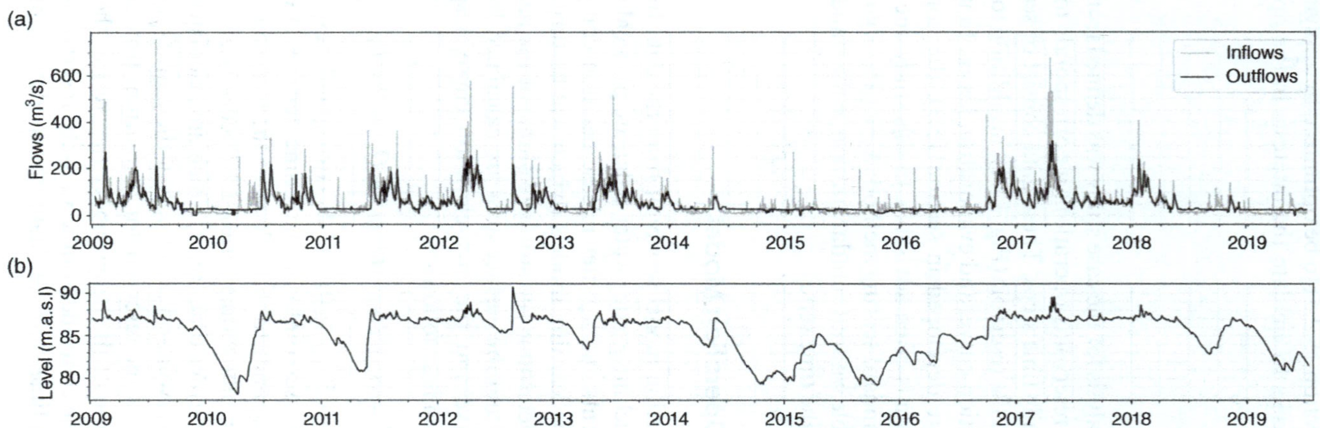

Figure 15.4 Real Data of the analysis period: (a) Inflows and outflows of the reservoir, (b) levels of the reservoir.

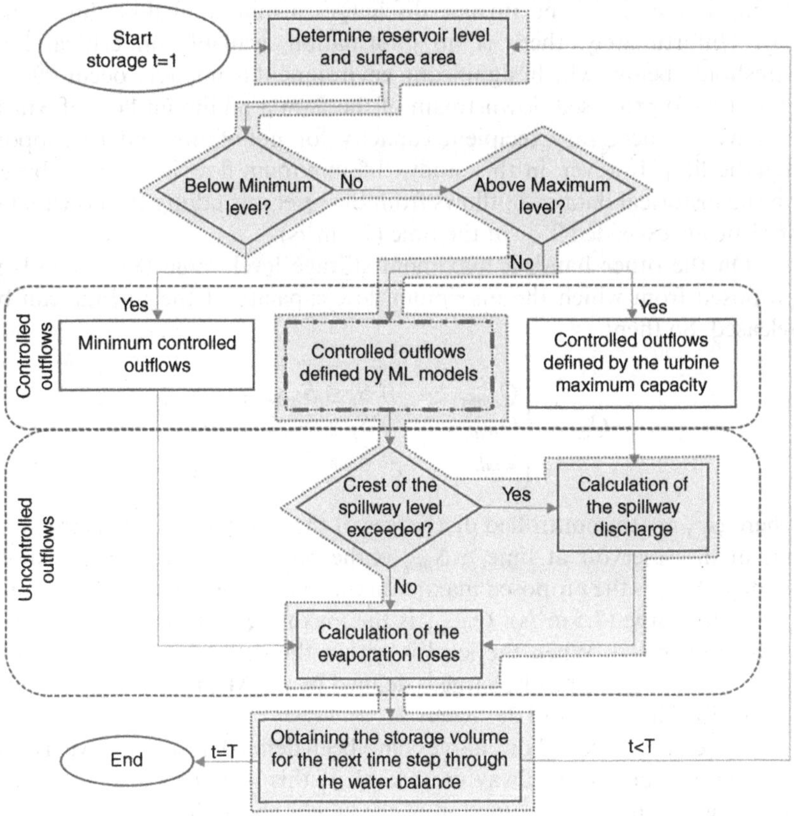

Figure 15.5 Flow diagram of the reservoir operation model built for the Hatillo reservoir.

Figure 15.5 summarizes the step by step of the reservoir operation model. Starting with the storage volume of the first day of analysis, reservoir level and surface areas are determined by applying the storage-elevation curve and the area-elevation curve.

The outflows of the reservoir can be controlled through the Francis turbine and the two hollow jet valves. The controlled discharges of the reservoir $Q_{c,t}$ are defined taking into account the respective capacity constraints (equation (15.2)). Therefore, conditional rules are used to limit minimum and maximum controlled outflows.

Limits of the reservoir levels are proposed. A minimum storage level (75 m.a.s.l) is considered to guarantee that when there are low levels during dry periods, the turbine discharges a minimum flow during the day. Unfortunately, there is no information available on critical flow thresholds below which significant environmental impacts occur. Berge et al. (2003) proposed downstream of the dam a minimum flow of $3\,m^3/s$ to have the necessary recipient capacity for pollutants and to support aquatic life. However, in this study, the minimum flow is assumed based on the historical data of outflows from the reservoir taking the flow that is statistically exceeded 95% of the time ($7.5\,m^3/s$).

On the other hand, a maximum storage level value (87.5 m.a.s.l) is proposed from which the maximum flow capacity of the turbine can be released. So then:

$$Q_{c,t} = \begin{cases} Q_{cmin} & if\ S_t \leq S_{min} \\ Q_{tbmax} & if\ S_t \geq S_{max} \\ Q_{ml,t} & if\ S_{min} < S_t < S_{max} \end{cases}, \tag{15.2}$$

where $Q_{c,t}$ are the controlled discharges of the reservoir at t; S_t is the storage in the reservoir at time t; S_{min} is the proposed minimum threshold storage; S_{max} is the proposed maximum threshold storage; Q_{cmin} is the minimum discharge ($7.5\,m^3/s$); Q_{tbmax} is the maximum discharge through the turbine ($30\,m^3/s$). When the level is within the defined storage threshold ($S_{min} < S_t < S_{max}$), the discharge is defined by the ML models $Q_{ml,t}$, which is described in section 15.3.

Once the controlled discharges have been defined, reservoir level is verified with respect to spillway crest level. If this is exceeded, the spillway release is calculated as a function of the height above the crest of the spillway. The specific rating curve of the spillway for Hatillo reservoir is available at Viparelli (1983). After all outflows (i.e., controlled and uncontrolled) are defined (including the evaporation losses EVP_t), reservoir storage volume for the next time step S_{t+1} is obtained by means of water balance (equation (15.1)), starting the process again and so the cycle continues until the analysis period is completed. The route shaded in the flow diagram (Fig. 15.5) corresponds to the main operating route in this proposed model (i.e., when the storage levels are within the defined threshold), where the segmented rectangle highlights the component of ML models, which will be described in section 15.3.

15.3. Methodology

Initially, section 15.3.1 provides a brief overview of the ML models used. The proposed methodology is divided into two general parts. In

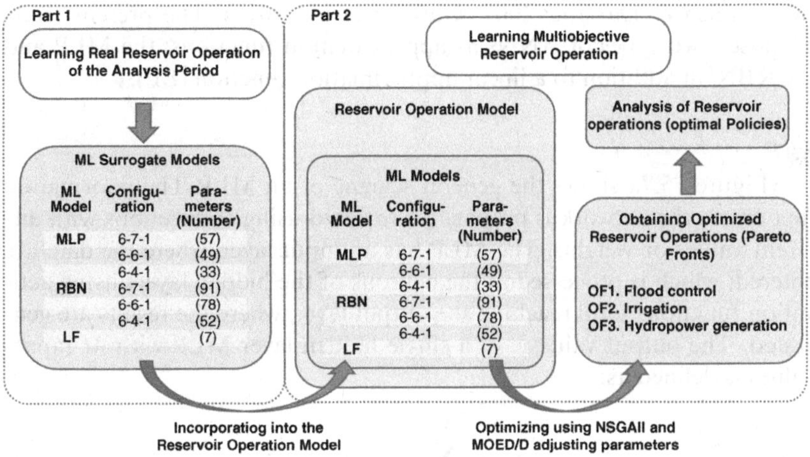

Figure 15.6 Scheme of the proposed methodology.

part 1 (section 15.3.2), ML models are built to simulate the operation of the reservoir during the analysis period (2009–2019), that is, to learn the real operation. In part 2 (sections 15.3.3–15.3.5), the ML models are incorporated into the reservoir operation model (which was described in section 15.2.3) and which, through multi-objective optimization, will learn optimal operations for the three objectives defined in section 15.3.4. Finally, section 15.3.5 describes the process of optimization and some computational details. Figure 15.6 summarizes the methodology proposed in this study, which will be developed in following sections.

15.3.1. Machine Learning (ML) Models

The reservoir operations in terms of daily release can be parameterized using mathematical functions adjusted to the patterns or shapes that define them. To this end, some artificial intelligence techniques or ML models, most notably ANNs, have proven to be efficient; some have even been classified as universal approximation functions (Tikk et al., 2003). Several authors have shown the benefits of these functions in the analysis of reservoir operation. Not only can they learn and replicate real operating behaviors (Ehsani et al., 2016; Zhang et al. 2018), they also adjust to artificial behaviors induced by purpose fulfillment in optimization problems

(Momtahen & Dariane, 2007; Giuliani et al., 2016). The present study proposes two types of ANNs as approximation functions: the MLP and the RBN, in addition to a linear approximation function (LF).

MLP

Figure 15.7a shows the general scheme of an MLP. The information that enters the network is propagated forward using connections with different values or weights. The MLP has an input layer, where the data are entered, which is processed in the neurons of the hidden layers using activation functions until reaching the output layer, where the results are generated. The output value y_t of a single hidden layer MLP with M inputs values is defined as:

$$y_t = c + \sum_{i=1}^{N} v_i f_i(I_t \cdot w_i + b_i), \tag{15.3}$$

where N is the number of neurons with an activation function $f_i(\bullet)$ (e.g., sigmoidal function); I_t is the M-dimensional inputs vector of the network; w_i is the M-dimensional vector that represents the connections between the input and the hidden layer; v_i is the N-dimensional vector for the connections between the hidden layer and the output layer; b_i is the N-dimensional vector of bias for the neurons of the hidden layer; and c is the bias value of the output layer. The number of parameters in this network is $p = N(M + 2) + 1$.

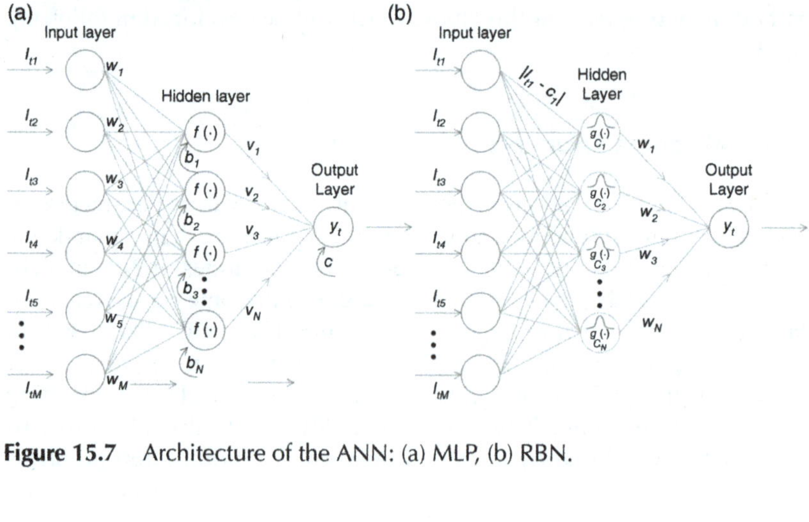

Figure 15.7 Architecture of the ANN: (a) MLP, (b) RBN.

A commonly used activation function in MLP is the sigmoid function, which is presented as

$$f_i = \frac{1}{[1 + exp[-(I_t \cdot w_i + b_i)]]}. \tag{15.4}$$

For this MLP function, the output layer does not have an activation function because the data are standardized using the standard scoring method (see section 15.2.2).

RBN

These networks use Gaussian functions for data processing, incorporating the center and radius parameters of processing units known as bases (instead of neurons, as in MLPs). Figure 15.7b shows the general schematic of an RBN. The function that the RBN represents is as follows:

$$y_t = \sum_{i=1}^{N} w_i g_i(I_t), \tag{15.5}$$

$$g_i(I_t) = exp\left[-\sum_{j=1}^{M} \frac{[(I_t)_j - c_{j,i}]^2}{r_{j,i}^2}\right], \tag{15.6}$$

where y_t is the output value; N is the number of bases with Gaussian functions g_i; M is the number of input variables; w_i is the N-dimensional weight vector between the hidden layer and the output layer; I_t is the M-dimensional inputs vector of the network; and $c_{j,i}$ and $r_{j,i}$ are the center and radial M-dimensional vectors of the ith base. The number of parameters for an RBN is $p = N(2M + 1)$.

LF

These are the mathematical functions used to find relationships between variables with linear trends. Unlike the previous functions, they do not have the ability to accurately replicate complex patterns. If y_t is a dependent variable and I_t is the independent variables (input values), the function can be expressed as follows:

$$y_t = b + \beta_1 I_{t,1} + \beta_2 I_{t,2} + \cdots + \beta_M I_{t,M}, \tag{15.7}$$

where y_t is the output value; $I_{t,1}, I_{t,2}, \ldots I_{t,M}$ are the input values; M is the number of input values; $\beta_1, \beta_2, \ldots \beta_M$ are the weight coefficients of the Input values; and b is the bias of the function. The number of parameters is $p = M + 1$.

15.3.2. Building ML Surrogate Models of the Real Reservoir Operation

As mentioned in section 15.2.1, the information about operating rules of the reservoir were not available for the case study. Therefore, ML surrogate models were used to simulate the operation of the reservoir. That is, to represent the amount of water that must be released at each time step through models that learn from the real data of the reservoir. This method in which the model learns from known output data or examples is known as *supervised learning* (Cunningham, 2008).

Initially, the parameterization of the reservoir's actual discharge Q_t was sought for the analysis period, in which three ML models were tested: MLP, RBN, and LF. The outputs Q_t correspond to total discharges (i.e., turbine discharge, valves discharge, and spills).

Considering that the correlation coefficient is commonly used to determine mathematical linear relations between two samples of random variables, or time series, correlation analysis was applied to select the inputs of the ML models, similar to Corzo (2009). Therefore, these correlations were performed between discharges at each time step t and the other time series (i.e., inflows, reservoir storages levels, and previous discharges). Correlation analysis revealed that for the inflows IN, the storages levels

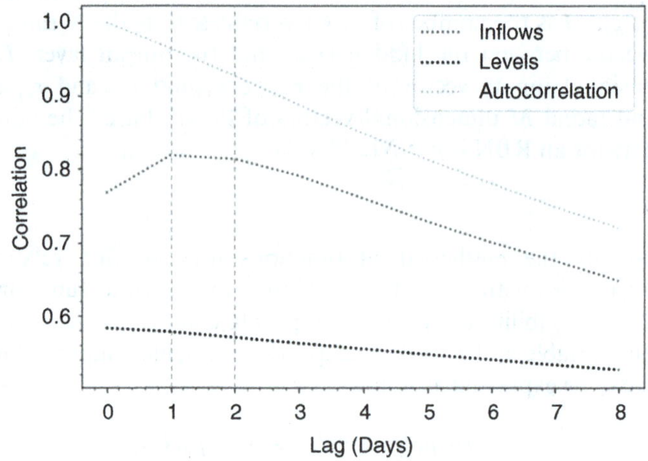

Figure 15.8 Correlation analyses for the selection of ML inputs: between inflows and discharges; levels and discharges; and autocorrelation of discharges.

S and the previous discharges Q, each using a lag time of 1 and 2 days (IN_{t-1}, IN_{t-2}; S_{t-1}, S_{t-2}; Q_{t-1}, Q_{t-2}), have the highest correlation values with respect to discharges Q_t (Fig. 15.8).

Using these six inputs for all ML models, different configurations for MLP and RBN were tested varying the number of neurons (MLP) or basis (RBN). The Scikit-learn Python library (Pedregosa et al., 2011) was used for MLP and LF models, while for RBN, Keras was used (Chollet, 2015). In all ML models, 80% of the data was used for training and 20% for the test. The final configurations of the ML models were determined by trial and error. In the case of MLP, configurations with several layers and different numbers of neurons were tested; however, in this case, by increasing the complexity of the network with more hidden layers and neurons, the results were not improved but the number of parameters and decision variables (weights and biases) was increased. Therefore, simple configurations of a single hidden layer and a maximum of 7 neurons in the hidden layer were chosen. Likewise, for the RBN, configurations of a maximum of 7 bases were used. The final configurations of the ML models are as follows:

1. Three MLP configurations: 6-7-1, 6-6-1, and 6-4-1 (i.e., 6 inputs - N neurons in a hidden layer - 1 output layer) using the parameters of equation (15.3) with the application of a sigmoid activation function (equation (15.4)).
2. Three RBN configurations: 6-7-1, 6-6-1 and 6-4-1. (i.e., 6 inputs - N Basis - 1 output layer) using the parameters of equations (15.5) and (15.6).
3. Finally, a LF is proposed using the parameters of equation (15.7).

The test runs for the 6-7-1 MLP configuration generated the best performance metrics (RMSE 5.17 and MAE 2.97).

Table 15.1 shows performance metrics for test runs for each of the ML models, which are compared against a simple forecast model. Note that the best performance for test runs were obtained for the MLP and RBN models in all the performance metrics used, mainly for the RMSE, which is an indicator that penalizes large errors. However, for the other measures, there are no very large differences with respect to the LF and naive models. This could be due to the inertia of the system, that is, for the actual operating condition of the reservoir, the daily discharges do not vary considerably from one day to the next. For example, when there are no spillway discharges, the daily outflows under normal conditions are 26 m^3/s as mentioned in section 15.2 and as shown in Figure 15.9 between June 2018 and

Table 15.1 Performance metrics for test runs of the ML models.

	MLP			RBN				
Measures	6-7-1	6-6-1	6-4-1	6-7-1	6-6-1	6-4-1	LF	Naive
RMSE	5.17	5.33	5.25	5.19	5.47	5.48	6.22	6.89
MAE	2.97	3.27	3.14	2.92	3.32	3.20	3.60	3.38
Correlation	0.988	0.987	0.988	0.988	0.987	0.987	0.982	0.978
Nash-Sutcliffe	0.976	0.974	0.975	0.976	0.973	0.973	0.965	0.957

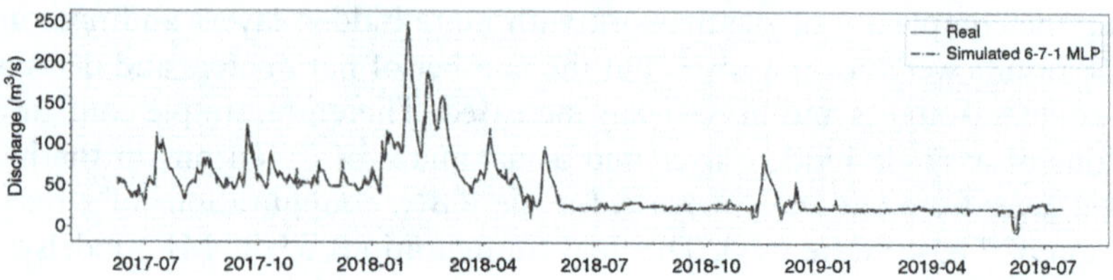

Figure 15.9 Test run for 6-7-1 MLP configuration for simulation of real reservoir operation.

July 2019. This means that the actual operation of the reservoir for that period can be very predictive for any model.

15.3.3. Incorporating ML Models Into the Reservoir Operation Model

Each of the ML models built in the previous section were coupled to the reservoir operation model described in section 15.2.3 to define the controlled discharges $Q_{c,t}$ at each time step t. The ML models were integrated into the water release decision component (see dash line segmented rectangle in Fig. 15.5). Therefore, inside the reservoir operation model, the outputs of the ML models Q_t are redefined as $Q_{ml,t}$, as indicated in equation (15.2). Table 15.2 and Figure 15.10 summarize the main components of the ML models.

The reservoir operation model described in section 15.2.3 is completed by coupling the ML models. In summary, seven different ML model configurations were built, each one elaborated in Python. The parameters of the ML models are the decision variables of the optimization process, described in section 15.3.5.

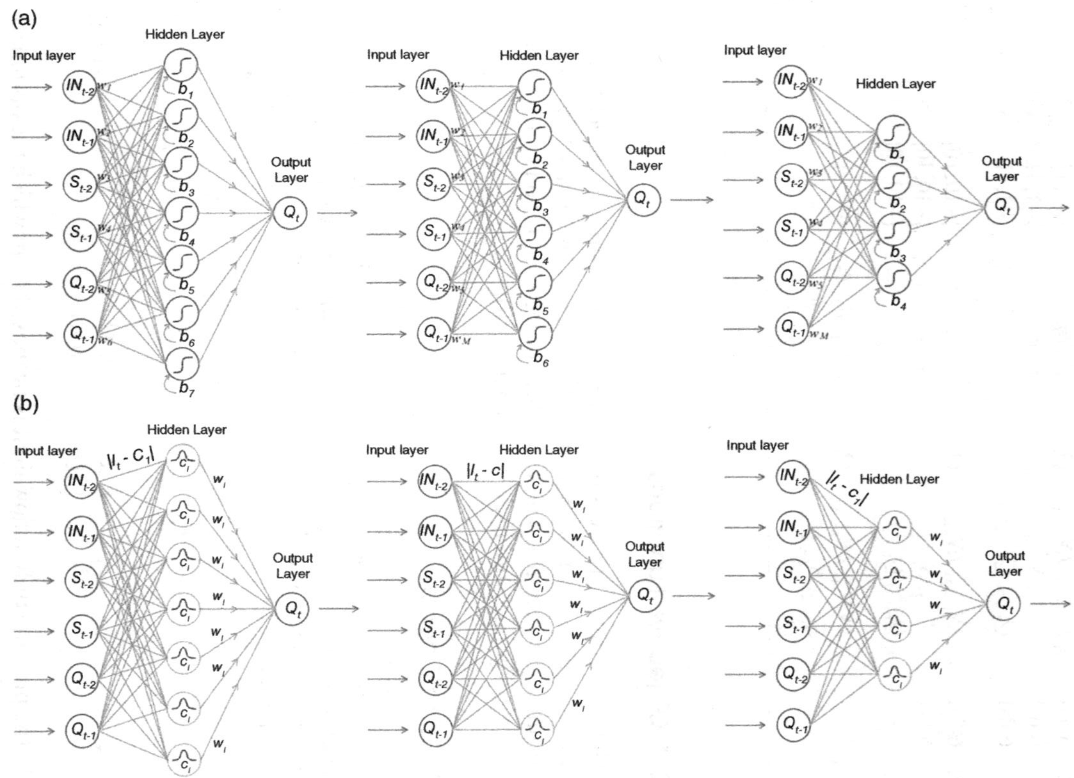

Figure 15.10 Configuration of the (a) MLP and (b) RBN proposed for the operation models.

Table 15.2 Summary of the parameters and configuration of the ML models.

	Configu-ration	Inputs	Output	Parameters	Range	Number of parameters
MLP	6-7-1	$IN_{t-2}, IN_{t-1},$	Q_t	$w_i, b_i,$	[−1, 1]	57
	6-6-1	$S_{t-2}, S_{t-1},$		v_i, c	[−1, 1]	49
	6-4-1	Q_{t-2}, Q_{t-1}			[−1, 1]	33
RBN	6-7-1	$IN_{t-2}, IN_{t-1},$	Q_t	$c_i, r_i,$	[−10, 10]	91
	6-6-1	$S_{t-2}, S_{t-1},$		w_i	[−10, 10]	78
	6-4-1	Q_{t-2}, Q_{t-1}			[−10, 10]	52
LF	–	$IN_{t-2}, IN_{t-1},$ $S_{t-2}, S_{t-1},$ Q_{t-2}, Q_{t-1}	Q_t	$b, \beta_1, \beta_2, \beta_3,$ $\beta_4, \beta_5, \beta_6$	[−5, 5]	7

15.3.4. Objective Functions

The optimization process uses the reservoir operation model to identify what parameters of the ML models represent the best solutions (i.e., nondominated solutions). There are three objectives: Flood control, water supply for irrigation, and hydropower generation.

OF1. Flood Control

The goal is to minimize $Q_{f,t}$, which is obtained in each time step t from the conditional equation (15.9), where $(Q_t - \bar{q})^2$ represents the square of the excess flows, being Q_t (m^3/s) the reservoir's daily discharge and $\bar{q}$ a proposed limit flood flow. The following function defines OF1 for the entire analysis period T:

$$OF1 = \sum_{t=1}^{T}(Q_{f,t}), \tag{15.8}$$

$$Q_{f,t} = \begin{cases} 0 & if\ Q_t \leq \bar{q} \\ (Q_t - \bar{q})^2 & otherwise \end{cases}. \tag{15.9}$$

In the present study, a flow limit is proposed from which the optimizer seeks to punish the discharges. Similar to Saavedra et al. (2010), this limit flow is set as the average of all the daily discharge values (Q_i) of the reservoir that exceeded the average discharge flow of the series (Q_{mean}). Therefore,

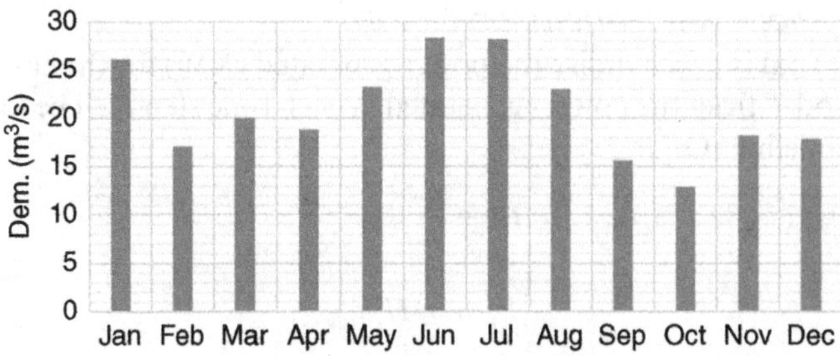

Figure 15.11 Demand for irrigation downstream of the reservoir.

using equation (15.10), the flow limit $\bar{q}$ is obtained, where N = number of records that exceeded the average discharge flow of the series.

$$\bar{q} = \sum_{i=1}^{N} \frac{Q_i}{N}; \quad Q_i > Q_{mean}. \tag{15.10}$$

OF2. Water Supply for Irrigation

The goal is to minimize deficits in the irrigation water supply $Q_{ir,\,t}$ in each time step t from the reservoir's daily discharge Q_t (m^3/s) with respect to the water demand for irrigation d_t. The following function defines OF2 for the entire period of analysis T:

$$OF2 = \sum_{t=1}^{T} (Q_{ir,t}) \tag{15.11}$$

$$Q_{ir,t} = \begin{cases} 0 & if\, Q_t > d_t \\ (d_t - Q_t)^2 & otherwise \end{cases} \tag{15.12}$$

Irrigation demands were obtained using the CROPWAT program (Smith, 1992). Taking into account that the cultivation of rice demands the greatest amount of water and covers the largest area of crops in the lower Yuna, irrigation demand calculations were made from the requirements and the cycle of the rice cultivation. Rice in the Dominican Republic has two sowing periods in the year: (1) December to April, and (2) the months of June, July, and August (Moquete, 2004). Climatic data (temperature, humidity, wind speed, hours of sunshine) and rainfall data were provided by INDRHI (taking monthly mean data for the analysis period). From the cultivation areas, the monthly average demands for the analysis period (2009–2019) were calculated, which are shown in Figure 15.11.

OF3. Hydropower Generation (Total)

The goal is to maximize the power generated (KW) for the entire analysis period T from the powers generated in each time step P_t. The following function defines OF3:

$$OF3 = \sum_{t=1}^{T} P_t, \tag{15.13}$$

$$P_t = \eta \gamma H_t q_{turb,t}, \tag{15.14}$$

$$\eta = f(q_{turb,t}), \tag{15.15}$$

where γ is the specific weight of water; H_t is the hydraulic head, which depends on the reservoir's water level and the elevation of the turbine axis; η is the efficiency of the turbine as a function of the discharge; and $q_{turb,\,t}$ is the discharge through turbine (m^3/s). It is important to mention that a conversion factor of -1 is applied to OF3 (equation 15.13) within the optimizer in this study, since the modules used by the program (JMETALPY) work only for minimization objectives (Benítez-Hidalgo et al., 2019).

In hydroelectric plants, there are other important indicators, such as firm energy, with which the energy security of a system can be evaluated. Due to the fact that Hatillo hydroelectric plant has a low installed capacity compared with the large hydroelectric plants in the Dominican Republic, total energy is the only objective considered in this study. This simplifies the optimization process; however, a firm energy indicator is evaluated in the results obtained (section 15.4).

15.3.5. Optimization and Computational Details

The optimization was performed on the reservoir operation model for the analysis period (2009–2019), using the parameters of the ML models as decision variables. This method belongs to the direct search policy (DPS) approach where the optimizer acts on the parameters of the functions that represent an operation policy, this being the ML models. Table 15.3 summarizes the number of optimized parameters for each model. The objective functions were defined in section 15.3.4.

Two EAs were used to optimize the operation model: the NSGA II and the MOEA / D. JMETALPY (Benítez-Hidalgo et al., 2019), a multi-objective optimization environment (developed in Python) that contains a set of modules with different multi-objective and single-objective optimization algorithms, was also used. With JMETALPY, a set of problems was used to evaluate the optimization algorithms. The operation model was inserted into JMETALPY as a subclass of floating problems.

Table 15.3 Parameters of the optimization process.

	MLP			RBN			FL
Configuration	6-7-1	6-6-1	6-4-1	6-7-1	6-6-1	6-4-1	
Weights (w_i, v_i)	42+7	36+6	24+4	7	6	4	–
Bias (b_i)	7+1	6+1	4+1	–	–	–	1
Radius ($r_{j,i}$)	–	–	–	42	36	24	–
Centers ($c_{j,i}$)	–	–	–	42	36	24	–
LF coefficients	–	–	–	–	–	–	6
Total	57	49	33	91	78	52	7

The values for population size and individual parameters in the algorithms largely depended on the optimization problem (Ritter, 2016). Trial and error tests were carried out using various population ranges to select the population size. Populations greater than 100 individuals did not improve the results and increased the computational time needed for the NSGA II. For the MOEA / D, the population was defined as 300 individuals. For the other parameters, trial and error tests were performed using small populations (up to 50 individuals) so that the program could converge rapidly. Default values were taken as a basis for the algorithms.

The optimization process was developed by a computer with 16 GB of RAM and eight 2.6 GHz cores. The codes were executed by implementing the Multiprocess Evaluator module, which allowed faster calculations and for different individuals within the same generation to be calculated in parallel. On average, the program reported convergence for each operating model after 8 hr in 10,000 evaluation processes.

Using the JMETALPY display module, the Pareto fronts were obtained for each of the program executions. For each model, a 3-D Pareto front was generated, having been optimized by both the NSGA II algorithm and the MOEA / D. A total of 14 models was simulated. As mentioned above, the decision variables were the parameters of the ML models. Therefore, the optimized parameters for MLP were weights (w_i, v_i) and biases (b_i). In RBN models, they were centers ($c_{j,i}$), radii ($r_{j,i}$), and weights w_i. For LF, the coefficients are $\beta_1, \beta_2, \dots \beta_M$ and the bias is (b_i). The parameters are summarized in Table 15.3.

15.4. Results

After compiling the Pareto fronts generated by the optimizers, non-dominated solutions were obtained. These represent different reservoir

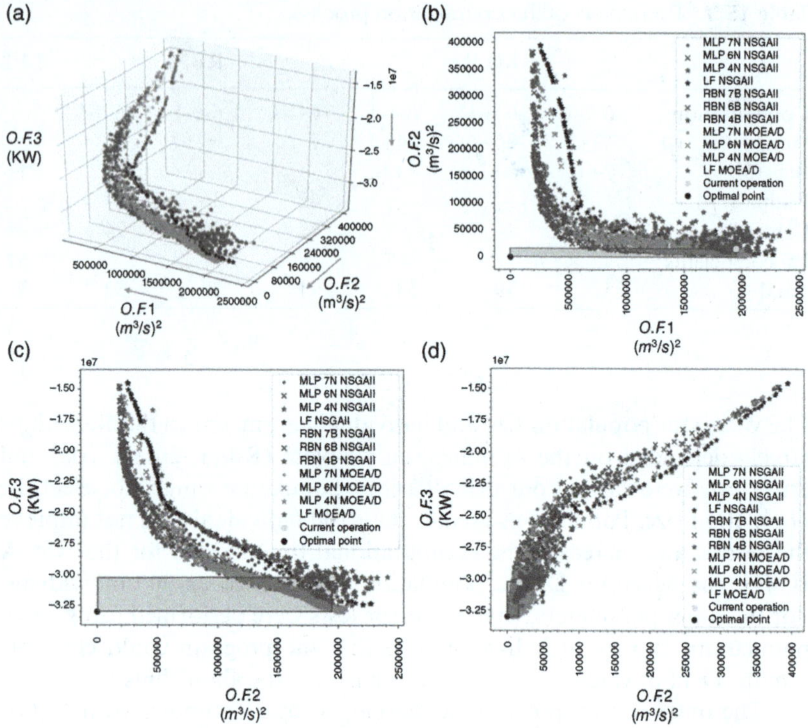

Figure 15.12 Unified Pareto fronts: (a) All the operation models; (b) operation models OF1-OF2; (c) operation models OF1-OF3; (d) operation models OF2-OF3.

operations in terms of daily releases. Figure 15.12 shows the 3-D Pareto front that compiles all the optimal solutions from the reservoir operation model. Each graph point corresponds to a different reservoir operation (operating policy) obtained by the optimizers. The solutions constructed by the MLP are the closest to the ideal optimum solution (black point). The real reservoir operation from the analysis period is represented by the light grey point; the shaded rectangles represent decisions space regarding reservoir operations that equaled or improved the real operation per the three objectives. For the objective function OF3 (a maximization function), a minimization function with a conversion factor of -1 was used; since the modules used by JMETALPY in the present study work only for minimization objectives.

Figure 15.13 shows the envelope of hydrographs generated by simulating all the obtained reservoir operations reflected in Figure 15.12. The

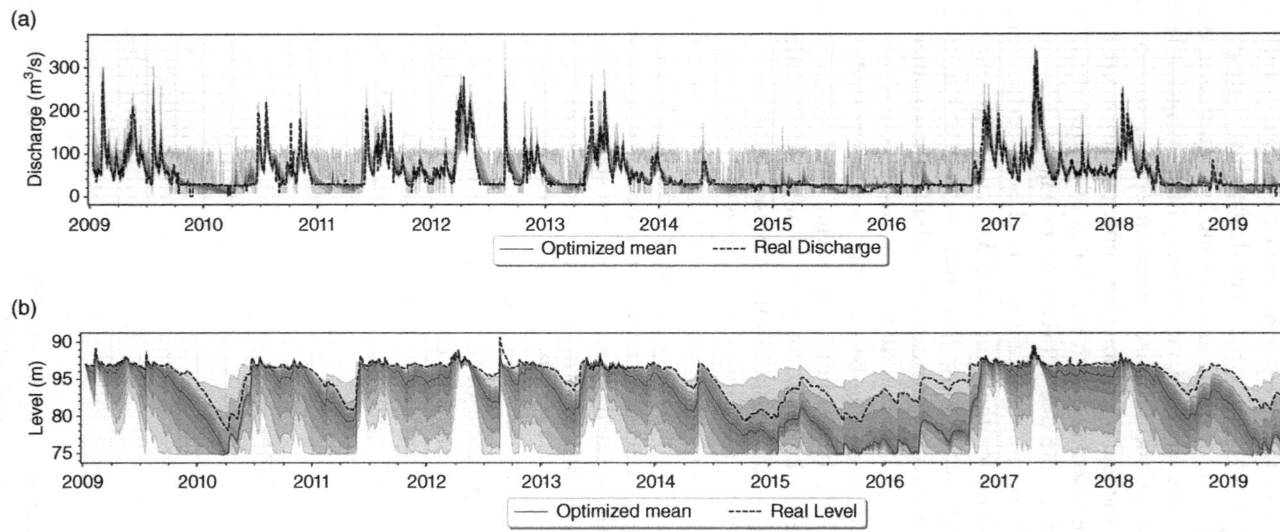

Figure 15.13 Envelope for hydrographs generated by all reservoir operations obtained from optimization: (a) Flow rates and (b) levels. The mean values of the reservoir operations for the entire population are shown in dark gray, and the gradient color shows the quintiles.

envelope of levels is presented in the same way. The mean values of the reservoir operations for the entire population are shown in dark gray, and the gradient color shows the quintiles. The dashed line represents the real operation of the system; here, the peaks of the hydrographs approach the maximum values of the envelope, maintaining high operation levels compared with the mean of the optimized reservoir operations. This could be due to the fact that the real operation gives more priority to OF2 and OF3, which may be more associated with high levels of the reservoir. Note that in dry periods such as the one between the middle of 2014 and the end of 2016, most of the solutions obtained were close at certain times to the minimum level defined for the reservoir (75 m.a.s.l). This can be understood as the consequence of water discharges to meet irrigation demands (OF2), in addition to having more capacity to buffer the volumes of the following wet season (OF1). In this study, the analysis period was evaluated globally, without distinguishing between dry and wet periods. Therefore, optimizers look for the best solutions by taking stock of the entire analysis period. Future works could incorporate different operations for the reservoir depending on the seasons of the analysis period.

Figure 15.14 shows the comparative values (Parallel graphic) of the different solutions for each of the three objectives. Additionally, the maximum peak flow and a firm energy indicator generated by each reservoir operation for the entire time series have been added. The firm energy ($F.E$) is evaluated considering the daily power that can statistically be exceeded 95% of the time, or in other words, the fifth percentile of all daily power generation: $F.E = PCTL_5(P_{t,1}, P_{t,2}, \dots, P_{t,T})$. The segmented black line reflects the evaluation of the three objectives plus firm energy and peak flow for the actual operation of the reservoir in the analysis period. As shown in the hydrograph and level envelope (Fig. 15.13), and also reflected in Figure 15.14, the real reservoir operation aims to meet the objectives OF2 and OF3 (i.e., water for irrigation and hydroelectric generation), but leaves

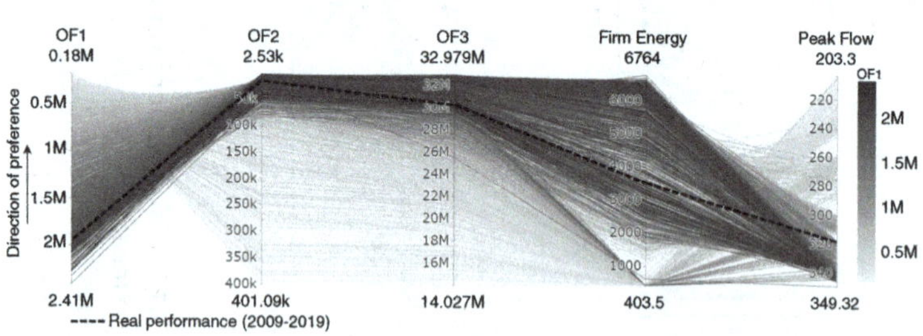

Figure 15.14 Comparison of results of all the solutions to reservoir operations for the three objectives, firm energy, and peak flow.

considerable room for improvement in flood control OF1. Note that the firm energy is greater as the generated power increases (OF3), presenting correlation; however, certain reservoir operations are observed that considerably reduce the firm energy. This can be associated with the fact that these solutions generate minimal discharges when there are low levels in the reservoir at certain times (dry periods).

Figure 15.15 shows different groups of solutions: Figure 15.15a represents all the reservoir operations that equal or improve each of the three proposed objectives with respect to the real reservoir operation. As can be seen, these solutions also improve firm energy; however, some present peak flows above the peak of the real operation. Therefore, Figure 15.15b illustrates the reservoir operations that improve the three objectives, firm energy, and peak flows.

Given that for the real reservoir operation, flood control is the objective that has the greatest margin for improvement, as can be seen in Figure 15.14, in another group of solutions, it is observed that by reducing hydropower production, greater benefits are obtained for the OF1. Figure 15.15c shows reservoir operations where energy production is decreased by 2 million kW. Note that this decrease allows for considerable improvements in flood control without affecting firm power and irrigation. Finally, for another group of solutions, it is observed that by affecting both irrigation and hydropower production, even better operations are obtained for flood control without affecting firm energy as can be seen in Figure 15.15d. The above, increases the OF2 to 50,000 $(m^3/s)^2$ and reduces the OF3 to 26 million kW.

Based on the hypervolume indicator (Table 15.4) and the compiled Pareto fronts (Fig. 15.12), the ANN models offer the best results overall. It is evident that the MLP models tend to generate the best results for

Table 15.4 Hypervolume indicator for optimized models.

Operation model		Hypervolume (10^{19})	
		NSGA II	MOEA/D
Linear function		1.06	1.03
MLP	4N	1.33	1.26
	6N	1.31	1.26
	7N	1.34	1.26
RBN	4B	1.26	–
	6B	1.27	–
	7B	1.19	–

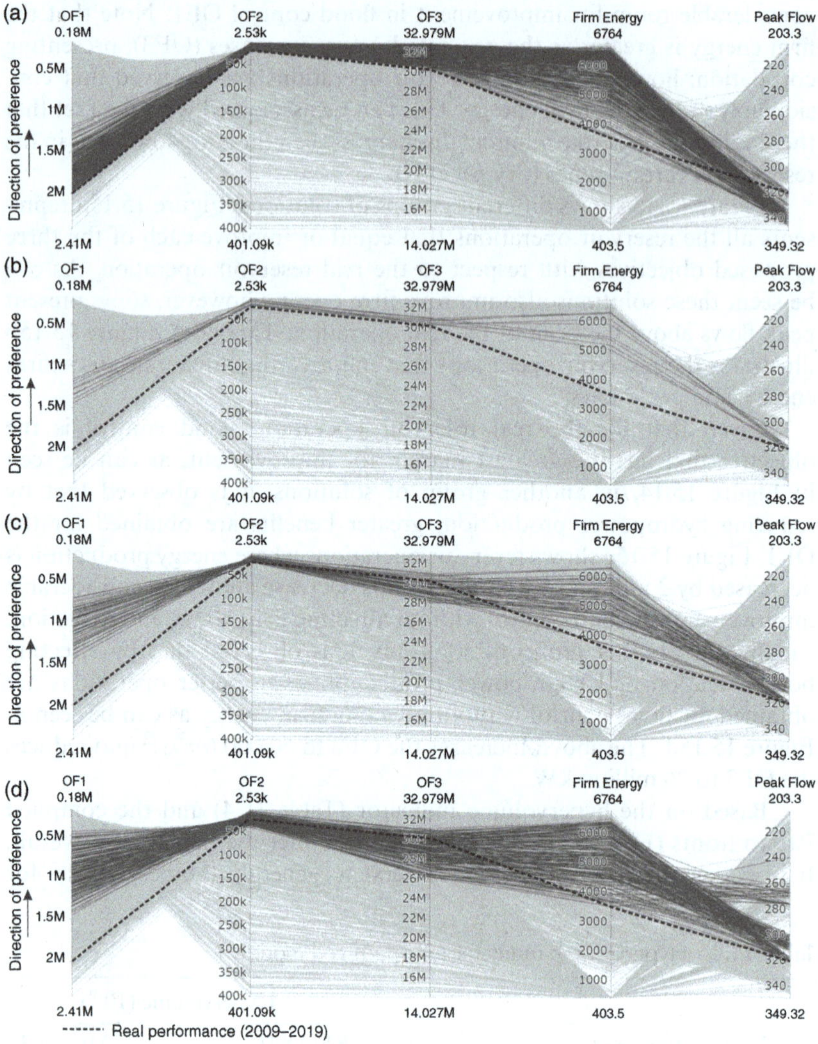

Figure 15.15 Parallel graphic: (a) Reservoir operations that improve the three objectives proposed for the real operation; (b) reservoir operations that improve the three objectives, firm energy, and peak flow for the real operation; (c) reservoir operations that improve OF1 by reducing hydropower production; (d) reservoir operations that improve OF1 by affecting both irrigation and hydropower production.

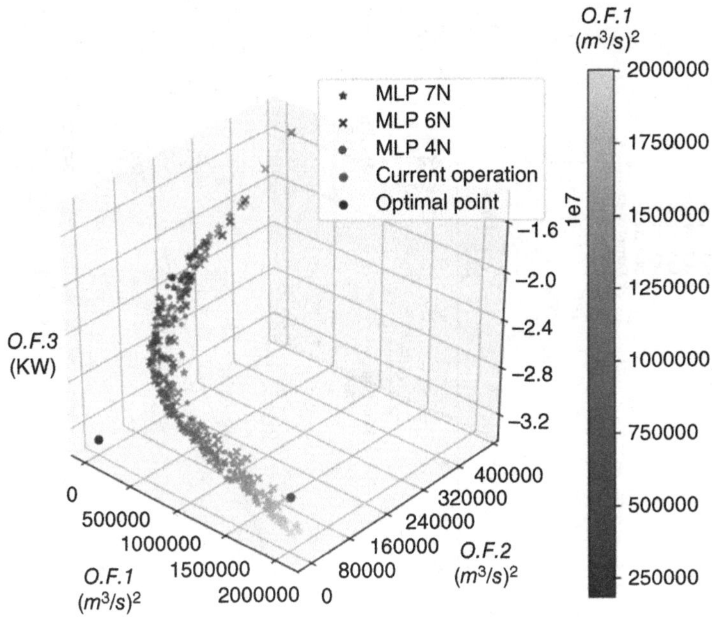

Figure 15.16 Pareto front for models with MLP and NSGA II optimizers.

this case study (Fig. 15.16). This may be due to the fact that RBN models use a greater number of parameters (decision variables), which implies greater difficulty for optimization. The results were similar regardless of the settings used for each ML model. The 6-7-1 MLP configuration had the best results overall, being better when using the NSGA II algorithm. For the RBN, the 6-6-1 configuration (using the same algorithm) yielded the best results. When comparing the two optimizers, similar results were observed between the MLP functions, with better outcomes from the NSGA II. This may be because the decision space in this study was not oriented toward a specific area of preference; however, the MOEA / D results may be improved by narrowing the decision space (Ma et al., 2015). Biasing the Pareto front toward a specific area, better convergence could be achieved by covering an area of interest. Regarding the RBN functions, the MOEA / D optimizer did not converge. In the case of the LF, the results were considerably lower according to the hypervolume indicator, which shows that this function does not simulate reservoir operations in the best way.

Some of the solutions obtained for the reservoir operations were selected and simulated in order to evaluate their performance. Figure 15.17

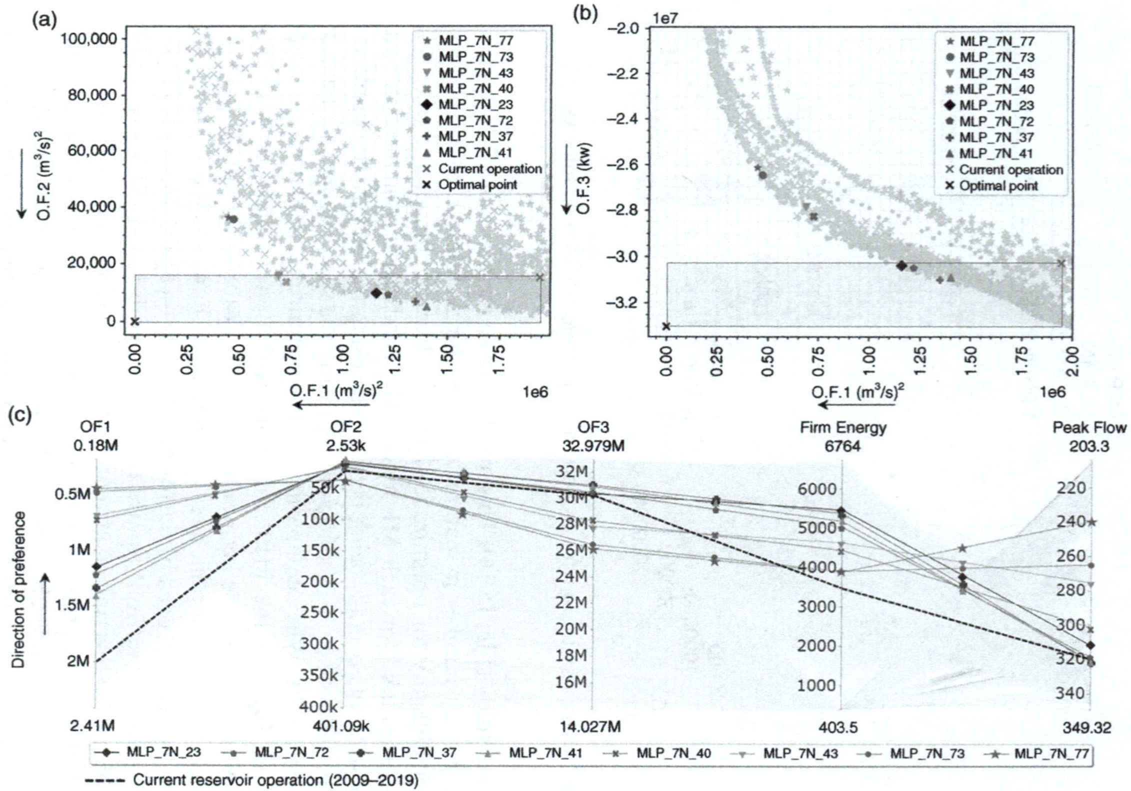

Figure 15.17 Zoomed-in Pareto front of the selected optimal reservoir operations: (a) OF1-OF2; (b) OF1-OF3; (c) parallel graphic of the selected solutions.

shows in detail the Pareto front OF1-OF2 and OF1-OF3, where shaded rectangles indicate the decision space of the solutions that equal or exceed the reservoir's actual operations toward the three objectives. Some points located inside and outside the decision space are simulated and compared with the real operation: four reservoir operations shown in Figure 15.15b are simulated (MLP_7N_23; MLP_7N_37; MLP_7N_41; MLP_7N_72), as well as two solutions shown in Figure 15.15c (MLP_7N_40; MLP_7N_43), and two shown in Figure 15.15d (MLP_7N_40; MLP_7N_43). Figure 15.17c shows the parallel graphic of the selected solutions.

Results are shown in Figure 15.18, which confirms that it is possible to reduce the peaks of the hydrograph of the real operation, as they are more to the left in the Pareto front. The MLP_7N_77 reservoir operation, as expected, had the best result in reducing all the peaks of the reference hydrograph (real operation), followed by MLP_7N_73. However, the effect negatively affected irrigation and power generation, keeping firm energy. Reservoir operations such as MLP_7N_40 and RN_7N_43, located within the shaded rectangle for the OF1 and OF2 (Fig. 15.17), allow one to reduce peak flows by slightly affecting the total power (OF3) but improving the irrigation objective and keeping the firm energy. Finally, the MLP_7N_23, MLP_7N_37, MLP_7N_41, MLP_7N_72 solutions, located within the shaded rectangle for the three objectives despite not having the same peak flow reduction capacity as the previous ones, managed to reduce the reference hydrograph while increasing power generation and guaranteeing the supply of water for irrigation. Figure 15.19 shows the hydrographs of the reservoir operations selected for the wettest portion of the analysis period, which spanned from November 2016 to May 2017.

Note that all the selected operations improved the real operation in terms of flood control and firm energy, however, as expected, the operations with the best performance in flood control (i.e., MLP_7N_77 and MLP_7N_73) present some intermittence (i.e., presence of minimum values) in power generation and irrigation in dry periods, as can be seen in Figure 15.18, for example, in the months of February to March 2010 and in the months of July to September 2014. The MLP_7N_40 and MLP_7N_43 solutions present less intermittence than the previous solutions in these terms, even improving the irrigation objective and having a considerable improvement in flood control (see Fig. 15.19). Regarding the solutions located within the shaded rectangles in Figure 15.17a and b, they considerably improved the real operation in terms of intermittence in power generation and irrigation, and also improving the flood control objective (although not in the same magnitude as the previous solutions: MLP_7N_40 and MLP_7N_43).

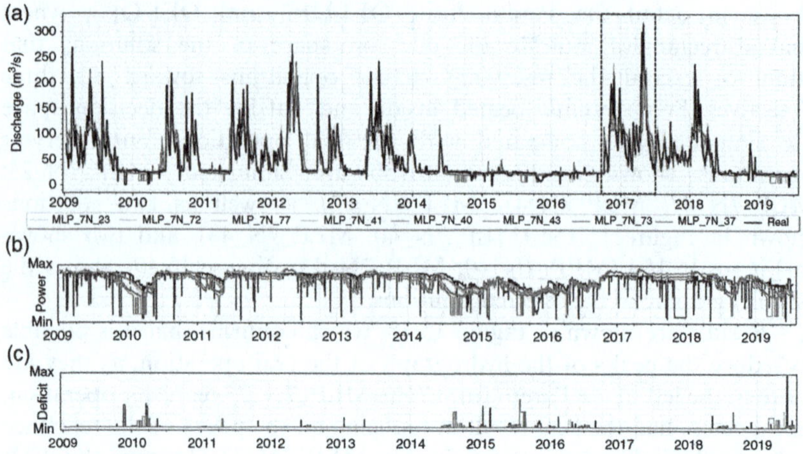

Figure 15.18 Simulation of the selected reservoir operations for the analysis period (2009–2019): (a) Hydrographs of output flow; (b) generated power; (c) deficit in irrigation water supply.

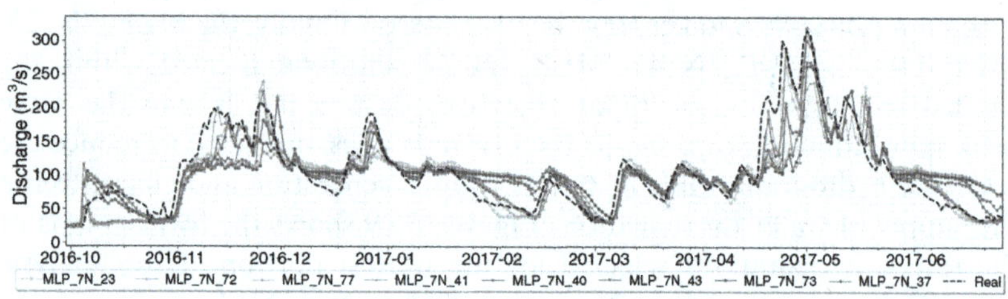

Figure 15.19 Hydrograph of the selected reservoir operations for the wet period (2016–2018).

15.5. Conclusions

This study presented a methodology using artificial intelligence techniques and EAs for obtaining optimal operational policies for multipurpose reservoirs. The approach was applied to a case study on the Hatillo reservoir in the Dominican Republic, which has three purposes: hydroelectric generation, irrigation, and flood control. Through this study, reservoir operations policies were obtained that improved the real operation of the reservoir for each of the three proposed objectives. Through the Pareto fronts obtained, the wide variety of operations for the Hatillo reservoir can be evidenced, which could even serve as a reference for decision making toward a balance of interests, thus improving the integral use of the reservoir. This methodology could be applied in multipurpose reservoirs in general, though the operation model must be adapted to each case.

Notably, the flood control objective of the reservoir's real operation was found to have wide room for improvement. Several reservoir operations policies were obtained that reduced the discharge peaks of the reservoir (as evidenced in the wet period from November 2016 to May 2017), equaling or even improving the objectives for irrigation and hydroelectric generation with respect to the real operation (e.g., MLP_7N_23; MLP_7N_37; MLP_7N_41; MLP_7N_72). Other reservoir operations policies, such as MLP_7N_77, considerably reduced the discharge peaks of the dam and only slightly affected the other purposes.

Overall, these results demonstrate that reservoir operation can be parameterized using ML models such as ANNs, which can accurately replicate the patterns and policies that govern reservoir operation. The 6-7-1 MLP configuration using the NSGA II optimizer had the best results overall. The MLP models tend to generate the best results for this case study. This may be due to the fact that RBN models use a greater number of parameters, making the optimization process more difficult. Results were similar regardless of the settings used for each ML model (i.e., varying the number of neurons). When the two optimizers were compared, similar results were observed for the MLP functions, with the NSGA II results being better. This may be because of the decision space in the present study was not oriented toward a specific area of preference; however, the MOEA/D results could be improved by narrowing the decision space (Ma et al., 2015). In the case of the LF, the results were considerably lower according to the hypervolume indicator, which showed that the reservoir operations policies did not adequately adjust to linear patterns.

The lack of information was one of the main limitations for the development of this research, although it was possible to have data from the reservoir for the period 2009–2019. For models based on data, such as those of the present study, it is desirable to have longer time series. Unfortunately, the reservoir operation rules as well as target levels, power production targets, and irrigation discharges were not available, which would have served as a reference for the elaboration of the reservoir operation model and for the definition of the objectives of optimization, reducing some assumptions.

Future works in this area are recommended to evaluate different multi-objective optimizers and ML models. For the Hatillo reservoir, new objectives can be included, such as firm energy or the flow regime alteration downstream of the dam. Further, if the necessary information were available, the models could be expanded to include economic variables that allow the reservoir's operation to be evaluated directly in terms of decision-making costs. Likewise, the performance of the operation models incorporating forecasting reservoir inflows, as well as the incorporation of rainfall data as input variables, should be evaluated. It would also be contemplated to evaluate different scenarios including possible changes in the components of the system, such as the increase in its installed hydropower capacity. Finally, future research could evaluate different optimization formulation for different seasons (i.e., dry seasons or wet seasons).

Acknowledgments

The authors acknowledge the information provided by Dominican Institute of Water Resources (INDHRI) and Empresa de Generación Eléctrica Dominicana (EGEHID) who supplied the reservoir operation data. The authors also acknowledge MESCYT in the Dominican Republic, for the research grant awarded to PUCMM (FONDOCYT # 2015-1H2-107) that led to the project in the Yuna River basin.

References

Basu, M. (2004). Goal-attainment method based on simulated annealing technique for economic-environmental-dispatch of hydrothermal power systems with cascaded reservoirs. *Electric Power Components and Systems, 32*(12), 1269–1286. https://doi.org/10.1080/15325000490446692

Benítez-Hidalgo, A., Nebro, A., García-Nieto, J., & Oregi, I. (2019). jMetalPy: A Python framework for multi-objective optimization with metaheuristics. *Swarm and Evolutionary Computation, 51,* 100598. https://doi.org/10.1016/j.swevo.2019.100598

Berge, D., Almanzar, L., & Bækken, T. (2003). *Environmental impact assessment of the Alto Yuna*. Oslo, Norway: NIVA.

Cámara, D. (2015). Evolution and evolutionary algorithms. In *Bio-inspired networking*. Elsevier. https://doi.org/10.1016/B978-1-78548-021-8.50001-6

Castelletti, A., Pianosi, F., Quach, X., & Soncini-Sessa, R. (2012). Assessing water resources management and development in northern Vietnam. *Hydrology and Earth System Sciences Discussions, 16*, 189–199. https://doi.org/10.5194/hess-16-189-2012

CEPAL. Naciones Unidas. (2008). *Evolución del impacto de la tormenta Noel en República Dominicana*. México: LC/MEX7L.853.

CEPAL (2020). *Análisis espacial de datos históricos y escenarios de cambio climático en México, Centroamérica, Cuba, Haití y la República Dominicana*. México: LC/MEX/TS.2020/43.

Chollet, F. K. (2015). *Keras*. GitHub. https://github.com/fchollet/keras

Corzo Perez, G. A. (2009). *Hybrid models for hydrological forecasting: Integration of data-driven and conceptual modeling techniques*. IHE Delft Institute for Water Education.

Cunningham, P., Cord, M., & Delany, S. (2008). Supervised Learning. In Cord, M., Cunningham, P. (eds) *Machine learning techniques for multimedia. cognitive technologies* (pp. 21–49). Berlin, Heidelberg: Springer. https://doi.org/10.1007/978-3-540-75171-7_2

Deb, K., Pratap, A., Agarwal, S., & Meyarivan. (2002). A fast and elitist multiobjective genetic algorithm: NSGA-II. *IEEE Transactions on Evolutionary Computation, 6*(2), 182–197. https://doi.org/10.1109/4235.996017

Ehsani, N., Fekete, B. M., Vörösmarty , C. J., & Tessler, Z. D. (2016). A neural network based general reservoir operation scheme. *Stochastic Environmental Research and Risk Assessment, 30*(4), 1151–1166. https://doi.org/10.1007/s00477-015-1147-9

Gembicki, F., & Haimes, Y. (1975). Approach to performance and sensitivity multiobjective optimization: The goal attainment method. *IEEE Transactions on Automatic Control, 20*(6), 769–771. https://doi.org/10.1109/TAC.1975.1101105

Giuliani, M., Castelletti, A., Pianosi , F., Mason, E., & Reed, P. M. (2016). Curses, tradeoffs, and scalable management: Advancing evolutionary multiobjective direct policy search to improve water reservoir operations. *ASCE, Journal of Water Resources Planning and Management, 142*, 04015050-1 - 04015050-17. https://doi.org/10.1061/(ASCE)WR.1943-5452.0000570

Giuliani, M., Herman, J. D., Castelletti, A., & Reed, P. (2014). Many-objective reservoir policy identification and refinement to reduce policy inertia and myopia in water management. *Water Resources Research, 50*, 3355–3377. https://doi.org/10.1002/2013Wc154700

Haimes, Y. Y., Lasdon L. S, & Wismer, D. A. (1971). On a bicriterion formulation of the problems of integrated system identification and system optimization. *IEEE Transactions on Systems, Man, and Cybernetics, 1*(3), 296–297. https://doi.org/10.1109/TSMC.1971.4308298

Heydari, M., Othman, F., & Taghieh, M. (2016). Optimization of multiple and multipurpose reservoir system operations by using matrix structure (Case study: Karun and Dez Reservoir Dams). *PLOS ONE, 11*(6), e0156276. https://doi.org/org/10.1371/journal.pone.0156276

Jansen, L., Lariyah, S., Mohamed, D., & Pierre, J. (2013). Hydropower reservoir for flood control: A case study on Ringlet Reservoir, Cameron Highlands, *Malaysia. Journal of Flood Engineering, 4*(1), 87–109.

Jesús Encarnación, A. (2019). *Planteamiento de plan de emergencia de las presas del complejo Nizao (República Dominicana), frente a los fenómenos climáticos extremos*. Valencia: Universitat Politècnica de València.

Jury, M., Malmgren, B., & Winter, A. (2007). Subregional precipitation climate of the Caribbean and relationships with ENSO and NAO. *Journal of Geophysical Research, 112*, D16107. https://doi.org/org/10.1029/2006JD007541

Khattab, M., & Al-Mohseen, K. (2020). Planning and Decision Making Under Uncertainty (Mosul Reservoir Optimal Operating Policy, Case Study).

Al-Rafidain Engineering Journal, 25, 85–96. https://doi.org/10.33899/rengj.2020.126885.1032

Ko, S.-K., Fontane, D. G., & Labadie, J. W. (1992). Multiobjective Optimization of Reservoir Systems Operation. *Water Resources Bulletin, 28*(1), 111–27v, https://doi.org/org/10.1111/j.1752-1688.1992.tb03158.x

Kollat, J., Reed, P., & Maxwell, R. (2011). Many-objective groundwater monitoring network design using bias-aware ensemble Kalman filtering, evolutionary optimization, and visual analytics. *Water Resources Research, 47*, W02529. https://doi.org/10.1029/2010WR009194

Koutsoyiannis, D., & Economou, A. (2003). Evaluation of the parameterization-simulation-optimization approach for the control of reservoir systems. *Water Resources Research, 39*(6), 1170. https://doi.org/10.1029/2003WR002148

Labadie, J. W. (2004). Optimal operation of multireservoir systems: State-of-the-art review. *Journal of Water Resources Planning and Management, 130*(2), 93–111. https://doi.org/10.1061/(ASCE)0733-9496(2004)130:2(93)

Loucks, D., & van Beek, E. (2017). Data-fitting, evolutionary, and qualitative modeling. In *Water resource systems planning and management. Gewerbestrasse*, Switzerland: Springer. https://doi.org/10.1007/978-3-319-44234-1_5

Ma, X., Liu, F., Qi, Y., Li, L., Jiao, L., Deng, et al. (2015). MOEA/D with biased weight adjustment inspired by user preference and its application on multi-objective reservoir flood control problem. *Soft Computing, 20*(12), 4999–5023. https://doi.org/10.1007/s00500-015-1789-z

MEPyD (2017). *Pérdidas provocadas por los fenómenos naturales entre noviembre 2016 y septiembre 2017*. Santo Domingo: Ministerio de Economía, Planificación y Desarrollo.

Miettinen , K. (1999). *Nonlinear multiobjective optimization*. Springer, Boston, MA.

Momtahen, S., & Dariane, A. (2007). Direct search approaches using genetic algorithms for optimization of water reservoir operating policies. *ASCE, Journal of Water Resources Planning and Management, 133*(3), 202–209. https://doi.org/10.1061/(ASCE)0733-9496(2007)133:3(202)

Moquete, C. (2004). *Generalidades del cultivo de arroz*. Santo Domingo: IDIAF.

Myo Lin, N., & Rutten, M. (2016). Optimal operation of a network of multi-purpose reservoir: A review. *Procedia Engineering, 154*, 1376–1384. https://doi.org/10.1016/j.proeng.2016.07.504

Myo Lin, N., Tian, X., Rutten, M., Abraham, E., Maestre, J. M., & van de Giesen, N. (2020). Multi-objective model predictive control for real-time operation of a multi-reservoir system. *Water, 12*(7), 1898. https://doi.org/10.3390/w12071898

Pedregosa, F., Varoquaux, G., Gramfort, A., Michel , V., Thirion, B., Grisel, O., et al. (2011). Scikit-learn: Machine learning in Python. *Journal of Machine Learning Research, 12*(2011), 2825–2830.

Ritter, J. (2016). *Optimization of hydropower dam cascade operations with respect to energy generation, flood hazard and flow regime alteration, using operational modeling: A case study in the Nechí catchment*. Delf: UNESCO-IHE.

Ritter, J., Corzo, G., Solomatine, D. P., & Angarita, H. (2020). Multiobjective direct policy search using physically based operating rules in multireservoir systems.

Journal of Water Resources Planning and Management, 146(4), 05020002. https://doi.org/10.1061/(ASCE)WR.1943-5452.0001159

Rosenstein, M., & Barto, A. (2001). *Robot weightlifting by direct policy search. Seventeenth International Joint Conference on Artificial Intelligence, 2*, 839–844.

Ruijie, Z., Cai, X., Ringler, C., & Zhu, T. (2017). Hydropower versus irrigation: An analysis of global patterns. *Enviromental Research Letters, 12*, 034006. https://doi.org/org/10.1088/1748-9326/aa5f3f

Saavedra, O. C., Koike, T., Yang, K., & Yang, D. (2010). Optimal dam operation during flood season using a distributed hydrological model and a heuristic algorithm. *Journal of Hydrologic Engineering ASCE, 15*(7), 580–586. https://doi.org/10.1061/(ASCE)HE.1943-5584.0000212

Scola, L., Takahashi, R. H., & Cerqueira, S. (2014). Multipurpose water reservoir management: An evolutionary multiobjective optimization approach. *Mathematical Problems in Engineering, 2014*(638259), 14. https://doi.org/org/10.1155/2014/638259

Smith, M. (1992). *CROPWAT: A computer program for irrigation planning and management*. Rome: Food and Agriculture Organization of the United Nations.

Sun, X., Luo, J., & Xie, J. (2018). Multi-objective optimization for reservoir operation considering water diversion and power generation objectives. *Water, 10*, 1540. https://doi.org/org/10.3390/w10111540

Tami, C. (2020). *Identificación de reglas de operación óptimas de embalses para el control de inundaciones a partir de modelos de operación. Caso de estudio: Cuenca del Río Yuna en República Dominicana*. Bogotá D.C: Escuela Colombiana de Ingeniería. https://repositorio.escuelaing.edu.co/handle/001/1290

Tikk, D., Kóczy, L., & Gedeon, T. (2003). A survey on universal approximation and its limits in soft computing techniques. *International Journal of Approximate Reasoning, 33*(2), 185–202. https://doi.org/10.1016/S0888-613X(03)00021-5

Viparelli, R. (1983). *Spillways for multipurpose reservoirs, to regulate the runoff available at the dam section and for the control of floods. International Conference on the Hydraulics Aspects of Floods & Flood Controls*. London, England: City, University of London.

Zhang, D., Lin, J., Peng, Q., Wang, D., Yang, T., Sorooshian, S., et al. (2018). Modeling and simulating of reservoir operation using the artificial neural network, support vector regression, deep learning algorithm. *Journal of Hydrology, 565*, 720–736. https://doi.org/10.1016/j.jhydrol.2018.08.050

Zhang, Q., & Li, H. (2007). MOEA/D: A multiobjective evolutionary algorithm based on decomposition. *IEEE Transactions on Evolutionary Computation, 11*(6), 712–731. https://doi.org/10.1109/TEVC.2007.892759

Zhou, C., Sun, N., Chen, L., Ding, Y., Zhou, J., Zha, G., et al. (2018). Optimal operation of cascade reservoirs for flood control of multiple areas downstream: A case study in the upper Yangtze River Basin. *Water, 10*, 1250. https://doi.org/10.3390/w10091250

INDEX

Advanced Hydroinformatics: Machine Learning and Optimization for Water Resources, Special Publications 78, First Edition. Edited by Gerald A. Corzo Perez and Dimitri P. Solomantine.

DOI: 10.1002/9781119639268.bindex

D

O

Q

R

Printed in the USA/Agawam, MA
May 19, 2026

905303.005